PROTEOMICS: BIOMEDICAL AND PHARMACEUTICAL APPLICATIONS

Proteomics: Biomedical and Pharmaceutical Applications

Edited by

HUBERT HONDERMARCK

University of Sciences and Technologies,
Lille, France

KLUWER ACADEMIC PUBLISHERS
DORDRECHT / BOSTON / LONDON

A C.I.P. Catalogue record for this book is available from the Library of Congress.

ISBN 1-4020-2322-7 (HB)
ISBN 1-4020-2323-5 (e-book)

Published by Kluwer Academic Publishers,
P.O. Box 17, 3300 AA Dordrecht, The Netherlands.

Sold and distributed in North, Central and South America
by Kluwer Academic Publishers,
101 Philip Drive, Norwell, MA 02061, U.S.A.

In all other countries, sold and distributed
by Kluwer Academic Publishers,
P.O. Box 322, 3300 AH Dordrecht, The Netherlands.

Cover illustration: Breast cancer cells observed by electron scanning microscopy
(reproduced with the permission of Molecular and Cellular Proteomics, USA),
two-dimensional electrophoresis gel and mass spectrometry spectrum used in protein identification.

Printed on acid-free paper

CONTENTS

PREFACE

HUBERT HONDERMARCK
* *INSERM-ESPRI, University of Sciences and Technologies Lille, France.*

A constancy in human history is that new discoveries have systematically been used to improve health and increase the duration and quality of life. From the use of flints to perform primitive surgery during prehistory, to the preparation of plant extracts to cure diseases throughout the world and historic times, the will to turn new basic knowledge into practical tools for medicine is probably a shared value of all civilizations and cultures. During the first part the 20th century, shortly after discovering the radioactive compounds, radium and polonium, the physicist Marie Curie created in Paris the first center for radiotherapy, demonstrating that discovery in basic science can have a rapid and significant impact in medicine. The second half of the last century was intellectually dominated by the discovery of the genetic code and the description of the structure of its carrying molecule, the DNA, with the end of this era essentially marked by the complete sequencing of the human genome. This major achievement was based on considerable technological progress in biochemistry and bioinformatics, but a significant impetus was the promise of a better understanding of human health through the discovery of genes involved in pathology. However, during the course of human genome sequencing, the scientific community and the public progressively realized that the accomplishment of this huge project was not going to revolutionize, on its own, the field of biomedicine. In large part, this was due to the presence of a significant number of genes for which function was (and remains) unknown, and because biology is much more complex than the structure and functioning of nucleic acids. In a way, the genome project somehow contributed, for a while, to shorten our vision of biology.

The beginning of the 21st century has seen the 'rediscovery' of the complexity of life at the molecular and cellular levels. While genes are the support for transmission of genetic information, most of the chemical reactions inside cells are carried out by proteins, which are the functional products of gene expression. This has given way to the realization that proteomics is the most relevant way to identify markers of pathologies and therapeutic targets. Whereas it is now clear that the number of human genes is around 35,000, there is no consensus on the number of proteins. Taking under consideration alternative splicing of mRNA and post-translational modifications, most of the current projections of protein number range from 5×10^5 to 5×10^6, but interestingly it cannot be excluded that the number of proteins might in fact be closer to infinity. To add further complexity, proteins interact one with another and the proteome is intrinsically dynamic as it varies between cell types and from physiological condition to condition as occurs in living cells. There is no doubt that defining the human proteome is going to be a much more difficult task than the sequencing of genome.

Together, it might not be an exaggeration to say that the major outcome of the human genome sequencing has finally been to open the way to the exploration of the proteome, transferring the goals and hopes in terms of biomedical and pharmaceutical applications. This is clearly a challenging, but also a promising heritage that this book explores through a series of ongoing experiences and projects representative of the new era in which biology and medicine have now entered.

CONTRIBUTORS

ADRIAENSSENS, Eric
INSERM-ESPRI, IFR-118, University of Sciences and Technologies Lille,
Villeneuve d'Ascq, France.

AEBISCHER, Toni
Department of Molecular Biology, Max-Planck-Institute for Infection Biology,
Schumannstrasse 21/22, D-10117 Berlin, Germany.

BECK, Alain
Centre d'Immunologie Pierre Fabre, 5 Avenue Napoléon III, 74160 Saint-Julien en
Genevois, France.

BERTRAND, Eric
Novartis Pharma Research, Functional Genomics, WJS 88.701, CH4002 Basel
Switzerland.

BRADSHAW, Ralph A.
Department of Physiology & Biophysics, and Neurobiology & Anatomy.
University of California, Irvine. Irvine, CA 92697, USA.

BUMANN, Dirk
Department of Molecular Biology, Max-Planck-Institute for Infection Biology,
Schumannstrasse 21/22, D-10117 Berlin, Germany.

CASH, Phillip
Department of Medical Microbiology, University of Aberdeen, Foresterhill,
Aberdeen AB25 2ZD, Scotland.

CHUNG, Maxey C.M.
Department of Biochemistry, Faculty of Medicine , National University of
Singapore, Singapore 119260, Republic of Singapore.

COLLINS, Mark O.
Division of Neuroscience, University of Edinburgh, 1 George Square, Edinburgh,
EH8 9JZ, UK.

DRUMMELSMITH, Jolyne
Centre de Recherche en Infectiologie du Centre de Recherche du CHUL and
Division de Microbiologie, Faculté de Médecine, Université Laval, Québec,
Canada.

DUNN, Michael J.
Department of Neuroscience, Institute of Psychiatry, King's College London,
London, UK.

EL YAZIDI-BELKOURA, Ikram
INSERM-ESPRI, IFR-118, University of Sciences and Technologies Lille,
Villeneuve d'Ascq, France.

FAUPEL, Michel
Novartis Pharma Research, Functional Genomics, WJS 88.701, CH4002 Basel
Switzerland.

GRANT, Seth G. N.
Wellcome Trust Sanger Institute, Hinxton, CAMBS, CB10 1SA, UK.

HAEUW, Jean-François
Centre d'Immunologie Pierre Fabre, 5 Avenue Napoléon III, 74160 Saint-Julien en
Genevois, France.

HONDERMARCK, Hubert
INSERM-ESPRI, IFR-118, University of Sciences and Technologies Lille,
Villeneuve d'Ascq, France.

HUSI, Holger
Division of Neuroscience, University of Edinburgh, 1 George Square, Edinburgh,
EH8 9JZ, UK.

JAIN, Kewal K.
Jain PharmaBiotech, Blaesiring 7, CH-4057 Basel, Switzerland.

JUNGBLUT, Peter R.
Proteomics Unit, Max-Planck-Institute for Infection Biology, Schumannstrasse
21/22, D-10117 Berlin, Germany.

KRAH, Alexander
Department of Molecular Biology and Proteomics Unit, Max-Planck-Institute for
Infection Biology, Schumannstrasse 21/22, D-10117 Berlin, Germany.

LEMOINE, Jerome
UMR-8576, IFR-118, University of Sciences and Technologies Lille, Villeneuve
d'Ascq, France.

LO, Siaw L.
Department of Biological Science, Faculty of Science, National University of
Singapore, Singapore 119260, Republic of Singapore.

LIANG, Rosa C.M.Y.
Department of Biological Science, Faculty of Science, National University of
Singapore, Singapore 119260, Republic of Singapore.

MCGREGOR, Emma
Proteome Sciences plc, Institute of Psychiatry, King's College London, London,
UK.

MEYER, Thomas F.
Department of Molecular Biology, Max-Planck-Institute for Infection Biology,
Schumannstrasse 21/22, D-10117 Berlin, Germany.

MISUMI, Shogo
Department of Biochemistry, Faculty of Pharmaceutical Sciences, Kumamoto
University, 5-1 Oe-Honmachi, Kumamoto 862-0973, Japan.

NEO, Jason C.H.
Department of Biological Science, Faculty of Science, National University of
Singapore, Singapore 119260, Republic of Singapore.

OUELLETTE, Marc
Centre de Recherche en Infectiologie du Centre de Recherche du CHUL and
Division de Microbiologie, Faculté de Médecine, Université Laval, Québec,
Canada.

PAGE, Nigel M.
School of Animal and Microbial Sciences, The University of Reading, Reading,
RG6 6AJ, UK.

PAPADOPOULOU, Barbara
Centre de Recherche en Infectiologie du Centre de Recherche du CHUL and
Division de Microbiologie, Faculté de Médecine, Université Laval, Québec,
Canada.

SCHNEIDER, Philippe
Service Régional Vaudois de Transfusion Sanguine, Lausanne, rue du Bugnon 27,
CH-1005 Lausanne, Switzerland.

SEOW, Teck K.
Department of Biological Science, Faculty of Science, National University of
Singapore, Singapore 119260, Republic of Singapore.

SHOJI, Shozo
Department of Biochemistry, Faculty of Pharmaceutical Sciences, Kumamoto
University, 5-1 Oe-Honmachi, Kumamoto 862-0973, Japan.

TAKAMUNE, Nobutoki
Department of Biochemistry, Faculty of Pharmaceutical Sciences, Kumamoto
University, 5-1 Oe-Honmachi, Kumamoto 862-0973, Japan.

TAN, Gek S.
Department of Biological Science, Faculty of Science, National University of
Singapore, Singapore 119260, Republic of Singapore.

TISSOT, Jean-Daniel
Service Régional Vaudois de Transfusion Sanguine, Lausanne, rue du Bugnon 27,
CH-1005 Lausanne, Switzerland.

VANDERMOERE, Franck
INSERM-ESPRI, IFR-118, University of Sciences and Technologies Lille,
Villeneuve d'Ascq, France.

VAN OOSTRUM, Jan
Novartis Pharma Research, Functional Genomics, WJS 88.701, CH4002 Basel
Switzerland.

WILKINS, Martin R.
Experimental Medicine and Toxicology, Imperial College London, Hammersmith
Hospital, Du Cane Road, London W12 0NN, UK.

Chapter 1

Proteomics Today, Proteomics Tomorrow

RALPH A. BRADSHAW
Department of Physiology & Biophysics, and Neurobiology & Anatomy. University of California, Irvine. Irvine, CA 92697, USA

1.   INTRODUCTION

The basic mantra 'DNA makes RNA makes protein' is the cornerstone of molecular and cellular biology and has interestingly enough now devolved into three 'sub-sciences': genomics, transcriptomics and proteomics. Each field is driven by its own technologies and each offers its own potential in providing new knowledge that will be useful in diagnosing and treating human and animal diseases. However, the boundaries separating these fields are diffuse and indeed there are no exact definitions for any of them. This is particularly true of proteomics, which clearly encompasses the greatest territory, because there are probably as many views of what proteomics encompasses (and therefore expectations) as there are interested parties (Huber, 2003).

In general, genomics is the study of the genome of an organism, which includes identifying all the genes and regulatory elements, and de facto the sequence of all of the DNA making up the genetic material. As the number of determined genomes rises (it is presently around 50), comparative analyses also become valuable. The expression of the genome leads to the formation of the transcriptome, which is the compliment of mRNAs that reflects the structural genes of an organism; however, this is more complex primarily due to the fact that the transcription of the genome is sensitive to a variety of factors related to the context of the cell and its environment. Indeed, at any given moment, only a subset of the full set of genes will be expressed and this is true whether it is a single cell organism or a cell from the organ of a higher eukaryote. As conditions change the expressed transcriptome will change too. Thus, the major confounding factor is time, a

1

H. Hondermarck (ed.), Proteomics: Biomedical and Pharmaceutical Applications, 1–17.
© 2004 *Kluwer Academic Publishers. Printed in the Netherlands.*

dimension that is largely absent in genomic studies. In eukaryotes, splicing events can produce several mRNAs from a single gene, which introduces yet another element of diversity. Clearly, transcriptomics (the description of the transcriptome and its expression) is potentially far more complex than genomics.

The conversion of the transcriptome to its protein products produces the corresponding proteome. This term was introduced by Marc Wilkins to describe "all proteins expressed by a genome, cell or tissue" in work for his doctoral degree (Cohen, 2001, Huber, 2003) and has been essentially universally adopted. In parallel with genomics and transcriptomics, proteomics is thus the description of a proteome. Since in cells and whole organisms the transcriptome is temporally dependent, it follows that the corresponding proteome is as well. However, the proteome is distinguished from both the genome and transcriptome in another very important way: whereas in the first two the information is essentially linear (sequential), the proteome is defined by both sequential and three-dimensional structures. Along with the many co-/posttranslational modifications that also occur, the complexity of the proteome is clearly much, much greater and presents a problem that is of undefined magnitude. In fact, it is this realization that really distinguishes proteomics from either genomics or transcriptomics. While both the genome and transcriptome (albeit that this has yet to be achieved in the latter case) can be seen as finite and thus definable, even for a entity as large as the human genome, the description of the complete proteome of a single cell or even a simple organism, let alone higher eukaryotes, is not presently imaginable and may well not be achievable, i.e. it may be effectively infinite in size (Cohen, 2001, Huber, 2003).

It must be noted that everyone does not accept this position and where the term proteome is applied to the third potential source included by Wilkins, i.e. tissue, in contradistinction to genome (organism) or cell, something approaching a complete description might be possible. The plasma proteome is a case in point (see below). In the ensuing remarks, a more detailed definition of proteomics is developed along with an outline of how this has produced a 'working' agenda for today's proteomic studies. Illustrations of some of what this has yielded and is likely to yield in the foreseeable future - tomorrow's proteomics - will be contrasted with some of the more unlikely 'promises' that have been made.

2.   PROTEOMICS - A UNIVERSAL DEFINITION

Table 1 provides a four-part definition for proteomics that outlines the scope and nature of the field and thus the information required to fully describe a proteome. Each part represents a different aspect of proteomic research and some have even acquired names for all or part of what they encompass. A brief description of these areas (and sub areas) follows.

Table 1. Proteomics - A Universal Definition.

Determination of the structure and function of the complete set of proteins produced by the genome of an organism, including co- and posttranslational modifications

Determination of the all of the interactions of these proteins with small and large molecules of all types

Determination of the expression of these proteins as a function of time and physiological condition

The coordination of this information into a unified and consistent description of the organism at the cellular, organ and whole animal levels

The first part deals primarily with the identification of all the protein components that make up the proteome and can be viewed in some respects as a giant catalog. The simplest form of this list would be the one to one match up of proteins with genes. This can be most easily accomplished in prokaryotes where mRNA splicing does not occur, but still is a realizable, if much larger and more elusive, goal in eukaryotes. It is this aspect of proteomics that most closely parallels genomics and transcriptomics and what undoubtedly gives workers from those fields the highly questionable sense that the proteome is a finite and therefore determinable entity. The match up, however, extends beyond the sequence level as one must also add to the catalog the three dimensional structure of each protein and all covalent modifications that accompany the formation and maturation of that protein, beginning with the processing of the N-terminus and ending with its ultimate proteolytic demise. This is a considerably larger challenge and one that has not been achieved for any proteome and likely will not be for some time. Indeed, the structural aspects of the proteome, which have also been referred to as 'structural genomics' and 'structural proteomics' (the difference between these terms is obscure at best), lag well behind the identification of sequences and covalent modifications (Sali *et al.*, 2003).

The final component of this part of the definition is the determination of function for each protein and it is here that one must begin to seriously depart from the cataloging concept. While one can usually connect one or more functions (catalytic, recognitive, structural, etc.) to most proteins, ascribing single functions to proteins is at best a gross simplification of the situation. As taken up in part three, protein-protein interactions, with their resultant effects on activity, are enormously widespread and since the presence or absence of these interactions, along with many co-/posttranslational modifications are time (expression) dependent, simple listings are at best a draconian description of a proteome and are really inadequate to describe function in the cellular context.

While the concept of a 'protein catalog' undoubtedly existed, at least in the minds of some people, prior to the coining of the term proteome, the realization that protein-protein interactions were a dominant feature of proteome function was certainly considerably less clear. However, such is the case and the second part of the definition requires a detailed knowledge of all of these recognitive events, both stable and transient. The elucidation of these interactions and the underlying networks that they describe have been termed cell-mapping proteomics (Blackstock & Weir, 1999). This in fact has been a quite productive area of proteomic research, as described below.

The third part of the definition clearly represents the greatest challenge in terms of data collection. If the first two parts can be said to loosely represent the 'who, what and how' of proteomics, then the third part is the 'when and where', and is by far the most dynamic of the three. Obtaining this information requires perturbation almost by definition and in essence represents what has been described as 'systems biology'. It is commonly known as 'expression proteomics' (Blackstock & Weir, 1999) and forms the heart of signal transduction studies, a field that has thrived independently and has been one of the major areas of molecular and cellular biology research for the past 10 years or more (Bradshaw & Dennis, 2003). However, expression proteomics encompasses more than just signal transduction events, even if these are not readily separated from them. These include all nature of metabolic pathways, cell cycling events and intracellular transport, to name only a few. For the most part, efforts in expression proteomics have, to date, been limited to 'snapshots' of the proteome, obtained following an appropriate cellular stimulus, and have concentrated on two kinds of measurements: changes in the level of proteins (up or down) and changes in post-ribosomal modifications, which are usually but not exclusively protein phosphorylations. The former are closely linked to studies using mRNA arrays although formerly these are really a part of transcriptomics.

The final part of the definition can, to complete the analogy, be considered the 'why' of proteomics. This part also encompasses the bioinformatics component of proteomics, another vaguely defined discipline that arose from the realization, as genomics moved into full swing, that the amount of information being obtained was increasing, and would continue to do so in an exponential manner and that 'managing' this inundation was a serious challenge in its own right. Proteomics, far more than genomics and transcriptomics, will accelerate this trend and hence data management must be an integral part of the definition. However, bioinformatics does cut across all of the 'omics' fields, or perhaps more appropriately, provides the fabric that links all of them together so it is not an exclusive part of proteomics either. This last part of the definition really also represents the ultimate goal of proteomics, if a field of study can be said to have a goal, which is to understand not only the nature of the components and how they change with time and condition but also how they integrate to produce a living entity. Some may argue that this is not an achievable goal either but that is a debate for another time and place.

3.   PROTEOMICS TODAY

3.1   Applications - An Overview

Proteomic research legitimately encompasses both 'classical' protein chemistry and the new more complex approaches that feature analyses designed to measure large numbers of proteins at the same time (Aebersold & Mann, 2003, Phizicky *et al.*, 2003), often in a high throughput fashion. Indeed, to many people the latter approaches, which feature such methodology as 2-dimensional polyacrylamide gel electrophoresis (2D PAGE or 2DE), various configurations of mass spectrometry (MS) (most notably applications of electrospray and matrix-assisted laser desorption ionizations [ESI and MALDI]) and the use of arrays (both protein and nucleic acid-based), are proteomics. This uncharitable view diminishes a large amount of important data that was collected primarily over the last half of the previous century and more importantly ignores the fact that the methods and techniques that grew out of that period are still making important contributions to achieving the definitions of Table 1. While new methods often supplant old ones as the one of choice, they rarely if ever eliminate them. Rather they simply expand the arsenal of tools available to answer the experimental question at hand. In this regard it is instructive to briefly compare the foci of the older protein analyses with the newer ones (Table 2). Clearly, the emphasis of earlier studies was on individual protein entities by first striving to obtain homogeneous preparations, initially by isolation of natural proteins and more recently recombinant ones, and then submitting them to detailed characterization including, but certainly not limited to, sequence and structure. These studies were closely coupled to functional analyses that examined kinetic, mechanistic or recognitive properties. This was the core of the reductionist research tradition that produced the basis on which most of modern molecular and cellular biology rests today. In the last few years, the emphasis has shifted in large part to understanding the 'sociology of proteins', i.e. how these individual entities form a cohesive, responsive (dynamic) network. To use common scientific jargon, this represents a paradigm shift, driven in part by the advances in genomics and transcriptomics and in part by the maturation of 2DE, MS and array technologies. However, it is still of great importance that we understand individual proteins in molecular detail and such studies should not be considered to be outside the scope of proteomics but simply important additional parts of it.

Table 2. A Comparison of Protein Chemical vs. Proteomic Experiments.

Protein Chemistry	Proteomics
Function/identity generally known	Function/identity often unknown (identification from sequences)
Single entity	Complexes of multiple entities (including protein machines)
Single function	Coordinated functions
Free standing (studied as an isolated entity)	Multiple interactions
Focused studies	High throughput

Given this perspective that proteomics is not so much a new field as the substantial expansion of an old one, it is germane to briefly consider where this synthesis of old and new technologies has brought us in terms of some of the more active areas of research before discussing some of the future prospects. Table 3 lists four such areas, which not surprisingly correlate quite closely to the definitions described in Table 1. There is necessarily some overlap between these. For example, both the determination of co-/posttranslational modifications and the identification of protein-protein interactions are also major components of expression profiling in addition to the many modifications and interactions that are not (as yet) directly linked to changes in cell response. Two of these topics are particularly associated with proteomics (protein identification and protein-protein interactions) perhaps because they have provided (or promise to) more 'new' insights than the other areas for which there was already substantial knowledge from independent studies. The ensuing two sections provide brief summaries of some of the advances to date.

1.1   3.2   Protein Identification and Database Mining

From the point of view of one interested in proteomics, there was nothing quite as interesting in the solution of the human (or for that matter any other) genome as the determination that a substantial amount of the putative revealed genes were unknown in both function and three dimensional structure. Although the latter remains true, a number of 'associative' analyses have at least provided a substantial number of indications as to what the general function may be, e.g. DNA repair, even if the exact activity remains to be determined. These have ranged from quite inventive 'in silico' analyses (Eisenberg *et al.*, 2000) that variously take advantage of evolutionary relationships, as revealed by genomic

Table 3. Major Proteomic Applications.

Area	Description
Proteome Mining	Identifying known and unknown proteins in complex mixtures
Protein Modifications	Identifying co/posttranslational changes
Expression Profiling	Measuring proteome changes in response to physiologic stimuli
Protein Complexes	Identifying protein-protein interactions

comparisons, to networks established through protein-protein interactions (see Section 3.3). There nonetheless remains quite large numbers of proteins that still must be identified and characterized. Prof. Sydney Brenner, in a plenary lecture entitled "From Genes to Organisms" at the 'Keystone 2000' meeting in Keystone, CO that showcased the leaders in molecular and cellular biology at the turn of the century, noted that if the human genome is composed of 50,000 genes (the number of human genes at that juncture was still quite undecided), it would simply require 50,000 biochemists, each working on one protein product, to define the human proteome. Since he has also been quoted as saying that proteomics 'is not about new knowledge but about amassing data' and 'that it (proteomics) will prove to be irrelevant' (Cohen, 2001), it isn't clear that he is actually in favor of this. Nonetheless, even though this clearly would not provide all the information required to satisfy the definition given in Table 1, it is not entirely an exaggeration either. We will not make serious headway into human (or other species) proteomics until these unknown gene products are defined and it is hard to see how this wouldn't provide new knowledge.

Aside from determining the nature of the players in the human proteome drama, there is also the considerably larger task of identifying them in real samples. The first major effort to do this was by 2 DE, which was introduced by O'Farrell (O'Farrell, 1975) and Klose (Klose, 1975) in the mid 1970's, and arguably marked the beginning of proteomic research. This technique has continued to improve both in reproducibility and sensitivity and has provided a great deal of useful information during the nearly 30 years that it has been available. However, its value is pretty much directly proportional to the ability of the investigator to identify the spots. This has been done in a number of ways but elution and coupling to mass spectrometry has clearly been the most powerful and there are now available well-annotated 2D maps for a number of species and conditions. However, despite the improvements, 2D gels have built in limitations in terms of resolving power and dynamic range that will always limit their usefulness particularly as the focus moves to the lower end of the concentration scale. Thus various chromatographic approaches, such as MudPIT (Wolters *et al.*, 2001) have been introduced to broaden the size and complexity of sample that can be effectively analyzed. These liquid chromatographic systems can be directly linked to the mass spectrometry and in theory can greatly increase the amount of data that can be generated and analyzed. However, only about a third (~1800) of the proteins in a yeast extract have been identified by this approach to date.

There are two basic applications of mass spectrometry used for the identification of proteins; both are dependent on the analysis of generated peptides, usually produced by trypsin (Yates, 1998). In the first, MALDI time-of-flight (TOF) spectra are generated for a mixture of peptides from a pure protein (or relatively simple mixture) and a table of masses is generated that is then matched against similar tables produced from theoretical digests of all proteins listed in the database. For relatively pure samples or simple mixtures, such as would typically be the case with a spot excised from a 2D gel, and where the sample is derived from an organism whose genome is known, this peptide mapping methodology is quite an effective method. However, in cases where the protein sample is derived from a species where the genome may be at best only partially known, its effectiveness falls off rapidly. Of course, co-/posttranslational modifications or unexpected enzyme cleavages will also complicate the picture. Any peptide that contains such alterations or arise from spurious proteolysis will fall outside the theoretical values of the peptides that are computer generated from the database and the coverage (percentage of the complete sequence accounted for in the experimental data set) of the protein will drop. Indeed, one must recover a sizeable portion of the peptides in order that the fit will be statistically significant and this becomes increasingly demanding as databases expand because the number of possible candidates also rises. At the same time, sequence errors in the database will also produce damaging mismatches that will affect the accuracy of such identifications, which unfortunately is a significant problem.

The second method, which generally employs tandem MS (MS/MS) in various configurations, depends on collusion-induced dissociation (CID) events to produce a series of staggered ions that correspond to derivatives of a selected peptide (separated in the initial MS run) that are systematically foreshortened from one end or the other (and then identified in the second MS run). The most useful ions, designated b and y, result from cleavages at the peptide bonds (although the other two types of bonds found in the polypeptide backbone can also be broken as well) and their masses differ successively by the loss of the previous amino acid residue (from one side or the other). Thus a set of b- or y-ions can be directly 'translated' into the sequence of the original peptide. This is clearly far more useful information in making an identification since in theory a sequence of ~10 amino acids should allow a positive identification, particularly if there is added knowledge such as the species of origin. However, in practice, one needs at least two or more sequences from the same protein to be certain of the identification because of a variety of complicating factors (related to converting real data, which is often incomplete, into a theoretical match). This technology is rapidly becoming the method of choice for protein identifications, both for the level of information that it provides and the fact that it is, as noted above, highly adaptable to on-line sampling through direct coupling of liquid chromatography (LC) to the MS/MS instrument.

Both of these methods require facile interaction with and interrogation of protein databases. In the first place, the conversion of the MS/MS data, called *de novo* sequence interpretation, realistically requires computerized algorithms and search programs since the amount of data that a typical LC MS/MS experiment can generate can run to thousands of spectra. One can interpret MS/MS spectra by hand and then run a BLAST search to make the identification but this is a reasonable approach only if a few spectra are involved, as might be the case for identifying a spot on a 2D gel. For dealing with the large-scale experiments, there are several programs, some of which are publicly available and some of which are not. Some were developed for different MS applications but essentially all can now interpret MS/MS data. There is also related programs available to search for specific features in MS/MS data, such as tyrosine phosphorylation, that are useful, for example, in expression proteomics (Liebler, 2002).

Protein identification permeates all of proteomics and will continue to do so for some time to come. The technology continues to improve and the magnitude and concentration range of samples that can be effectively analyzed has constantly increased for the past decade. There is, however, a long way to go. Consider for example the plasma proteome, long considered a potentially valuable and quite underutilized source of diagnostic information for human disease (Anderson & Anderson, 2002). Although not without its own dynamic aspects and evolving composition, its content is certainly more static than, for instance, the corresponding lysate of a cell. As such, it is clearly a more tractable target. At the same time it is derived from an all-pervasive fluid and as such likely contains products, remnants and detritus indicative of normal and abnormal situations and conditions. The bad news is that that creates an enormous concentration range, and probably number, of analytes. Thus, while the ability to completely define the plasma protein and to facilely assay it would potentially allow physicians to diagnose a far broader spectrum of pathologies than is currently possible, such tests remain a presently unrealized dream. This challenge, among many others like it, is a substantial impetus to continue to develop protein identification methodologies (Petricoin *et al.*, 2002).

3.3   Protein-protein Interactions

Although the presence of such a substantial number of undefined proteins (in terms of both structure and function) in both simple and complex organismal proteomes was not entirely unexpected, it certainly stimulated (and still does) the imagination and enthusiasm of life scientists. If anything, the revelation of the extent to which intracellular protein-protein interactions occur has done so to an even greater extent. It is perhaps the greatest surprise that proteomics has produced to date.

Protein chemists have long been aware that most intracellular enzymes are oligomeric (dimers and tetramers being strongly favored) but there was little evidence from earlier studies for extensive interactions beyond that. However, to be fair, the types of experiments and the questions being asked then did not lend themselves, for the most part, to detecting them. With the availability of recombinant proteins in the 1980's, this scenario started to change and in the early 1990's, for example, signal transduction experiments began to show the formation of substantial complexes induced by and associated with activated receptors. Similar observations were made regarding the transcriptional machinery as well as many other cellular complexes. These experiments often utilized antibodies to precipitate a putative member of a complex and were then analyzed, after separation on gels, by additional immuno-reagents in Western blot format. Increasingly MS was also used to identify the partners as new entities, for which antibodies were not available at the time, were discovered. These "pulldowns" were quite instructive and a goodly number of the known signal transducers (scaffolds, adaptors and effectors), for example, were so identified. Immunological recognition still remains a powerful tool in many proteomic applications. However, it was not until more non-specific methodology was employed that the vast scope and range of intracellular protein interactions became evident. This began with yeast 2-hybrid analyses (Fields & Song, 1989) and then branched out into MS-tag methods (where a member of the complex is labeled or "tagged" with an entity that can be detected by an antibody or other affinity reagent) (Aebersold & Mann, 2003); both were soon adapted to high-throughput (HTP) capability that substantially ratcheted up the number of interactions identified. The results have been impressive although in no case are they complete.

Although these kinds of studies have been applied to a number of research paradigms, the most extensive analyses have been made with yeast. The resultant development of the yeast protein interaction map is well illustrated by four seminal articles (Gavin *et al.*, 2002, Ho *et al.*, 2002, Ito *et al.*, 2001, Uetz *et al.*, 2000) (and a fifth that summarizes the collected findings of these and related studies (von Mering *et al.*, 2002)). The first two (Ito *et al.*, 2001, Uetz *et al.*, 2000) described the application of scaled up 2-hybrid analyses, accounting collectively for over 5000 binary interactions, while the latter two used two forms of MS-tag technology that, when reduced to binary interactions, accounted for about 18,000 and 33,000 assignments (von Mering *et al.*, 2002). The first of these (Gavin *et al.*, 2002) used tandem-affinity purification (TAP); the second (Ho *et al.*, 2002) was based on the expression of a large number of tagged 'bait' proteins expressed in yeast and was called high-throughput mass spectrometric protein complex identification (HMS-PCI). It is important to note that the latter two methods are meant to trap and identify larger complexes containing multiple partners while the 2-hybrid method is designed to identify only interacting pairs. In comparing these results, von Mering et al. (von Mering *et al.*, 2002) also included interactions detected indirectly, i.e. from gene expression ("chip") and genetic lethality data, and predicted from various genomic analyses (Eisenberg *et al.*, 2000). They were all compared against a reference standard of some 10,900 interactions culled from 'manually curated catalogues' that were derived from the Munich Information Center for Protein Sequences and the Yeast Proteome Database. They conclude that all the data sets are incomplete and that there is substantial numbers of false positives, particularly among the identifications that link proteins in different functional categories and/or are identified by only one method. Importantly the interactions that are common to the various high-throughput data sets, which they regard as reliable positives, represent only about a third of the known (reference set). Thus, they conclude that a minimum number of interactions in yeast would be three times the known number or about 30,000 and that the actual number may be many times this due to external stimuli or other physiological change.

These values and conclusions are both exciting and disappointing. On the one hand, they suggest that the copious number of interactions already observed is representative of a detailed intracellular protein network through which metabolic and informational fluxes flow to support the essential life processes.   However, on the other hand, they clearly indicate the inadequacies of the present methodologies. Although these HTP methods have produced masses of data, large parts of it are apparently spurious or at least questionable and, perhaps even more troubling, they are not detecting even the majority of presumably relevant interactions. Clearly there is a long way to go in this area as well.

4.   ARRAYS, CHIPS AND HIGH-THROUGHPUT - IS BIGGER BETTER ?

Regardless of the nature of the proteome, proteomic analyses are inherently large in scope (Table 2) but at the same time offer the opportunity to look at biology (or biological systems) on a different and more complex level (Phizicky *et al.*, 2003). However, size also offers challenges and these have been met, at least initially, by the introduction of HTP methods (or ones that hold the promise to be adaptable to HTP). Chief among these methods are arrays, ensembles of molecules of different types that can be built with robots and interrogated with large-scale query sets of various types. Indeed, some consider that arrays are proteomics (Petach & Gold, 2002) and apparently leave every thing else for (modern?) protein chemistry.

Intellectually, protein arrays are a direct extension of nucleic acid arrays (or chips) that display cDNAs (or corresponding oligonucleotides) derived from a germane library that can be probed with mRNA prepared from cells that have been stimulated in some fashion. This has become the premier way to examine gene expression (Shalon *et al.*, 1996) and has been useful in proteomics to compare transcript and protein levels and to identify protein-protein interactions among other things. In the main its value is to reveal patterns that can be related to the causative stimulus or underlying pathology of the test cell and it has shown some diagnostic potential. Protein arrays are, however, more complex. They can be built by immobilizing either the test molecules, sometimes designated analyte specific reagents (ASR) (Petach & Gold, 2002), or the proteins to be tested. The means of attachment also vary tremendously reflecting the greater difficulty in attaching native proteins and maintaining their three-dimensional structure throughout the analyses. ASRs include such moieties as antibodies (Hanash, 2003), peptide and nucleic acid aptamers (Geyer & Brent, 2000, Petach & Gold, 2002), and protein domains (Nollau & Mayer, 2001), e.g., the SH2 domain that recognizes phosphotyrosine-containing sequences in signal transduction systems. Protein arrays reflecting a (nearly) whole proteome have also been prepared and these can be used to identify function, protein-protein interactions, etc. (Grayhack & Phizicky, 2001, Zhu *et al.*, 2001). New and inventive protein arrays appear with regularity and their use will be a major part of tomorrow's proteomics.

As exciting as these approaches are, and the volume of new information they produce notwithstanding, they are neither comprehensive nor error free. Indeed, as with the TAP and HMS-PCI data (see Section 3.3), it is not until the data have been substantially filtered and reconfirmed (by independent means) do they begin to become really reliable. That seems to be something of a norm for all HTP experiments. Thus, one might reasonably ask if trying to examine too bigger picture may not be at least in part counter productive. The response to this really depends on the objective of the experiment. If large-scale screening to find new drug targets is the purpose, any identification will ultimately be checked and cross-checked during the target validation phase. However, how are the vast numbers of protein-protein interactions in say a HMS-PCI experiment to be checked? Ultimately each will have to be verified (and undoubtedly will be) but almost certainly by a process that is slower, more exact but more time consuming. Large-scale array experiments have and will continue to provide ever broader views into biology but their contribution to the fine details will, at least for the moment, likely be considerable less.

5.   WITHER PROTEOMICS ?

Few fields have emerged with as much promise as proteomics, and with the possible exception of genomics, as much hype (Cohen, 2001). Even casual assessments have suggested that neither biology nor medicine will ever be quite the same again (as proteomics takes root) and that is pretty strong stuff. However, it is not necessarily wrong or even exaggerated. The problem of course is not in the field, for it is easy to realize that if the proteomic definitions given in Table 1 are even modestly achieved the effects on animal and human health care, food production and world ecology will be enormous, but rather in our ability to make the needed measurements. The problems are partly technological and partly intellectual. Chief among the technical problems is the dynamic (concentration) range of samples, quantification (or lack thereof), and information (overload) management. However, it may be expected that in time these and other technological hurdles will be over come; one cannot be quite so sure about the intellectual barriers.

Some insight into this latter issue comes from a comparison of the manned space program of NASA, which first placed a human on the moon in 1969, and the determination of the human genome, which was completed in the last year (or so). When the decision to reach both of these goals was taken, the only significant obstacles were technological and in both cases highly inventive minds produced the needed breakthroughs. As a result, both stand as monumental human achievements. However, the promises of the manned space program have not continued to live up to the initial event in terms of accomplishment or return of new knowledge. That is not to say that space studies have not yielded nor will continue to yield important advances; it is only the gap between expectation and reality that is disappointing. Perhaps colonies on the moon or manned exploration of Mars were not really realistic so quickly but such thoughts were certainly entertained and, more importantly, expressed after the first lunar landing. Since in this analogy functional genomics represents the 'lunar colonies', i.e. the long-term promise, and since proteomics is substantially synonymous with functional genomics, it follows that the burden of capitalizing on the human genome project falls largely to proteomics. And this is where the 'intellectual' side appears. NASA has suffered enormously from all manner of over-analyses, priority changes and shifting political winds and agendas, in addition to continued daunting technical challenges, and these quite human, but largely non-scientific activities have sapped its resources and opportunities. The same could certainly happen to proteomics. A few concerns are illustrative.

Much of proteomics, and indeed much of its lure, lies in the fact that so much of the proteome - any proteome - is unknown. However, mapping unknown functions and elucidating new protein-protein interactions is not hypothesis driven research and funding agencies, particularly in the US, have pretty much eschewed anything that is not. Perhaps because discovery research has been largely the province of industry, particularly the pharmaceutical industry, grant reviewers deem it somehow an inappropriate approach. Just how these same people think that these data, which are crucial for all proteomic research, and arguably all biological research, are going to be obtained remains a mystery although it seems likely that at least some of the information will be derived from industrial research (in part because academic laboratories will be shut out of this activity, particularly smaller ones) (Cohen, 2001). This will contribute to a related problem, to wit the control of intellectual property and the need to place proteomic data in the public domain. There are no easy solutions to this complex problem and it promises to be a significant deterrent. An internationally funded public human proteome project would be a very worthy investment for the world's governments. It would have to have more tightly defined objectives than the global definitions (Table 1) given above but just coordinating the various activities underway would be a good first step (Tyers & Mann, 2003).

Nowhere is the promise of proteomics greater than in clinical applications (Hanash, 2003). The ensuing chapters of this volume discuss these at length, illustrating the degree to which proteomics is already contributing to medical care and providing signposts for the future directions. As basic research, drug discovery and diagnostic and therapeutic applications coalesce, the true potential of proteomics will become clear. Let us hope that it is of the magnitude and significance predicted in the following pages.

ACKNOWLEDGEMENTS

The author wishes to thank Ms. Erin Schalles for reading the manuscript and providing expert editorial advice.

REFERENCES

Aebersold, R., and Mann, M., 2003, Mass spectrometry-based proteomics. *Nature* **422:** 198-207.

Anderson, N. L., and Anderson, N. G., 2002, The human plasma proteome: history, character, and diagnostic prospects. *Mol. Cell. Proteomics* **1:** 845-67.

Blackstock, W. P., and Weir, M. P., 1999, Proteomics: quantitative and physical mapping of cellular proteins. *Trends Biotechnol.* **17:** 121-7.

Bradshaw, R. A., and Dennis, E. A., 2003, *Handbook of Cell Signaling*, Elsevier Academic Press, San Diego.Vol. 1-3.

Cohen, J., 2001, The Proteomics Payoff. *Tech. Rev.* **104:** 55-63.

Eisenberg, D., Marcotte, E. M., Xenarios, I., and Yeates, T. O., 2000, Protein function in the post-genomic era. *Nature* **405:** 823-6.

Fields, S., and Song, O., 1989, A novel genetic system to detect protein-protein interactions. *Nature* **340:** 245-246.

Gavin, A. C., Bosche, M., Krause, R., Grandi, P., Marzioch, M., Bauer, A., Schultz, J., Rick, J. M., Michon, A. M., Cruciat, C. M., Remor, M., Hofert, C., Schelder, M., Brajenovic, M., Ruffner, H., Merino, A., Klein, K., Hudak, M., Dickson, D., Rudi, T., Gnau, V., Bauch, A., Bastuck, S., Huhse, B., Leutwein, C., Heurtier, M. A., Copley, R. R., Edelmann, A., Querfurth, E., Rybin, V., Drewes, G., Raida, M., Bouwmeester, T., Bork, P., Seraphin, B., Kuster, B., Neubauer, G., and Superti_Furga, G., 2002, Functional organization of the yeast proteome by systematic analysis of protein complexes. *Nature* **415:** 141-7.

Geyer, C. R., and Brent, R., 2000, Selection of genetic agents from random peptide aptamer expression libraries. *Meth. Enzymol.* **328:** 171-208.

Grayhack, E., and Phizicky, E., 2001, Genomic analysis of biochemical function. *Curr. Opin. Chem. Biol.* **5:** 34-39.

Hanash, S., 2003, Disease proteomics. *Nature* **422:** 226-32.

Hanash, S., 2003, Harnessing immunity for cancer marker discovery. *Nat. Biotechnol.* **21:** 37-8.

Ho, Y., Gruhler, A., Heilbut, A., Bader, G. D., Moore, L., Adams, S. L., Millar, A., Taylor, P., Bennett, K., Boutilier, K., Yang, L., Wolting, C., Donaldson, I., Schandorff, S., Shewnarane, J., Vo, M., Taggart, J., Goudreault, M., Muskat, B., Alfarano, C., Dewar, D., Lin, Z., Michalickova, K., Willems, A. R., Sassi, H., Nielsen, P. A., Rasmussen, K. J., Andersen, J. R., Johansen, L. E., Hansen, L. H., Jespersen, H., Podtelejnikov, A., Nielsen, E., Crawford, J., Poulsen, V., Sorensen, B. D., Matthiesen, J., Hendrickson, R. C., Gleeson, F., Pawson, T., Moran, M. F., Durocher, D., Mann, M., Hogue, C. W., Figeys, D., and Tyers, M., 2002, Systematic identification of protein complexes in Saccharomyces cerevisiae by mass spectrometry. *Nature* **415:** 180-3.

Huber, L. A., 2003, Is proteomics heading in the wrong direction? *Nat. Rev. Mol. Cell. Biol.* **4:** 74-80.

Ito, T., Chiba, T., Ozawa, R., Yoshida, M., Hattori, M., and Sakaki, Y., 2001, A comprehensive two-hybrid analysis to explore the yeast protein interactome. *Proc. Natl. Acad. Sci. U.S.A.* **98:** 4569-74.

Klose, J., 1975, Protein mapping by combined isoelectric focusing and electrophoresis of mouse tissues. A novel approach to testing for induced point mutations in mammals. *Humangenetik* **26:** 231-43.

Liebler, D. C., 2002, Introduction to Proteomics, Humana Press, Totowa, NJ.

Nollau, P., and Mayer, B. J., 2001, Profiling the global tyrosine phosphorylation state by Src homology 2 domain binding. *Proc. Natl. Acad. Sci. U. S. A.* **98:** 13531-36.

O'Farrell, P. H., 1975, High resolution two-dimensional electrophoresis of proteins. *J. Biol. Chem.* **250:** 4007-21.

Petach, H., and Gold, L., 2002, Dimensionality is the issue: use of photoaptamers in protein microarrays. *Curr. Opin. Biotech.* **13:** 309-314.

Petricoin, E. F., Zoon, K. C., Kohn, E. C., Barrett, J. C., and Liotta, L. A., 2002, Clinical proteomics: translating benchside promise into bedside reality. *Nat. Rev. Drug Discov.* **1:** 683-95.

Phizicky, E., Bastiaens, P. I., Zhu, H., Snyder, M., and Fields, S., 2003, Protein analysis on a proteomic scale. *Nature* **422**: 208-15.

Sali, A., Glaeser, R., Earnest, T., and Baumeister, W., 2003, From words to literature in structural proteomics. *Nature* **422**: 216-25.

Shalon, D., Smith, S. J., and Brown, P. O., 1996, A DNA microarray system for analyzing complex DNA samples using two-color fluorescent probe hybridization. *Genome Res.* **6**: 639-645.

Tyers, M., and Mann, M., 2003, From genomics to proteomics. *Nature* **422**: 193-7.

Uetz, P., Giot, L., Cagney, G., Mansfield, T. A., Judson, R. S., Knight, J. R., Lockshon, D., Narayan, V., Srinivasan, M., Pochart, P., Qureshi-Emili, A., Li, Y., Godwin, B., Conover, D., Kalbfleisch, T., Vijayadamodar, G., Yang, M., Johnston, M., Fields, S., and Rothberg, J. M., 2000, A comprehensive analysis of protein-protein interactions in Saccharomyces cerevisiae. *Nature* **403**: 623-7.

von Mering, C., Krause, R., Snel, B., Cornell, M., Oliver, S. G., Fields, S., and Bork, P., 2002, Comparative assessment of large-scale data sets of protein-protein interactions. *Nature* **417**: 399-403.

Wolters, D. A., Washburn, M. P., and Yates, J. R., III, 2001, An automated multidimensional protein identification technology for shotgun proteomics. *Anal. Chem.* **73**: 5683 -5690.

Yates, J. R., 1998, Mass spectrometry and the age of the proteome. *J. Mass. Spect.* **33**: 1-19.

Zhu, H., Bilgin, M., Bangham, R., Hall, D., Casamayor, A., Bertone, P., Lan, N., Jansen, R., Bidlingmaier, S., Houfek, T., Mitchell, T., Miller, P., Dean, R. A., Gerstein, M., and Snyder, M., 2001, Global analysis of protein activities using proteome chips. *Science* **293**: 2101-5.

Chapter 2

Proteomic Technologies and Application to the Study of Heart Disease

EMMA MCGREGOR[*], and MICHAEL J. DUNN[#]
[*]*Proteome Sciences plc, Institute of Psychiatry, King's College London, London, UK;*
[#]*Department of Neuroscience, Institute of Psychiatry, King's College London, London, UK*

1. INTRODUCTION

In April 2003 we celebrated the 50th anniversary of the publication of the paper by Watson and Crick (Watson and Crick, 1974) describing the double helix of DNA. However, more than 40 years were to pass before the first complete genome sequence, that of the bacterium *Haemophilus influenzae* (Fleischmann *et al.*, 1995), was published. Since then there has been intensive effort in genomic sequencing initiatives which at the time of writing (June 2003) has resulted in the publication of 141 complete genome sequences (16 archael, 106 bacterial, 19 eukaryotic) with more than 700 other genome sequencing projects in progress (GOLD, Genomes OnLine Database, http://igweb.integratedgenomics.com/GOLD/). A major milestone was reached in 2001 with the publication of the draft sequence of the human genome (Lander *et al.*, 2001; Venter *et al.*, 2001). Perhaps surprisingly the human genome contains fewer open reading frames (around 30,000 ORFs) encoding functional proteins than was generally predicted and, like all other completed genomes, contains many "novel" genes with no ascribed functions. Moreover, it is now apparent that one gene does not encode a single protein, because of processes such as alternative mRNA splicing, RNA editing and post-translational protein modification. Therefore the functional complexity of an organism far exceeds that indicated by its genome sequence alone. It is, therefore, clear that the "omic" approaches to the global study of the products of gene expression, including

H. Hondermarck (ed.), Proteomics: Biomedical and Pharmaceutical Applications, 19–55.

transcriptomics, proteomics and metabolomics, will play a major role in elucidating the functional role of the many novel genes and their products and in understanding their involvement in biologically relevant phenotypes in health and disease.

Powerful techniques such as DNA micro-arrays and serial analysis of gene expression (SAGE) facilitate rapid, global transcriptomic profiling of mRNA expression. However, there is often a poor correlation between mRNA abundance and the quantity of the corresponding functional protein present within a cell (Anderson and Seilhamer, 1997; Gygi *et al.*, 1999). In addition concomitant co- and post-translational modification (PTM) events can result in a diversity of protein products from a single open reading frame (Gooley and Packer, 1997). These modifications can include phosphorylation, sulphation, glycosylation, hydroxylation, *N*-methylation, carboxymethylation, acetylation, prenylation and *N*-myristolation. Moreover, protein maturation and degradation are dynamic processes that can control the amount of functionally active protein within a cell.

These arguments have been generally accepted as providing a compelling justification for direct and large scale analysis of proteins. The concept of mapping the human complement of protein expression was first proposed more than twenty years ago (Anderson *et al.*, 2001; Anderson and Anderson, 1982) with the development of a technique where large numbers of proteins could be separated simultaneously by two-dimensional polyacrylamide gel electrophoresis (2-DE) (Klose, 1975; O'Farrell, 1975). However, it was not until 1995, that the term 'proteome', defined as the protein complement of a genome, was first coined by Wilkins working as part of a collaborative team at Macquarie (Australia) and Sydney Universities (Australia) (Wasinger *et al.*, 1995; Wilkins *et al.*, 1996).

In this chapter we will review proteomic investigations of cardiac proteins and focus on their application to the study of heart disease in the human and in animal models of cardiac dysfunction. The majority of these studies of the cardiac proteome have involved protein separation, visualisation and quantitation using the traditional 2-DE approach combined with protein identification by mass spectrometry. These essential technologies will be briefly described. However, there is increasing interest in using alternative gel-free techniques based on mass spectrometry or protein arrays for high throughput proteomics. These alternative approaches will be introduced, but further details can be found in Chapter 2 of this volume by Michel Faupel.

2. PROTEOMIC TECHNOLOGIES

2.1 Sample Preparation

Sample preparation is perhaps the most important step in a proteomics experiment as artefacts introduced at this stage can often be magnified with the potential to impair the validity of the results. No single method for sample preparation can be applied universally due to the diverse nature of samples that are analysed by 2-DE (Dunn and Gorg, 2001), but some general considerations can be mentioned. The high resolution capacity of 2-DE is exquisitely able to detect subtle post-translational modifications such as phosphorylation, but it will also readily reveal artefactual modifications such as protein carbamylation that can be induced by heating of samples in the presence of urea. In addition, proteases present within samples can readily result in artefactual spots, so that samples should be subjected to minimal handling and kept cold at all times, and it is possible to add cocktails of protease inhibitors.

2.2 Solubilisation

The ideal solubilisation procedure for 2-DE would result in the disruption of all non-covalently bound protein complexes and aggregates into a solution of individual polypeptides (Dunn and Gorg, 2001). If this is not achieved, persistent protein complexes in the sample are likely to result in new spots in the 2-D profile, with a concomitant reduction in the intensity of those spots representing the single polypeptides. In addition, the method of solubilisation must allow the removal of substances, such as salts, lipids, polysachharides and nucleic acids that can interfere with the 2-DE separation. Finally, the sample proteins must remain soluble during the 2-DE process. For the foregoing reasons sample solubility is one of the most critical factors for successful protein separation by 2-DE.

The most popular method for protein solubilisation for 2-DE remains that originally described by O'Farrell (O'Farrell, 1975), using a mixture of 9.5 M urea, 4% w/v of the non-ionic detergent NP-40 or the zwitterionic detergent CHAPS (3[(cholamidopropyl)dimethylammonio]-1-propane sulphonate), 1% w/v of the reducing agent dithiothreitol (DTT) and 2% w/v of synthetic carrier ampholyte of the appropriate pH range (so-called "lysis buffer"). While this method works well for many types of sample, it is not universally applicable, with membrane proteins representing a particular challenge (Santoni *et al.*, 2000). Variations in solubilization buffer constituents using newly developed detergents such as sulfobetaines (Luche *et al.*, 2003), additional denaturing agents such as thiourea (Rabilloud, 1998), and alternative reducing agents, e.g. trubutyl phosphine (Herbert *et al.*, 1998), can help to improve protein solubilization and hence the concentration of extracted protein for certain sample types. It must be stressed that the choice of solubilisation buffer must be optimised for each sample type to be analysed by 2-DE; see Stanley et al, 2003 (Stanley *et al.*, 2003) for an example of optimising the solubilisation of human myocardium.

2.3   Two-Dimensional Gel Electrophoresis (2-DE)

2-DE involves the separation of solubilised proteins in the first dimension according to their charge properties (isoelectric point, p*I*) by isoelectric focusing (IEF) under denaturing conditions, followed by their separation in the second dimension according to their relative molecular mass (M$_r$) by sodium dodecyl sulphate polyacrylamide gel electrophoresis (SDS-PAGE). As the charge and mass properties of proteins are essentially independent parameters, this orthogonal combination of charge (p*I*) and size (M$_r$) separations results in the sample proteins being distributed across the two-dimensional gel profile (Fig. 1).

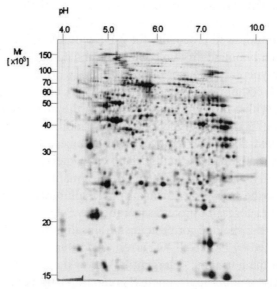

Figure 1. A two-dimensional electrophoresis (2DE) separation of 80 μg of heart (ventricle) proteins. The first dimension comprised an 18 cm non-linear pH 3-10 immobilised pH gradient (IPG) subjected to isoelectric focusing. The second dimension was a 21 cm 12% SDS-PAGE (sodium dodecylsulphate polyacrylamide gel electrophoresis) gel. Proteins were detected by silver staining. The non-linear pH range of the first-dimension IPG strip is indicated along the top of the gel, acidic pH to the left. The M$_r$ (relative molecular mass) scale can be used to estimate the molecular weights of the separated proteins.

Although the basic technique of 2-DE has a history spanning more than 25 years (Klose, 1975; O'Farrell, 1975; Scheele, 1975), it remains the core technology of choice for separating complex protein mixtures in the majority of proteome projects (Rabilloud, 2002). This is due to its unrivalled power to separate simultaneously thousands of proteins, to indicate post-translational modifications that result in alterations in protein p*I* and/or M_r, the subsequent high-sensitivity visualisation of the resulting 2-D separations that are amenable to quantitative computer analysis to detect differentially regulated proteins, and the relative ease with which proteins from 2-D gels can be identified and characterised using highly sensitive microchemical methods. Developments over the last few years that have resulted in the current 2-DE method that combines increased resolving power and high reproducibility with relative simplicity of use will be briefly described here. Further details can be found in recent reviews (Dunn and Gorg, 2001; Gorg *et al.*, 2000; Patton *et al.*, 2002; Rabilloud, 2002).

To be effective and accurate 2-DE must produce highly reproducible protein separations and have high resolution. This has been achieved in recent years by the use of immobilized pH gradients (IPG) (Amersham Biosciences) generated using immobiline reagents (Bjellqvist *et al.*, 1982) to replace the synthetic carrier ampholytes (SCA) previously used to generate the pH gradients required for IEF. The immobiline reagents consist of a series of eight acrylamide derivatives with the structure $CH_2=CH-CO-NH-R$, where R contains either a carboxyl or tertiary amino group, giving a series of buffers with different p*K* values distributed throughout the pH 3 to 10 range. The appropriate IPG reagents can be added to the gel polymerization mixture according to published recipes. During the polymerization process the buffering groups that form the pH gradient are covalently attached via vinyl bonds to the polyacrylamide backbone. IPG generated in this way are immune to the effects of electroendosmosis that results in cathodic drift with the consequent loss of basic proteins from 2-D gel profiles generated using SCA IEF. IPG IEF separations are extremely stable and generate highly reproducible 2-D protein profiles (Blomberg *et al.*, 1995; Corbett *et al.*, 1994).

The IPG IEF dimension of 2-DE is performed on individual gel strips, 3-5mm wide, cast on a plastic support. IEF is carried out generally using either an IPGphor (Amersham Biosciences) or Multiphor (Amersham Biosciences). The advantages associated with both pieces of equipment when performing IEF are controversial. By applying a low voltage during sample application by in-gel rehydration (Rabilloud *et al.*, 1994; Sanchez *et al.*, 1997), improved protein entry, especially of high molecular weight proteins has been reported using the IPGphor (Gorg *et al*, 2000). However, an increased number of detectable protein spots in 2D gels has been reported following IEF using the Multiphor compared to the same samples subjected to IEF using the IPGphor (Choe and Lee, 2000). Recently the IPGphor was used for sample loading under a low voltage and the strips were then transferred to the Multiphor for focusing, with an increased sample entry compared to using the Multiphor alone being reported (Craven *et al.*, 2002).

Following steady-state IEF, strips are equilibrated and then applied to the surface of either vertical or horizontal slab SDS-PAGE gels (Gorg *et al.*, 2000). Using standard format SDS-gels (20 x 20 cm) with 18 cm IPG strips it is possible to routinely separate 2000 proteins from whole-cell and tissue extracts. Resolution can be significantly enhanced (separation of 5000-10,000 proteins) using large format (40 x 30 cm) 2D gels (Klose and Kobalz, 1995). However, gels of this size are very rarely used due to the handling problems associated with such large gels. The longest commercial IPG IEF gels have a length of 24 cm (Gorg *et al.*, 1999). In contrast, mini-gels (7x7 cm) can be run using 7 cm IPG strips. Whilst these gels will only separate a few hundred proteins they can be very useful for rapid screening purposes. Second-dimension SDS-PAGE is usually carried out using apparatus capable of running simultaneously multiple large-format 2-D gels (*e.g.* Ettan DALT 2, 12 gels, Amersham Biosciences; Protean Plus Dodeca Cell,12 gels, Bio-Rad). While such equipment allows large numbers of 2-D protein separations to be performed, the procedure is still very time consuming and labour intensive. Automating the process of 2-DE has proved a formidable task and has largely been the province of commercial proteomics companies (*e.g.* the ProGEx™ platform developed by Large Scale Biology Corporation). Recently, however, two commercial systems that will be available to the general proteomic community, or at least those that can afford such a solution, have been announced (ProTeam Suite™, Tecan; a2DE™, NextGen Sciences).

2.4   Increasing Proteomic Coverage by 2-DE

While large format 2-D gels using wide pH 3-10 gradients are able to separate around 2,000 proteins from a complex sample such as a total myocardial protein lysate (Fig. 1), it is clear that this gives incomplete proteomic coverage for tissues such as the human heart which may be expressing more than 10,000 proteins. This inability to cope with the enormous diversity of cellular proteins often results in several protein species co-migrating in the same spot on a 2-D gel (Gygi *et al.*, 2000). However, intermediate (*e.g.* pH 4-7, 6-9) and narrow (*e.g.* pH 4.0-5.0, 4.5-5.5, 5.0-6.0, 5.5-6.7) range IPG IEF gels are now available and have the capability of 'pulling apart' this protein profile and increasing the resolution in particular regions (Fig. 2).

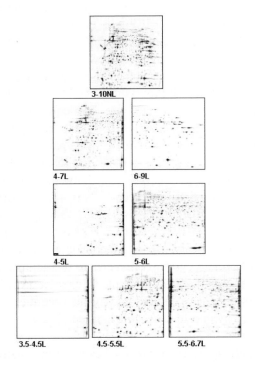

Figure 2. Zoom gels: increasing proteomic coverage using narrow-range pH gradients. Solubilised human heart (ventricle) proteins were separated in the first dimension using an 18 cm non-linear pH 3-10 immobilised pH gradient (IPG). In order to separate this sample still further, very narrow-range pH gradients have been used (e.g. pH 4-7L, 6-9L, 4-5L, 5-6L, 3.5-4.5L, 5.5-6.7L). The second dimension for all gels was a 21 cm 12% SDS-PAGE (sodium dodecylsulphate polyacrylamide gel electrophoresis) gel. Proteins were detected by silver staining.

This 'zoom gel' approach results in enhanced proteomic coverage (Westbrook *et al.*, 2001; Wildgruber *et al.*, 2000), but at the expense of increased workload if large numbers of samples are to be investigated. An important additional advantage of narrow range IPGs is that they can tolerate higher protein loading for micro-preparative purposes.

2.5   2-DE of basic proteins

Basic proteins have always represented a particular challenge to satisfactory separation by 2-DE. Cathodic drift in SCA IEF generally results in loss of most proteins with p*I* values greater than pH 7. Loss of these proteins is minimised using IPG IEF, but there is still a tendency for these basic proteins to form streaks rather than discrete spots. This is evident using wide range pH gradients (*e.g.* see the pH 7-10 region of Fig. 1), but is exacerbated if narrow range IPG IEF gels in the range from pH 6 to pH 12 are used. Streaking is a consequence of the loss of reducing agent from the basic region of the strip with consequent oxidation of the protein thiol groups resulting in inter- and intra-chain S-S bonds. Reducing agents such as dithiothreitol (DTT) and dithioerythritol (DTE) are weak acids with pK values in the region pH 8-9 and are consequently electrophoretically transported out of basic region of the IPG strip during IEF. This can be partially overcome using an additional electrode paper wick at the cathode as a source of DTT during IEF and can be combined with the inclusion of glycerol and isopropanol in the IPG rehydration solution to suppress the electroendosmotic flow that exacerbates streaking (Gorg *et al.*, 1995; Hoving *et al.*, 2002). Recently it has been found that streaking can be minimised by running IPG IEF gels in an excess of hydroxyethyldisulphide (HED, available as DeStreak^{TM} from Amersham Biosciences) which results in the oxidation of protein cysteinyl residues to mixed disulphides (Olsson et al., 2002). Streaking of basic proteins is even more pronounced when preparative protein loads are applied and in this case it is recommended that samples are applied by cup-loading at the anode.

2.6   Increasing Proteomic Coverage by Cellular, Sub-Cellular and Protein Fractionation

A major limitation to the capacity of 2-DE to display complete proteomes is the very high dynamic range of protein abundance, estimated at 10^6 for cells and tissues (Corthals *et al.*, 2000) and 10^{12} for plasma (Anderson and Anderson, 2002). This is beyond the dynamic range of 2-DE, with an estimated maximum dynamic range of 10^4 (Rabilloud, 2002). Clearly, the development of reliable and reproducible prefractionation strategies will be essential to access and retrieve quantitative data for the less abundant proteins present in biological samples such as the heart.

Methods for cell and sub-cellular fractionation include immuno-isolation, electromigration (*e.g.* free flow electrophoresis), flow cytometry, density gradient isolation of organelles and sequential extraction. The use of some of these approaches to investigate the mitochondrial and myofibrillar sub-proteomes of the heart will be described in Sections 3.4 and 3.6 respectively. A particular problem in proteomic analysis of the heart is the diversity of cell types that are present. Proteomic profiles of total myocardial lysates are dominated by the proteins present in cardiac myocytes, but such samples will also contain lower amounts of proteins derived other cell types including fibroblasts, smooth muscle and endothelial cells. Recently we have started to apply the technique of laser capture microdissection (LCM) in which a laser beam is used to isolate specific regions of interest from microscope sections of tissue. While this technique generally results in the isolation of relatively small amounts of material, it has been shown to be possible to perform proteomic studies of the resulting protein samples (Banks *et al.*, 1999; Craven and Banks, 2002; Jain, 2002). In preliminary studies we have been able to generate sufficient material by LCM of human cardiac tissue sections to produce large-format 2-D gels of proteins from isolated cardiac myocytes and microvessels (Fig. 3) (unpublished data) (De Souza, 2003).

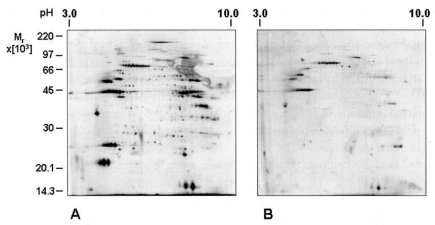

Figure 3. Laser capture microdissection (LCM) of human cardiac tissue. *A.)* Cardiac myocytes; *B.)* Blood vessels.

The alternative approach is to subfractionate the sample at the level of the solubilised proteins. These approaches include electrophoretic separation of the proteins in solution by techniques including continuous free flow electrophoresis (Hoffmann *et al.*, 2001), recycling IEF (Hesse *et al.*, 2001) and IEF using multi-compartment electrolysers (Zuo and Speicher, 2000). The latter approach seems particularly promising for use in conjunction with narrow range pH gradient IPG IEF for 2-DE (Zuo *et al.*, 2001). Protein mixtures can also be subfractionated by traditional chromatographic techniques such as ion exchange, size exclusion, chromtofocusing and affinity chromtograpy. The latter approach has recently been developed as a powerful immunoaffinity subtraction method to deplete human plasma of 14 of the most abundant proteins (Pieper *et al.*, 2003). This then allowed new proteins to be detected by 2-DE, with 3,000 spots representing 300 proteins being identified (Pieper, 2003). Finally, interacting protein complexes can be isolated if one or more components of the complex is known (or suspected) and can be used in conjunction with a specific antibody to isolate the "bait" protein together with the proteins forming the complex (Figeys *et al.*, 2001). These proteins can then be separated by any suitable method such as SDS-PAGE or 2-DE and subsequently identified.

2.7 Alternatives to 2-DE

The limitations inherent in the 2-DE approach have resulted in significant efforts to develop alternative approaches that do not depend on 2-DE, and in particular avoid the use of IEF with its attendant problems of protein solubility. The simplest approach is the use of one-dimensional SDS-PAGE in conjunction with protein identification by MS/MS, so that several proteins co-migrating in a single band can be identified. However, this method is limited by the complexity of the protein mixture that can be analysed and has been most successfully applied to the study of protein complexes that can be isolated by techniques such as immunoprecipitation (Figeys and Pinto, 2001). Other approaches have been designed to dispense with the need for a gel-based separation and to take advantage of the automation, throughput and sensitivity that can be provided by techniques based on a combination of liquid chromatography (LC) and MS. In these so-called "shotgun" approaches, the complex protein sample is initially digested with an enzyme (typically trypsin). The resulting peptide fragments are then separated by one or more dimensions of LC to reduce the complexity of peptide fractions that are then introduced (either on- or off-line) into a tandem mass spectrometer for sequence-based identification. For example, the so-called "MudPIT" approach of Yates (Wolters *et al.*, 2001) combining multidimensional liquid chromatography, tandem mass spectrometry, and database searching has been shown to be capable of detecting and identifying around 1,500 yeast proteins in a single analysis (Washburn *et al.*, 2001). A single dimension LC separation, combined with the very high resolution of a Fourier-transform ion cyclotron resonance (FTICR) MS was able to identify nearly 1,900 proteins (>61% of the predicted proteome) of the bacterium, *Deinococcus radiourans* (Smith *et al.*, 2002). While these approaches are certainly very powerful, it must be realised that they generate raw lists of proteins present in a sample. They do not give any information on the relative quantitative abundance of the individual protein components and give no indication as to whether the proteins identified are subject to post-translational modification.

This problem is currently being addressed by the development of MS-based techniques in which stable isotopes are used to differentiate between two populations of proteins. This approach consists of four steps: (1) differential isotopic labelling of the two protein mixtures, (2) digestion of the combined labelled samples with a protease such as trypsin or Lys-C, (3) separation of the peptides by multidimensional LC, and (4) quantitative analysis and identification of the peptides by MS/MS. The most widely used method which is now commercially available is the isotope-coded affinity tag (ICAT) method based on the labelling of cysteine residues (Gygi *et al.*, 1999), limiting its application to proteins containing such residues. Recently there has been a proliferation of different isotopic labelling methods for MS-based quantitative proteomics (Flory *et al.*, 2002). It seems likely that this group of methods will be increasingly used in proteomic investigations, but their quantitative reproducibility remains to be established. Moreover, the dynamic range of the ICAT technique is no better than 2-DE (Arnott *et al.*, 2002) and there is evidence that it can be complementary to a 2-DE approach in identifying a different subset of proteins from a given sample (Patton *et al.*, 2002). Finally, there is much interest in the development of antibody and protein arrays for quantitative expression profiling (Jenkins and Pennington, 2001; Lal *et al.*, 2002; Mitchell, 2002), but considerable work remains to be carried out before this approach can be routinely used in proteomic investigations.

It is therefore clear that until these alternative approaches mature into robust techniques for quantitative protein expression profiling, 2-DE will remain the separation work-horse in many proteomic investigations. This technique has the capacity to support the simultaneous analysis of the changes in expression of hundreds to thousands of proteins and as such it remains unrivalled as an "open'" protein expression profiling approach.

2.8   Protein Detection and Visualisation

Following separation by electrophoresis, proteins must be visualised at high sensitivity. Appropriate detection methods should therefore ideally combine the properties of a high dynamic range (*i.e.* the ability to detect proteins present in the gel at a wide range of relative abundance), linearity of staining response (to facilitate rigorous quantitative analysis), and if possible compatibility with subsequent protein identification by mass spectrometry. Staining by Coomassie brilliant blue (CBB) has for many years been a standard method for protein detection following gel electrophoresis, but its limited sensitivity (around 100 ng protein) stimulate the development of a more sensitive (around 10 ng protein) method utilising CBB in a colloidal form (Neuhoff *et al.*, 1988). Since its first description in 1979 (Switzer, III *et al.*, 1979), silver staining has often been the method of choice for protein detection on 2-D gels due to its high sensitivity (around 0.1 ng protein). However, silver staining suffers from significant inherent disadvantages; it has a limited dynamic range, it is susceptible to saturation and negative staining effects that compromise quantitation, and most protocols are not compatible with subsequent protein identification by mass spectrometry. This is because glutaraldehyde, included in many protocols as a sensitising reagent, causes extensive cross-linking through reaction with both ε-amino and α-amino groups. To achieve compatibility with MS, glutaraldehyde must be omitted (Shevchenko *et al.*, 1996; Yan *et al.*, 2002b), but at the expense of increased background and reduced sensitivity.

Detection methods based on the post-electrophoretic staining of proteins with fluorescent compounds have the potential of increased sensitivity combined with an extended dynamic range for improved quantitation. The most commonly used reagents are the SYPRO series of dyes from Molecular Probes (Patton, 2000). Extensive studies have been carried out to evaluate the sensitivity of these fluorescent dyes compared with silver staining (Berggren *et al.*, 2002; Lopez *et al.*, 2000). In our laboratory (Yan *et al.*, 2000) we found that SYPRO Ruby had a similar sensitivity to silver staining, but SYPRO Orange and Red were significantly less sensitive, requiring four and eight times more protein to be loaded to achieve equivalent 2-DE protein profiles. In addition, SYPRO resulted in linearity of staining response over a far greater dynamic range than was achieved by silver staining (Yan *et al.*, 2000). Importantly it has been shown that these fluorescent staining methods do not interfere with subsequent protein identification by MS (Lauber *et al.*, 2001; Patton, 2000).

Whilst fluorescent electrophoretic stains have the potential to enhance the sensitivity of detection and provide improved quantitation of proteins on 2-D gels, many pairs of gels are still required in order to establish biologically statistically significant differences between for example a control and disease state. However, the development of two-dimensional difference gel electrophoresis (DIGE) by Unlu et al (Unlu *et al.*, 1997) has now made it possible to detect and quantitate differences between experimentally paired protein samples resolved on the same 2-D gel. DIGE uses two mass- and charge-matched *N*-hydroxy succinimidyl ester derivatives of the fluorescent cyanine dyes Cy3 and Cy5, which possess distinct excitation and emission spectra. Cy3 and Cy5 covalently bind to lysine residues. The two samples that are to be compared are labelled with one of the dyes and equivalent aliquots of each are mixed and the resulting mixture of both samples is then subjected to 2DE (Fig. 4). The resulting 2-D gel is imaged twice using an appropriate fluorescent imaging device (once for the Cy3 dye and once for the Cy5 dye) and then both images are superimposed. The staining process for DIGE takes only 45 minutes which is much faster than other staining methods with similar detection sensitivities such as silver (>2 hours) and SYPRO Ruby (>3 hours). The main advantage to DIGE is the ability to compare two samples on the same gel. This reduces inter-gel variability, decreases the number of gels required by 50% and allows for more accurate and rapid analysis of differences. Improved accuracy of quantitation when multiple pairs of samples are to be compared can be achieved using a pooled internal standard containing all experimental samples labelled with a third dye, Cy2. This internal standard is included with every pair of experimental samples run on each gel (Alban *et al.*, 2003; Yan *et al.*, 2002a) (Fig. 4).

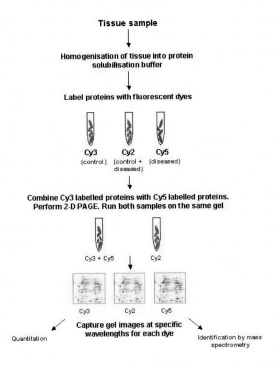

Figure 4. DIGE (differential gel electrophoresis) analysis. A flow diagram illustrating the use of Cy3, Cy5 and Cy2 dyes in the analysis of control versus diseased samples. The Cy3 dye is used to label one protein sample (e.g. control sample), whilst the other is labeled with Cy5 (e.g. diseased/experimental sample). The labeling reaction is terminated and equal amounts of each labeled sample are combined. In parallel, equal concentrations of both control and diseased samples are pooled in one tube and labeled with Cy2. This labeled sample is used for normalisation purposes and acts as an internal standard. Both the mixture of Cy3 and Cy5 labeled proteins and the Cy2 labeled pooled sample are separated on the same gel. Each 2DE protein profile associated with each individual dye is visualized at a specific wavelength.

DIGE does not interfere with subsequent analysis by mass spectrometry, however there are some disadvantages associated with DIGE. The method depends on the fluorescent dye for quantitation. Therefore, carefully controlled labelling is critical for accurate quantitation. The number of lysine residues contained within a particular protein influences the labelling efficiency of that protein; proteins with a high percentage of lysine residues may be labelled more efficiently than proteins with few or no lysine residues. Hence, DIGE cannot detect those proteins containing no lysine residues. Therefore, a protein spot can appear to be highly abundant when silver stained but appear significantly reduced using DIGE.

DIGE can affect protein spot patterns compared to patterns obtained with conventional systems. Proteins are physically labelled, *i.e.*. Cy3 and Cy5 molecules are covalently linked to lysine residues prior to electrophoresis. This has the potential to modify the location of a protein spot within a 2-D gel. The dyes used do not change the pI values of proteins because the dye molecules are positively charged, thereby negating the charge on the lysine group. However, the dye molecule has a molecular weight of 0.5 kDa. An additional 0.5 kDa will not significantly affect the migration of large proteins because only a small percentage of the protein becomes labelled. However this small increase in molecular weight becomes more evident as the molecular weight of proteins decreases.

The labeling reaction is carried out at low efficiency (protein:dye, 5:1) such that at most one lysine residue per protein molecule is conjugated with dye, This minimizes any alteration to the M_r of the protein. Since only a small percentage (typically 2-5%) of a protein becomes labeled with Cy3 or Cy5, most of the protein molecules cannot be visualaised in the Cy3 and Cy5 images. To locate unlabelled protein the 2-D gel can be stained post-DIGE using Sypro Ruby (Gharbi *et al.*, 2002; Zhou *et al.*, 2002). The protein of interest is identified by comparing the Sypro Ruby stained image and the Cy3 and Cy5 images.

2.9   Protein Identification

From the foregoing discussion it is clear that we have a panel of methods that facilitate the detection of proteins that show quantitative changes in relative abundance between the different types of sample being analysed. However, unless we can go on to unequivocally identify these proteins we will not be able to investigate the functional significance of these changes. Conventional methods of identifying proteins from 2-D gels have relied on Western blotting, microsequencing by automated Edman degradation (Patterson, 1994) and amino acid compositional analysis (Yan *et al.*, 1996).

The disadvantage of these techniques is that they are very time consuming with limited sensitivity.

A major breakthrough in the analysis of proteins and peptides came about with the development of sensitive methods based on the use of mass spectrometry. The importance of these developments was recognised in 2002 with award of the Nobel Prize in Chemistry to John Fenn (electrospray (ESI) ionization) and Koichi Tanaka (matrix-assisted laser desorption/ionization (MALDI) ionization) pioneers in the development of methods of ionisation that made protein and peptide MS a practicable procedure. These developments have led to MS being the method of choice for protein identification and characterisation of protein and peptides (Fig. 5).

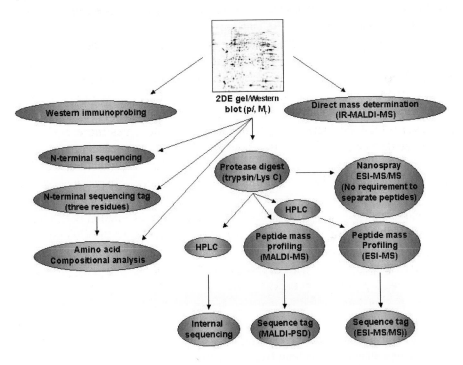

Figure 5. Methods employed for identifiying and characterizing proteins separated by 2DE. Abbreviations used: 2DE, two-dimensional electrophoresis; ESI, electrospray ionization; HPLC, high performance liquid chromatography; IR, infrared; M_r, relative molecular mass; MALDI, matrix-assisted laser desorption ionization; MS, mass spectrometry; MS/MS tandem mass spectrometry; p*I*, isoelectric point; PSD, post-source decay.

These methods are very sensitive, require small amounts of sample (femtomole to attomole concentrations) and have the capacity for high sample throughput (Patterson and Aebersold, 1995; Yates, III, 1998). Both

MALDI and ESI are capable of ionizing very large molecules with little or no fragmentation. The type of analyzer used with each can vary. MALDI sources are usually coupled to time-of-flight (TOF) analyzers, whilst ESI sources can be coupled to analyzers such as quadropole, ion-trap and hybrid quadropole time-of-flight (Q-TOF).

Peptide mass fingerprinting (PMF) is the primary tool for MS identification of proteins in proteomic studies. In this method protein spots of interest are excised and the gel plug containing the spot is digested with a protease, typically trypsin which cleaves polypeptide chains at basic amino acids (arginine/lysine residues).   In this way the intact protein is broken down into a mixture of peptides. The masses of the resulting peptides, or more strictly their mass to charge ratios (m/z), are then measured by mass spectrometry to produce a characteristic mass profile or 'fingerprint' of that protein (Fig. 6).

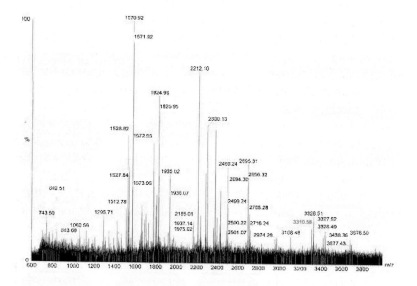

Figure 6. Peptide mass profiling of a silver stained protein spot from a narrow range, pH 4-7, 2DE separation of human heart (ventricle) proteins. A MALDI-TOF mass spectrum of tryptic peptides is shown, analyzed using a Micromass Tofspec 2E spectrometer (Manchester, UK) operated in the positive ion reflectron mode at 20 kV accelerating voltage with "time-lag focusing" enabled. The protein spot of interest was identified as human vimentin, (courtesy of J.A.Westbrook and R. Wait) (unpublished data). 2DE gels were silver stained using a modified Amersham Biosciences kit.

The mass profile is then compared with peptide masses predicted from theoretical digestion of known protein sequences contained within current protein databases or predicted from nucleotide sequence databases. This approach proves very effective when trying to identify proteins from species whose genomes are completely sequenced, but is not so reliable for organisms whose genomes have not been completed. This has been a problem in the past for proteomic studies of some animal models of heart disease, for example those involving rats, dogs, pigs and cows. This problem has been shown to be overcome effectively by improving PMF by adopting an orthogonal approach combined with amino acid compositional analysis (Wheeler *et al.*, 1996).

Nevertheless, it is sometimes impossible to assign an unequivocal identity to a protein based on PMF alone. Peptide sequence information is then required to confirm an identity. This can be generated by conventional automated chemical Edman microsequencing but is most readily accomplished using tandem mass spectrometry (MS/MS). MS/MS is a two-stage process, by means of MALDI-MS with post-source decay (PSD), MALDI-TOF-TOF-MS/MS or ESI-MS/MS triple-quadropole, ion-trap, or Q-TOF machines, to induce fragmentation of peptide bonds. One approach, termed peptide sequence tagging, is based on the interpretation of a portion of the MS/MS or PSD fragmentation data to generate a short partial sequence or 'tag'. Using the 'tag' in combination with the mass of the intact parent peptide ion provides significant additional information for the homology search (Mann and Wilm, 1994). A second approach uses a database searching algorithm SEQUEST (Eng, 1994). This matches uninterpreted experimental MS/MS spectra with predicted fragment patterns generated *in silico* from sequences in protein and nucleotide databases. These techniques are highly sensitive as stated above. A nano-electrospray ion source that has a spraying time of more than 30 minutes from only 1µl of sample, can achieve sensitivities of low femtomole range. This allows the sequencing of protein spots from silver stained 2-D gels containing as little as 5 ng protein (Wilm *et al.*, 1996). This approach can also be combined with single or multi-dimensional LC coupled either on-line (ESI-MS/MS) or off-line (MALDI-TOF/TOF) to MS as described in Section 2.6 to fractionate the peptide mixture thereby facilitating the identification of multiple protein species that can co-migrate in a single 2-D gel spot.

2.10   Bioinformatics

Bioinformatics plays a central role in proteomics (Dowsey *et al.*, 2003). At present the greatest bottleneck faced by all proteomic laboratories is image analysis of 2-D gels. This is a very time consuming process involving a certain amount of technical expertise if accurate conclusions are to be derived from the data. There are a variety of specialized software packages commercially available for quantitative analysis and databasing of electrophoretic separations and a range of bioinformatic tools for identifying proteins based on data from micro-chemical analyses. In addition, special bioinformatic tools have been developed for the further characterization of proteins, ranging from the calculation of their physicochemical properties (e.g. pI, M$_r$) to the prediction of their potential post-translational modifications and their three-dimensional structure. Most of these tools with their associated databases are available on the Internet through the World Wide Web (WWW), and can be accessed through the ExPASy proteomics server (http://www.expasy.ch/tools/).

Once image analysis is complete and proteins have been identified and characterized it is essential that annotated and curated databases are constructed to store all the data. These databases must allow effective interrogation by the user and, where possible, make data available to other scientists worldwide. Currently this is best achieved using the WWW. To maintain optimal inter-connectivity between these 2DE gels protein databases and other databases of related information it has been suggested that such 2-DE databases are constructed according to a set of fundamental rules (Appel *et al.*, 1996). Databases conforming to these rules are said to be 'federated 2-DE databases', while many of the other databases conform to at least some of the rules. An index to these 2-DE protein databases can be accessed via WORLD-2DPAGE (http://www.expasy.ch/ch2d/2d-index.html).

Finally, a major problem in proteomics will be integrating the output from large-scale proteomic studies. This is essential if we are to exploit the information on changes in the expression of what can be large numbers of proteins of diverse functions to understand their role in biological systems and in dysfunction in disease. This can be driven in an hypothesis-driven way by investigating individual (or groups) proteins in more detail. However, it is unlikely that a full understanding will be generated at the level of individual proteins and that this will only be achieved by understanding how proteins act within the context of sub-cellular, cellular, tissue, organ, and whole animal structures. Thus it will be essential to compute these interactions and networks to determine the functional system involved in health and disease. This has been termed "systems biology" and a recent article (Noble, 2002) illustrates the development of this approach in the case of the heart.

3. CARDIAC PROTEOMICS

3.1 Heart 2-DE Protein Databases

Four gel protein databases of human cardiac proteins, established by three independent groups, are accessible through the WWW. These databases known as HSC-2DPAGE (Evans *et al.*, 1997), HEART-2DPAGE (Pleissner *et al.*, 1996), and HP-2DPAGE (Muller *et al.*, 1996), conform to the rules for federated 2DE protein databases (Appel *et al.*, 1996). These databases facilitate proteomic research into heart diseases containing information on several hundred cardiac proteins that have been identified by protein chemical methods. In addition, 2DE protein databases for other mammals, such as the mouse, rat (Li *et al.*, 1999), dog (Dunn *et al.*, 1997), pig and cow, are also under construction to support work on animal models of heart disease and heart failure.

Table 1. 2DE Heart Protein Databases Accessible via the World Wide Web

Database	Institute	Web address	Organ
HEART-2DPAGE	German Heart Institute, Berlin	http://userpage.chemie.fu-berlin.de/~pleiss/dhzb.html	Human heart (ventricle)
			Human heart (atrium)
HP-2DPAGE	Heart 2-DE Database, MDC, Berlin	http://www.mdc-berlin.de/~emu/heart/	Human heart (ventricle)
HSC-2DPAGE	Heart Science Centre, Harefield Hospital	http://www.harefield.nthames.nhs.uk/nhli/protein/	Human heart (ventricle)
			Rat heart (ventricle)
			Dog heart (ventricle)
RAT HEART-2DPAGE	German Heart Institute, Berlin	http://www.mpiib-berlin.mpg.de/2D-PAGE/RAT-HEART/2d/	Rat heart

3.2 Dilated Cardiomyopathy

Dilated cardiomyopathy (DCM) is a disease of unknown aetiology. It is a severe disease characterized by impaired systolic function resulting in heart failure. To date proteomic investigations into human heart disease have centered around DCM. The known contributory factors of DCM are viral infections, cardiac-specific autoantibodies, toxic agents, genetic factors and sustained alcohol abuse. The expression of as many as 100 cardiac proteins have been observed to significantly alter in their expression in DCM, with the majority of these proteins being less abundant in the diseased heart. This has been reported in numerous studies (Corbett *et al.*, 1998; Dunn *et al.*, 1997; Knecht *et al.*, 1994b; Knecht *et al.*, 1994a; Li *et al.*, 1999; Pleissner *et al.*, 1995; Scheler *et al.*, 1999). Identification of many of these proteins has been via chemical methods such as mass spectrometry (Corbett *et al.*, 1998; Otto *et al.*, 1996; Pleissner *et al.*, 1997; Thiede *et al.*, 1996), classifying them into three broad functional classes:

1. Cytoskeletal and myofibrillar proteins
2. Proteins associated with mitochondria and energy production
3. Proteins associated with stress responses

Investigating the contribution of these changes to altered cellular function underlying cardiac dysfunction is now a major challenge. This has already begun. For example fifty-nine isoelectric isoforms of HSP27 have been observed to be present in human myocardium using traditional 2DE large format gels. Twelve of these protein spots in the p*I* range of 4.9-6.2 and mass range of 27,000-28,000 Da were significantly altered in intensity in myocardium taken from patients with DCM. Ten of these protein spots were significantly changed in myocardium taken from patients with ischaemic heart failure (Scheler *et al.*, 1999).

3.3   Sub-proteomics of the heart – PKC signal transduction pathways

Protein kinase signal transduction pathways have been extensively studied and characterised in the myocardium. Much work has focussed on studying the role of individual kinases involved in ischaemic preconditioning (IP). IP describes the reduction in susceptibility to myocardial infarction that follows brief periods of sublethal ischaemia (Murry *et al.*, 1986). This reduction can manifest itself as a 4-fold reduction in infarct size, this being secondary to a delay in the onset and rate of cell necrosis during the subsequent lethal ischaemia (Marber *et al.*, 1994).

The involvement of protein kinase C (PKC) in preconditioning was first suggested by Downey and colleagues (Ytrehus *et al.*, 1994). They demonstrated that pharmacological inhibition of PKC blocked ischaemic preconditioning and that the infarct-sparing effects of ischaemic preconditioning could be mimicked by activating PKC using phorbol esters. However, although many groups have repeated this experiment (Goto *et al.*,

1995; Kitakaze *et al.*, 1996; Liu *et al.*, 1995; Mitchell *et al.*, 1995) it is still uncertain which PKC isoform(s) is/are responsible.

PKC-ε activation has been shown to play a significant role in protection, using isoform-specific inhibitory peptides, which are able to abolish protection in response to IP (Gray *et al.*, 1997; Liu *et al.*, 1999). The binding of a specific PKC isoform is thus prevented from localizing with its substrate (s) and there is a loss of function. However, the selectivity of this approach has been brought into question. Another approach to validating PKC-ε as having a cardio-protective role is to use animal models where the gene of interest, in this case that encoding PKC-ε, is lacking (Saurin *et al.*, 2002).

A functional proteomic approach to investigate PKC-ε mediated cardioprotection has been adopted by Ping and colleagues (Vondriska *et al.*, 2001a). In their approach immunoprecipitation, using PKC-ε monoclonal antibodies, was performed to isolate PKC-ε complexes. This "sub-proteome" is then separated out using 1-D/2-D electrophoresis and putative candidate proteins, which associate with PKC-ε, are then identified using western blotting/mass spectrometry based techniques. In order to validate this proteomic data, the co-localisation of candidate proteins with the PKC-ε complex is established using PKC-ε-GST affinity pull-down assays. Expression of these candidiate proteins in cardiac cells is then confirmed using isolated mouse cardiac myocytes (Vondriska *et al.*, 2001a). Using this approach Ping and colleagues have found that, within the "PKC-ε subproteome", PKC-ε forms complexes with at least 93 proteins in the mouse heart (Baines *et al.*, 2002;Edmondson *et al.*, 2002;Li *et al.*, 2000;Ping *et al.*, 2001;Vondriska *et al.*, 2001b). The identified proteins can be separated into six different classes of molecule incorporating structural proteins, signalling molecules, metabolism-related proteins, stress-activated proteins, transcription- and translation-related proteins and PKC-ε binding domain containing proteins (Edmondson *et al.*, 2002; Ping *et al.*, 2001).

3.4   Sub-proteomics of the heart – mitochondria

Mitcochondria are involved in a variety of cellular processes. Their primary role is the production of ATP, but they are also involved in ionic homeostasis, apoptosis, oxidation of carbohydrates and fatty acids, and a variety of other catabolic and anabolic pathways. As a consequence of this functional diversity, mitochondrial dysfunction is involved in a variety of diseases including heart disease (Hirano *et al.*, 2001). Characterisation of the mitochondrial proteome could thus provide new insights into cardiac dysfunction in heart disease (Lopez and Melov, 2002).

It has been predicted that the human mitochondrial proteome comprises around 1500 distinct proteins (Taylor *et al.*, 2003). This level of complexity should be capable of being addressed using the currently available panel of proteomic technologies.   Providing that sufficient tissue is available, mitochondria can be purified relatively easily by differential centrifugation (Lopez and Melov, 2002), and the mitochondrial proteins separated by 2-DE and identified by PMF. The most comprehensive report using this approach is a study of rat liver mitochondria in which broad and narrow range IPG 2-D gels resulted in the identification of 192 different gene products from a total of around 1800 protein spots, of which approximately 70% were enzymes with a broad spectrum of catalytic activities (Fountoulakis *et al.*, 2002).   A similar study has identified 185 different gene products from around 600 protein spots in the mitochondrial proteome of the neuroblastoma cell line IMR-32 (Fountoulakis and Schlaeger, 2003). This approach has not been systematically applied to the study of the mitochondrial proteome of the heart, although the 2-DE/PMF approach has been applied in differential expression studies of hearts from knockout mouse strains deficient in creatine kinase (Kernec *et al.*, 2001) and mitochondrial superoxide dismutase (Lopez and Melov, 2002).

Proteomics based on the 2-DE approach, of course, suffers from problems associated with the analysis of membrane proteins, so that such mitochondrial proteins are poorly represented on 2-D profiles. In addition, many mitochondrial proteins are more basic than cytosolic proteins, mitochondria are rich in low molecular weight (<10 kDa) proteins, and mitochondrial proteins are poorly described in databases (Lescuyer *et al.*, 2003).   In an attempt to overcome some of these hurdles, Pflieger et al (Pflieger *et al.*, 2002) have separated the proteins from isolated yeast mitochondria by one-dimensional SDS-PAGE to overcome the problems associated with the IEF dimension of 2-DE. The SDS gel was then cut into 27 slices of around 2mm and tryptic digests of the proteins contained in these bands analysed by LC-MS/MS.   This approach resulted in the identification of 179 gene products, *i.e.* similar to the number identified from isolated rat liver mitochondria by 2-DE/MS (Fountoulakis *et al.*, 2002). However, these proteins represented a broader range of proteins than covered by the 2-DE-based approach, with their physicochemical properties spanning a wide range of pI, M_r and hydrophobicity.

An alternative approach to increasing proteomic coverage is based on the analysis of isolated intact mitochondrial protein complexes. One strategy is based on the use of sucrose density gradients to separate intact mitochondrial complexes solubilised with *n*-dodecyl-β-D-maltoside (Hanson *et al.*, 2001). Initially the proteins from the individual fractions were analysed by 2-DE. However, subsequently this approach has been coupled with 1-D SDS-PAGE and MALDI PMF analysis of tryptic digests of excised protein bands (Taylor *et al.*, 2003).   When applied to human heart mitochondria, this approach resulted in the identification of 615 *bona fide* or potential mitochondrial proteins, many of which had not been previously reported using 2-DE (Taylor *et al.*, 2003). Proteins with a wide range of pI, M_r and hydrophobicities were reported with a high coverage of the known subunits of the oxidative machinery of the inner mitochondrial membrane. A significant proportion of the identified proteins are associated with signalling, RNA, DNA, and protein synthesis, ion transport, and lipid metabolism (Taylor *et al.*, 2003). In a recent study of complex I purified from bovine heart mitochondria, three independent separation methods (1-D SDS-PAGE, 2-DE and reverse phase HPLC) combined with MALDI-PMF and ESI-MS/MS were employed and the intact enzyme was shown to be an assembly of 46 different proteins (Carroll *et al.*, 2003).

Another approach to overcoming the limitations inherent in the IEF dimension of 2-DE is to use alternative types of 2-D separations. 2-D blue native (BN) electrophoresis (Schagger and von Jagow, 1991) can be used to separate membrane and other functional protein complexes as intact, enzymatically active complexes in the first dimension. This is followed with a second-dimension separation by Tricine-SDS-PAGE to separate the complexes into their component subunits. This method, combined with protein identification by MALDI PMF, has been applied to several studies of the mitochondrial proteome (Brookes *et al.*, 2002; Kruft *et al.*, 2001). In a study of human heart mitochondria using BN/SDS-PAGE, the individual subunits of all five complexes of the oxidative phosphorylation system were represented and a novel variant of cytochrome *c* oxidase subunit Vic was reported (Devreese *et al.*, 2002).

3.5 Animal Models of Heart Disease

Investigations of human diseased tissue samples can be compromised by factors such as the disease stage, tissue heterogeneity, genetic variability and the patient's medical history/therapy. Avoiding any of the above complications when working with human samples can prove to be extremely difficult. An alternative approach is to apply proteomics to appropriate models of human disease, animal models being an attractive alternative.

There are several models of cardiac hypertrophy, heart disease and heart failure in small animals, particularly the rat. Proteomic analysis of these models has focussed on changes in cardiac proteins in response to alcohol (Patel *et al.*, 1997; Patel *et al.*, 2000) and lead (Toraason *et al.*, 1997) toxicity. Unfortunately the cardiac physiology of small animal models and their normal pattern of gene expression (e.g. isoforms of the major cardiac contractile proteins) differ from that in larger mammals such as humans. Therefore investigations have moved into higher mammals and recently two proteomic studies of heart failure in large animals have been published. One study investigated pacing-induced heat failure in the dog (Heinke *et al.*, 1998; Heinke *et al.*, 1999), whilst the second, based in our laboratory, investigated bovine DCM (Weekes *et al.*, 1999). Both studies demonstrated shared similarities with the proteome analysis of human DCM, with the majority of changes involving reduced protein abundance in the diseased heart.

Identifying altered canine and bovine proteins has proved to be particularly challenging since these species are poorly represented in current genomic databases. As a result of this, new bioinformatic tools (MultiIdent) have had to be developed to facilitate cross-species protein identification (Wilkins *et al.*, 1998). The most significant change observed for bovine DCM, was a seven-fold increase in the enzyme ubiquitin carboxyl-terminal hydrolase (UCH) (Weekes *et al.*, 1999). This could potentially facilitate increased protein ubiquitination in the diseased state, leading to proteolysis via the 26S proteosome pathway. Interestingly there is evidence to suggest that inappropriate ubiquitination of proteins could contribute to the development of heart failure (Field and Clark, 1997).

More recently we have investigated whether the ubiquitin-proteosome system is perturbed in the heart of human DCM patients (Weekes *et al.*, 2003). As in bovine DCM, expression of the enzyme UCH was more than 8-fold elevated at the protein level and more than 5-fold elevated at the mRNA level in human DCM. Moreover, this increased expression of UCH was shown by immunocytochemistry to be associated with the myocytes which do not exhibit detectable staining in control hearts. Overall protein ubiquitination was increased 5-fold in DCM relative to control hearts and using a selective affinity purification method we were able to demonstrate enhanced ubiquitination of a number of distinct proteins in DCM hearts. We have identified a number of these proteins by mass spectrometry. Interestingly many of these proteins were the same proteins that we have previously found to be present at reduced abundance in DCM hearts (Corbett *et al.*, 1998). This new evidence strengthens our hypothesis that inappropriate ubiquitin conjugation leads to proteolysis and depletion of certain proteins in the DCM heart and may contribute to loss of normal cellular function in the diseased heart.

3.6 Proteomics of Cultured cardiac Myocytes

Cell culture systems are attractive models for proteomic analysis as they should provide defined systems with much lower inherent variability between samples, particularly if established cell lines are used. Cell culture systems are therefore ideal for detailed proteomic investigations of responses in protein expression to controlled stimuli. However, it is clear that cells maintained in culture respond by alterations in their pattern of gene, and consequently protein, expression such that this can be quite different from that found *in vivo*. This process can occur quite rapidly in primary cultures of cells established from tissue samples and is even more profound in cells maintained long-term, particularly where transformation has been used to establish immortal cell lines. The situation is even worse in the case of cardiac myocytes. While neonatal cardiac myocytes can be maintained and grown *in vitro*, adult cells are terminally differentiated and can be maintained for relatively short times *in* vitro but are not capable of cell division.

In fact there are relatively few published proteomic investigations of isolated cardiac myocytes. In a study of beating neonatal rat cardiac myocytes, 2-DE was used to investigate the regulation of protein synthesis by catecholamines (He *et al.*, 1992a; He *et al.*, 1992b). Marked changes in protein expression were observed in response to treatment with norepinephrine (He *et al.*, 1992a). The use of the α-adrenoceptor blocker, prazosin, allowed a clear classification of α- and non-α (probably β) adrenoceptor-mediated catecholamine effects on protein expression (He et al., 1992b). Unfortunately none of the proteins could be identified at the time this studied was carried out.

Arnott et al (Arnott *et al.*, 1998) have used phenylephrine treated neonatal rat cardiac myocytes as a model of cardiac hypertrophy for proteomic analysis. In this 2-DE based study, 11 protein spots were found to display statistically significant changes in expression upon induction of hypertrophy. Of these, 8 showed higher expression and three were decreased in abundance in hypertrophied cells. All of these proteins were successfully identified by a combination of PMF by MALDI-TOF and partial sequencing by LC-MS/MS. The atrial isoforms of myosin light chains (MLC) 1 (2 spots) and 2 were increased, as was the ventricular isoforms of MLC2 (2 spots). Other proteins that increased were chaperonin cofactor a, nucleoside diphosphate kinase a and the 27 kDa heat shock protein (HSP27). The proteins that were decreased were identified as mitochondrial matrix protein p1 (2 spots) and NADH ubiquinone oxidoreductase 75 kDa subunit. The changes in expression of MLC isoforms are consistent with previous studies of the expression of MLC isoforms in cardiac hypertrophy both by Northern blot analysis of mRNA and by immunofluorescence. In contrast, the changes in expression of the other proteins had not previously been shown to be associated with cardiac hypertrophy (Arnott *et al.*, 1998).

In a similar study, endothelin (ET) was used to induce hypertrophy in neonatal rat cardiac myocytes. ET treatment was found by 2-DE to result in a two-fold decrease in 21 proteins compared to the levels in untreated cells (Macri *et al.*, 2000). The ET-induced hypertrophy was accompanied by a 30% increase in MLC1 and MLC2.

Isolated adult rabbit cardiac myocytes have been used in a proteomic study of myocardial ischaemic preconditioning, induced pharmacologically with adenosine (Arrell *et al.*, 2001). Here a subproteomic approach was used in which cytosolic and myofilament-enriched fractions were analysed by 2-DE. Various adenosine-mediated changes in protein expression were detected in the cytosolic fraction but these proteins were not subsequently identified. The most striking finding was novel post-translational modification of MLC1 that was shown to be due to phosphorylation at two sites (Arrell *et al.*, 2001). The functional significance of this finding to preconditioning remains to be established. In a recent study of adult human cardiac myocytes from patients with end stage heart failure, two protein spots associated with MLC1 were observed on 2-D gels (van, V *et al.*, 2003). These two spots may represent phosphorylated and non-phosphorylated MLC as described by Arrell et al (Arrell *et al.*, 2001), but the amount of the putatively phosphorylated form was not found to differ between failing and control (transplant donor) samples.

3.7   Proteomic Characterisation of Cardiac Antigens in heart Disease and Transplantation

Proteomics can be used to identify cardiac-specific antigens that elicit antibody responses in heart disease and following cardiac transplantation. This approach makes use of Western blot transfers of 2-D gel separations of cardiac proteins. These are probed with patient serum samples and developed using appropriately conjugated anti-human immunoglobulins. This strategy has revealed several cardiac antigens that are reactive with autoantibodies in DCM (Latif *et al.*, 1993; Pohlner *et al.*, 1997) and myocarditis (Pankuweit *et al.*, 1997). Cardiac antigens that are associated with antibody responses following cardiac transplantation have also been characterized and may be involved in acute (Latif *et al.*, 1995) and chronic (Wheeler *et al.*, 1995) rejection.

4.   CONCLUDING REMARKS

Established proteomic technologies, together with the new and alternative strategies currently under development, are now making it possible to address important issues in biomedicine. In this chapter we have attempted to illustrate how proteomics is being used to characterise protein expression in the human heart and to investigate changes in protein expression associated with cardiac dysfunction in disease. The proteomics studies that have been carried out on cardiac tissue from both human patients and appropriate animal models are providing new insights into the cellular mechanisms involved in cardiac dysfunction. In addition they should result in the discovery of new diagnostic and/or prognostic biomarkers and the identification of potential drug targets for the development of new therapeutic approaches for combating heart disease.

REFERENCES

Alban, A., Olu David, S., Bjorkesten, L., Andersson, C., Sloge, E., Lewis, S., and Currie, I. 2003, A novel experimental design for comparative two-dimensional gel analysis: two-dimensional difference gel electrophoresis incorporating a pooled internal standard. *Proteomics* 3:36-44

Anderson, L. and J. Seilhamer. 1997, A comparison of selected mRNA and protein abundances in human liver. *Electrophoresis* **18:** 533-537.

Anderson, N. G. and L. Anderson. 1982, The Human Protein Index. *Clin.Chem.* **28:** 739-748.

Anderson, N. G., A. Matheson, and N. L. Anderson. 2001, Back to the future: the human protein index (HPI) and the agenda for post-proteomic biology. *Proteomics* **1:** 3-12.

Anderson, N. L. and N. G. Anderson. 2002, The human plasma proteome: history, character, and diagnostic prospects. *Mol.Cell Proteomics.* **1:** 845-867.

Appel, R. D., A. Bairoch, J. C. Sanchez, J. R. Vargas, O. Golaz, C. Pasquali, and D. F. Hochstrasser. 1996, Federated two-dimensional electrophoresis database: a simple means of publishing two-dimensional electrophoresis data. *Electrophoresis* **17:** 540-546.

Arnott, D., A. Kishiyama, E. A. Luis, S. G. Ludlum, J. C. Marsters, Jr., and J. T. Stults. 2002, Selective detection of membrane proteins without antibodies: a mass spectrometric version of the Western blot. *Mol.Cell Proteomics* **1:** 148-156.

Arnott, D., K. L. O'Connell, K. L. King, and J. T. Stults. 1998, An integrated approach to proteome analysis: identification of proteins associated with cardiac hypertrophy. *Anal.Biochem.* **258:** 1-18.

Arrell, D. K., I. Neverova, H. Fraser, E. Marban, and J. E. Van Eyk. 2001, Proteomic analysis of pharmacologically preconditioned cardiomyocytes reveals novel phosphorylation of myosin light chain 1. *Circ.Res.* **89:** 480-487.

Baines, C. P., J. Zhang, G. W. Wang, Y. T. Zheng, J. X. Xiu, E. M. Cardwell, R. Bolli, and P. Ping. 2002, Mitochondrial PKCepsilon and MAPK form signaling modules in the murine heart: enhanced mitochondrial PKCepsilon-MAPK interactions and differential MAPK activation in PKCepsilon-induced cardioprotection. *Circ.Res.* **90:** 390-397.

Banks, R. E., M. J. Dunn, M. A. Forbes, A. Stanley, D. Pappin, T. Naven, M. Gough, P. Harnden, and P. J. Selby. 1999, The potential use of laser capture microdissection to selectively obtain distinct populations of cells for proteomic analysis--preliminary findings. *Electrophoresis* **20:** 689-700.

Berggren, K. N., B. Schulenberg, M. F. Lopez, T. H. Steinberg, A. Bogdanova, G. Smejkal, A. Wang, and W. F. Patton. 2002, An improved formulation of SYPRO Ruby protein gel stain: Comparison with the original formulation and with a ruthenium II tris (bathophenanthroline disulfonate) formulation. *Proteomics.* **2**: 486-498.

Bjellqvist, B., K. Ek, P. G. Righetti, E. Gianazza, A. Gorg, R. Westermeier, and W. Postel. 1982, Isoelectric focusing in immobilized pH gradients: principle, methodology and some applications. *J.Biochem.Biophys.Methods* **6**: 317-339.

Blomberg, A., L. Blomberg, J. Norbeck, S. J. Fey, P. M. Larsen, M. Larsen, P. Roepstorff, H. Degand, M. Boutry, A. Posch, and . 1995, Interlaboratory reproducibility of yeast protein patterns analyzed by immobilized pH gradient two-dimensional gel electrophoresis. *Electrophoresis* **16**: 1935-1945.

Brookes, P. S., A. Pinner, A. Ramachandran, L. Coward, S. Barnes, H. Kim, and V. M. Darley-Usmar. 2002, High throughput two-dimensional blue-native electrophoresis: a tool for functional proteomics of mitochondria and signaling complexes. *Proteomics* **2**: 969-977.

Carroll, J., I. M. Fearnley, R. J. Shannon, J. Hirst, and J. E. Walker. 2003, Analysis of the subunit composition of complex I from bovine heart mitochondria. *Mol.Cell. Proteomics* **2**: 117-126.

Choe, L. H. and K. H. Lee. 2000, A comparison of three commercially available isoelectric focusing units for proteome analysis: the multiphor, the IPGphor and the protean IEF cell. *Electrophoresis* **21**: 993-1000.

Corbett, J. M., M. J. Dunn, A. Posch, and A. Gorg. 1994, Positional reproducibility of protein spots in two-dimensional polyacrylamide gel electrophoresis using immobilised pH gradient isoelectric focusing in the first dimension: an interlaboratory comparison. *Electrophoresis* **15**: 1205-1211.

Corbett, J. M., H. J. Why, C. H. Wheeler, P. J. Richardson, L. C. Archard, M. H. Yacoub, and M. J. Dunn. 1998, Cardiac protein abnormalities in dilated cardiomyopathy detected by two-dimensional polyacrylamide gel electrophoresis. *Electrophoresis* **19**: 2031-2042.

Corthals, G. L., V. C. Wasinger, D. F. Hochstrasser, and J. C. Sanchez. 2000, The dynamic range of protein expression: a challenge for proteomic research. *Electrophoresis* **21**: 1104-1115.

Craven, R. A. and R. E. Banks. 2002, Use of laser capture microdissection to selectively obtain distinct populations of cells for proteomic analysis. *Methods Enzymol.* **356**: 33-49.

Craven, R. A., D. H. Jackson, P. J. Selby, and R. E. Banks. 2002, Increased protein entry together with improved focussing using a combined IPGphor/Multiphor approach. *Proteomics* **2**: 1061-1063.

De Souza, A. Unpublished data. 2003.

Devreese, B., F. Vanrobaeys, J. Smet, J. Van Beeumen, and R. Van Coster. 2002, Mass spectrometric identification of mitochondrial oxidative phosphorylation subunits separated by two-dimensional blue-native polyacrylamide gel electrophoresis. *Electrophoresis* **23**: 2525-2533.

Dowsey, A. W., M. J. Dunn, and GZ. Yang. *Proteomics*, in press.

Dunn, M. J., J. M. Corbett, and C. H. Wheeler. 1997, HSC-2DPAGE and the two-dimensional gel electrophoresis database of dog heart proteins. *Electrophoresis* **18**: 2795-2802.

Dunn, M. J. and A. Gorg. 2001, Two-dimensional Polyacrylamide Gel Electrophoresis for Proteome Analysis. In *Proteomics, From Protein Sequence to Function.*(S. R. Pennington and M. J. Dunn (eds.),), BIOS Scientific Publishers Ltd, pp. 43-63.

Edmondson, R. D., T. M. Vondriska, K. J. Biederman, J. Zhang, R. C. Jones, Y. Zheng, D. L. Allen, J. X. Xiu, E. M. Cardwell, M. R. Pisano, and P. Ping. 2002, Protein kinase C epsilon signaling complexes include metabolism- and transcription/translation-related proteins: complimentary separation techniques with LC/MS/MS. *Mol.Cell.Proteomics* **1**: 421-433.

Eng, JK. 1994, An approach to correlate tandem mass spectral data of peptides with amino acid sequenecs in a protein database. *J.Am.Soc.Mass Spec.* **5:** 976-989.

Evans, G., C. H. Wheeler, J. M. Corbett, and M. J. Dunn. 1997, Construction of HSC-2DPAGE: a two-dimensional gel electrophoresis database of heart proteins. *Electrophoresis* **18:** 471-479.

Field, M. L. and J. F. Clark. 1997, Inappropriate ubiquitin conjugation: a proposed mechanism contributing to heart failure. *Cardiovasc.Res.* **33:** 8-12.

Figeys, D., L. D. McBroom, and M. F. Moran. 2001, Mass spectrometry for the study of protein-protein interactions. *Methods* **24:** 230-239.

Figeys, D. and D. Pinto. 2001, Proteomics on a chip: promising developments. *Electrophoresis* **22:** 208-216.

Fleischmann, R. D., M. D. Adams, O. White, R. A. Clayton, E. F. Kirkness, A. R. Kerlavage, C. J. Bult, J. F. Tomb, B. A. Dougherty, J. M. Merrick, and . 1995, Whole-genome random sequencing and assembly of Haemophilus influenzae Rd. *Science* **269:** 496-512.

Flory, M. R., T. J. Griffin, D. Martin, and R. Aebersold. 2002, Advances in quantitative proteomics using stable isotope tags. *Trends Biotechnol.* **20:** S23-S29.

Fountoulakis, M., P. Berndt, H. Langen, and L. Suter. 2002, The rat liver mitochondrial proteins. *Electrophoresis* **23:** 311-328.

Fountoulakis, M. and E. J. Schlaeger. 2003, The mitochondrial proteins of the neuroblastoma cell line IMR-32. *Electrophoresis* **24:** 260-275.

Gharbi, S., P. Gaffney, A. Yang, M. J. Zvelebil, R. Cramer, M. D. Waterfield, and J. F. Timms. 2002, Evaluation of two-dimensional differential gel electrophoresis for proteomic expression analysis of a model breast cancer cell system. *Mol.Cell.Proteomics.* **1:** 91-98.

Gooley, A. A. and N. H. Packer. 1997, The Importance of Protein Co- and Post- Translational Modifications in Proteome Projects. In *Proteome Research: New Frontiers in Functional Genomics.*(M. R. Wilkins, K. L. Williams, R. D. Appel, and D. F. Hochstrasser (eds.),), Springer, pp. 65-91.

Gorg, A., G. Boguth, C. Obermaier, A. Posch, and W. Weiss. 1995, Two-dimensional polyacrylamide gel electrophoresis with immobilized pH gradients in the first dimension (IPG-Dalt): the state of the art and the controversy of vertical versus horizontal systems. *Electrophoresis* **16:** 1079-1086.

Gorg, A., C. Obermaier, G. Boguth, A. Harder, B. Scheibe, R. Wildgruber, and W. Weiss. 2000, The current state of two-dimensional electrophoresis with immobilized pH gradients. *Electrophoresis* **21:** 1037-1053.

Gorg, A., C. Obermaier, G. Boguth, and W. Weiss. 1999, Recent developments in two-dimensional gel electrophoresis with immobilized pH gradients: wide pH gradients up to pH 12, longer separation distances and simplified procedures. *Electrophoresis* **20:** 712-717.

Goto, M., Y. Liu, X. M. Yang, J. L. Ardell, M. V. Cohen, and J. M. Downey. 1995, Role of bradykinin in protection of ischemic preconditioning in rabbit hearts. *Circ.Res.* **77:** 611-621.

Gray, M. O., J. S. Karliner, and D. Mochly-Rosen. 1997, A selective epsilon-protein kinase C antagonist inhibits protection of cardiac myocytes from hypoxia-induced cell death. *J Biol.Chem.* **272:** 30945-30951.

Gygi, S. P., G. L. Corthals, Y. Zhang, Y. Rochon, and R. Aebersold. 2000, Evaluation of two-dimensional gel electrophoresis-based proteome analysis technology. *Proc.Natl.Acad.Sci.U.S.A* **97:** 9390-9395.

Gygi, S. P., Y. Rochon, B. R. Franza, and R. Aebersold. 1999, Correlation between protein and mRNA abundance in yeast. *Mol.Cell Biol.* **19:** 1720-1730.

Hanson, B. J., B. Schulenberg, W. F. Patton, and R. A. Capaldi. 2001, A novel subfractionation approach for mitochondrial proteins: a three-dimensional mitochondrial proteome map. *Electrophoresis* **22:** 950-959.

He, C., U. Muller, W. Oberthur, and K. Werdan. 1992a, Application of high resolution two-dimensional polyacrylamide gel electrophoresis of polypeptides from cultured neonatal rat cardiomyocytes: regulation of protein synthesis by catecholamines. *Electrophoresis* **13:** 748-754.

He, C., U. Muller, and K. Werdan. 1992b, Regulation of protein biosynthesis in neonatal rat cardiomyocytes by adrenoceptor-stimulation: investigations with high resolution two-dimensional polyacrylamide gel electrophoresis. *Electrophoresis* **13:** 755-756.

Heinke, M. Y., C. H. Wheeler, D. Chang, R. Einstein, A. Drake-Holland, M. J. Dunn, and C. G. dos Remedios. 1998, Protein changes observed in pacing-induced heart failure using two-dimensional electrophoresis. *Electrophoresis* **19:** 2021-2030.

Heinke, M. Y., C. H. Wheeler, J. X. Yan, V. Amin, D. Chang, R. Einstein, M. J. Dunn, and C. G. dos Remedios. 1999, Changes in myocardial protein expression in pacing-induced canine heart failure. *Electrophoresis* **20:** 2086-2093.

Herbert, B. R., M. P. Molloy, A. A. Gooley, B. J. Walsh, W. G. Bryson, and K. L. Williams. 1998, Improved protein solubility in two-dimensional electrophoresis using tributyl phosphine as reducing agent. *Electrophoresis* **19:** 845-851.

Hesse, C., C. L. Nilsson, K. Blennow, and P. Davidsson. 2001, Identification of the apolipoprotein E4 isoform in cerebrospinal fluid with preparative two-dimensional electrophoresis and matrix assisted laser desorption/ionization-time of flight-mass spectrometry. *Electrophoresis* **22:** 1834-1837.

Hirano, M., M. Davidson, and S. DiMauro. 2001, Mitochondria and the heart. *Curr.Opin.Cardiol.* **16:** 201-210.

Hoffmann, P., H. Ji, R. L. Moritz, L. M. Connolly, D. F. Frecklington, M. J. Layton, J. S. Eddes, and R. J. Simpson. 2001, Continuous free-flow electrophoresis separation of cytosolic proteins from the human colon carcinoma cell line LIM 1215: a non two-dimensional gel electrophoresis-based proteome analysis strategy. *Proteomics* **1:** 807-818.

Hoving, S., B. Gerrits, H. Voshol, D. Muller, R. C. Roberts, and J. van Oostrum. 2002, Preparative two-dimensional gel electrophoresis at alkaline pH using narrow range immobilized pH gradients. *Proteomics* **2:** 127-134.

Jain, K. K. 2002, Application of laser capture microdissection to proteomics. *Methods Enzymol.* **356:** 157-167.

Jenkins, R. E. and S. R. Pennington. 2001, Arrays for protein expression profiling: towards a viable alternative to two-dimensional gel electrophoresis? *Proteomics* **1:** 13-29.

Kernec, F., M. Unlu, W. Labeikovsky, J. S. Minden, and A. P. Koretsky. 2001, Changes in the mitochondrial proteome from mouse hearts deficient in creatine kinase. *Physiol Genomics* **6:** 117-128.

Kitakaze, M., K. Node, T. Minamino, K. Komamura, H. Funaya, Y. Shinozaki, M. Chujo, H. Mori, M. Inoue, M. Hori, and T. Kamada. 1996, Role of activation of protein kinase C in the infarct size-limiting effect of ischemic preconditioning through activation of ecto-5'-nucleotidase. *Circulation* **93:** 781-791.

Klose, J. 1975, Protein mapping by combined isoelectric focusing and electrophoresis of mouse tissues. A novel approach to testing for induced point mutations in mammals. *Humangenetik.* **26:** 231-243.

Klose, J. and U. Kobalz. 1995, Two-dimensional electrophoresis of proteins: an updated protocol and implications for a functional analysis of the genome. *Electrophoresis* **16:** 1034-1059.

Knecht, M., V. Regitz-Zagrosek, K. P. Pleissner, S. Emig, P. Jungblut, A. Hildebrandt, and E. Fleck. 1994a, Dilated cardiomyopathy: computer-assisted analysis of endomyocardial biopsy protein patterns by two-dimensional gel electrophoresis. *Eur.J.Clin.Chem.Clin.Biochem.* **32:** 615-624.

Knecht, M., V. Regitz-Zagrosek, K. P. Pleissner, P. Jungblut, C. Steffen, A. Hildebrandt, and E. Fleck. 1994b, Characterization of myocardial protein composition in dilated cardiomyopathy by two-dimensional gel electrophoresis. *Eur.Heart J.* **15 Suppl D:** 37-44.

Kruft, V., H. Eubel, L. Jansch, W. Werhahn, and H. P. Braun. 2001, Proteomic approach to identify novel mitochondrial proteins in Arabidopsis. *Plant Physiol* **127:** 1694-1710.

Lal, S. P., R. I. Christopherson, and C. G. dos Remedios. 2002, Antibody arrays: an embryonic but rapidly growing technology. *Drug Discov.Today* **7:** S143-S149.

Lander, E. S., L. M. Linton, B. Birren, C. Nusbaum, M. C. Zody, J. Baldwin, K. Devon, K. Dewar, M. Doyle, W. FitzHugh, R. Funke, D. Gage, K. Harris, A. Heaford, J. Howland, L. Kann, J. Lehoczky, R. LeVine, P. McEwan, K. McKernan, J. Meldrim, J. P. Mesirov, C. Miranda, W. Morris, J. Naylor, C. Raymond, M. Rosetti, R. Santos, A. Sheridan, C. Sougnez, N. Stange-Thomann, N. Stojanovic, A. Subramanian, D. Wyman, J. Rogers, J. Sulston, R. Ainscough, S. Beck, D. Bentley, J. Burton, C. Clee, N. Carter, A. Coulson, R. Deadman, P. Deloukas, A. Dunham, I. Dunham, R. Durbin, L. French, D. Grafham, S. Gregory, T. Hubbard, S. Humphray, A. Hunt, M. Jones, C. Lloyd, A. McMurray, L. Matthews, S. Mercer, S. Milne, J. C. Mullikin, A. Mungall, R. Plumb, M. Ross, R. Shownkeen, S. Sims, R. H. Waterston, R. K. Wilson, L. W. Hillier, J. D. McPherson, M. A. Marra, E. R. Mardis, L. A. Fulton, A. T. Chinwalla, K. H. Pepin, W. R. Gish, S. L. Chissoe, M. C. Wendl, K. D. Delehaunty, T. L. Miner, A. Delehaunty, J. B. Kramer, L. L. Cook, R. S. Fulton, D. L. Johnson, P. J. Minx, S. W. Clifton, T. Hawkins, E. Branscomb, P. Predki, P. Richardson, S. Wenning, T. Slezak, N. Doggett, J. F. Cheng, A. Olsen, S. Lucas, C. Elkin, E. Uberbacher, M. Frazier, R. A. Gibbs, D. M. Muzny, S. E. Scherer, J. B. Bouck, E. J. Sodergren, K. C. Worley, C. M. Rives, J. H. Gorrell, M. L. Metzker, S. L. Naylor, R. S. Kucherlapati, D. L. Nelson, G. M. Weinstock, Y. Sakaki, A. Fujiyama, M. Hattori, T. Yada, A. Toyoda, T. Itoh, C. Kawagoe, H. Watanabe, Y. Totoki, T. Taylor, J. Weissenbach, R. Heilig, W. Saurin, F. Artiguenave, P. Brottier, T. Bruls, E. Pelletier, C. Robert, P. Wincker, D. R. Smith, L. Doucette-Stamm, M. Rubenfield, K. Weinstock, H. M. Lee, J. Dubois, A. Rosenthal, M. Platzer, G. Nyakatura, S. Taudien, A. Rump, H. Yang, J. Yu, J. Wang, G. Huang, J. Gu, L. Hood, L. Rowen, A. Madan, S. Qin, R. W. Davis, N. A. Federspiel, A. P. Abola, M. J. Proctor, R. M. Myers, J. Schmutz, M. Dickson, J. Grimwood, D. R. Cox, M. V. Olson, R. Kaul, C. Raymond, N. Shimizu, K. Kawasaki, S. Minoshima, G. A. Evans, M. Athanasiou, R. Schultz, B. A. Roe, F. Chen, H. Pan, J. Ramser, H. Lehrach, R. Reinhardt, W. R. McCombie, B. M. de la, N. Dedhia, H. Blocker, K. Hornischer, G. Nordsiek, R. Agarwala, L. Aravind, J. A. Bailey, A. Bateman, S. Batzoglou, E. Birney, P. Bork, D. G. Brown, C. B. Burge, L. Cerutti, H. C. Chen, D. Church, M. Clamp, R. R. Copley, T. Doerks, S. R. Eddy, E. E. Eichler, T. S. Furey, J. Galagan, J. G. Gilbert, C. Harmon, Y. Hayashizaki, D. Haussler, H. Hermjakob, K. Hokamp, W. Jang, L. S. Johnson, T. A. Jones, S. Kasif, A. Kaspryzk, S. Kennedy, W. J. Kent, P. Kitts, E. V. Koonin, I. Korf, D. Kulp, D. Lancet, T. M. Lowe, A. McLysaght, T. Mikkelsen, J. V. Moran, N. Mulder, V. J. Pollara, C. P. Ponting, G. Schuler, J. Schultz, G. Slater, A. F. Smit, E. Stupka, J. Szustakowski, D. Thierry-Mieg, J. Thierry-Mieg, L. Wagner, J. Wallis, R. Wheeler, A. Williams, Y. I. Wolf, K. H. Wolfe, S. P. Yang, R. F. Yeh, F. Collins, M. S. Guyer, J. Peterson, A. Felsenfeld, K. A. Wetterstrand, A. Patrinos, M. J. Morgan, J. Szustakowski, P. de Jong, J. J. Catanese, K. Osoegawa, H. Shizuya, and S. Choi. 2001, Initial sequencing and analysis of the human genome. *Nature* **409:** 860-921.

Latif, N., C. S. Baker, M. J. Dunn, M. L. Rose, P. Brady, and M. H. Yacoub. 1993, Frequency and specificity of antiheart antibodies in patients with dilated cardiomyopathy detected using SDS-PAGE and western blotting. *J.Am.Coll.Cardiol.* **22:** 1378-1384.

Latif, N., M. L. Rose, M. H. Yacoub, and M. J. Dunn. 1995, Association of pretransplantation antiheart antibodies with clinical course after heart transplantation. *J.Heart Lung Transplant.* **14:** 119-126.

Lauber, W. M., J. A. Carroll, D. R. Dufield, J. R. Kiesel, M. R. Radabaugh, and J. P. Malone. 2001, Mass spectrometry compatibility of two-dimensional gel protein stains. *Electrophoresis* **22:** 906-918.

Lescuyer, P., J. M. Strub, S. Luche, H. Diemer, P. Martinez, A. van Dorsselaer, J. Lunardi, and T. Rabilloud. 2003, Progress in the definition of a reference human mitochondrial proteome. *Proteomics* **3:** 157-167.

Li, R. C., P. Ping, J. Zhang, W. B. Wead, X. Cao, J. Gao, Y. Zheng, S. Huang, J. Han, and R. Bolli. 2000, PKCepsilon modulates NF-kappaB and AP-1 via mitogen-activated protein kinases in adult rabbit cardiomyocytes. *Am.J.Physiol Heart Circ.Physiol* **279:** H1679-H1689.

Li, X. P., K. P. Pleissner, C. Scheler, V. Regitz-Zagrosek, J. Salnikow, and P. R. Jungblut. 1999, A two-dimensional electrophoresis database of rat heart proteins. *Electrophoresis* **20:** 891-897.

Liu, G. S., M. V. Cohen, D. Mochly-Rosen, and J. M. Downey. 1999, Protein kinase C-epsilon is responsible for the protection of preconditioning in rabbit cardiomyocytes. *J Mol Cell Cardiol* **31:** 1937-1948.

Liu, Y., A. Tsuchida, M. V. Cohen, and J. M. Downey. 1995, Pretreatment with angiotensin II activates protein kinase C and limits myocardial infarction in isolated rabbit hearts. *J.Mol.Cell Cardiol.* **27:** 883-892.

Lopez, M. F., K. Berggren, E. Chernokalskaya, A. Lazarev, M. Robinson, and W. F. Patton. 2000, A comparison of silver stain and SYPRO Ruby Protein Gel Stain with respect to protein detection in two-dimensional gels and identification by peptide mass profiling. *Electrophoresis* **21:** 3673-3683.

Lopez, M. F. and S. Melov. 2002, Applied proteomics: mitochondrial proteins and effect on function. *Circ.Res.* **90:** 380-389.

Luche, S., V. Santoni, and T. Rabilloud. 2003, Evaluation of nonionic and zwitterionic detergents as membrane protein solubilizers in two-dimensional electrophoresis. *Proteomics* **3:** 249-253.

Macri, J., Dubay, T., and Matteson, D. Characterization of the protein profile associated with endothelin-induced hypertrophy in neonatal rat myocytes. *J.Mol.Cell.Cardiol.* 32. 2000. Abstract.

Mann, M. and M. Wilm. 1994, Error-tolerant identification of peptides in sequence databases by peptide sequence tags. *Anal.Chem.* **66:** 4390-4399.

Marber, M., D. Walker, and D. Yellon. 1994, Ischaemic preconditioning. *BMJ* **308:** 1-2.

Mitchell, M. B., X. Meng, L. Ao, J. M. Brown, A. H. Harken, and A. Banerjee. 1995, Preconditioning of isolated rat heart is mediated by protein kinase C. *Circ.Res.* **76:** 73-81.

Mitchell, P. 2002, A perspective on protein microarrays. *Nat.Biotechnol.* **20:** 225-229.

Muller, E. C., B. Thiede, U. Zimny-Arndt, C. Scheler, J. Prehm, U. Muller-Werdan, B. Wittmann-Liebold, A. Otto, and P. Jungblut. 1996, High-performance human myocardial two-dimensional electrophoresis database: edition 1996. *Electrophoresis* **17:** 1700-1712.

Murry, C. E., R. B. Jennings, and K. A. Reimer. 1986, Preconditioning with ischemia: a delay of lethal cell injury in ischemic myocardium. *Circulation* **74:** 1124-1136.

Neuhoff, V., N. Arold, D. Taube, and W. Ehrhardt. 1988, Improved staining of proteins in polyacrylamide gels including isoelectric focusing gels with clear background at nanogram sensitivity using Coomassie Brilliant Blue G-250 and R-250. *Electrophoresis* **9**: 255-262.

Noble, D. 2002, Modeling the heart--from genes to cells to the whole organ. *Science* **295**: 1678-1682.

O'Farrell, P. H. 1975, High resolution two-dimensional electrophoresis of proteins. *J.Biol.Chem.* **250**: 4007-4021.

Olsson, I., K. Larsson, R. Palmgren, and B. Bjellqvist. 2002, Organic disulfides as a means to generate streak-free two-dimensional maps with narrow range basic immobilized pH gradient strips as first dimension. *Proteomics* **2**: 1630-1632.

Otto, A., B. Thiede, E. C. Muller, C. Scheler, B. Wittmann-Liebold, and P. Jungblut. 1996, Identification of human myocardial proteins separated by two-dimensional electrophoresis using an effective sample preparation for mass spectrometry. *Electrophoresis* **17**: 1643-1650.

Pankuweit, S., I. Portig, F. Lottspeich, and B. Maisch. 1997, Autoantibodies in sera of patients with myocarditis: characterization of the corresponding proteins by isoelectric focusing and N-terminal sequence analysis. *J.Mol.Cell Cardiol.* **29**: 77-84.

Patel, V. B., J. M. Corbett, M. J. Dunn, V. R. Winrow, B. Portmann, P. J. Richardson, and V. R. Preedy. 1997, Protein profiling in cardiac tissue in response to the chronic effects of alcohol. *Electrophoresis* **18**: 2788-2794.

Patel, V. B., G. Sandhu, J. M. Corbett, M. J. Dunn, L. M. Rodrigues, J. R. Griffiths, W. Wassif, R. A. Sherwood, P. J. Richardson, and V. R. Preedy. 2000, A comparative investigation into the effect of chronic alcohol feeding on the myocardium of normotensive and hypertensive rats: an electrophoretic and biochemical study. *Electrophoresis* **21**: 2454-2462.

Patterson, S. D. 1994, From electrophoretically separated protein to identification: strategies for sequence and mass analysis. *Anal.Biochem.* **221**: 1-15.

Patterson, S. D. and R. Aebersold. 1995, Mass spectrometric approaches for the identification of gel-separated proteins. *Electrophoresis* **16**: 1791-1814.

Patton, W. F. 2000, A thousand points of light: the application of fluorescence detection technologies to two-dimensional gel electrophoresis and proteomics. *Electrophoresis* **21**: 1123-1144.

Patton, W. F., B. Schulenberg, and T. H. Steinberg. 2002, Two-dimensional gel electrophoresis; better than a poke in the ICAT? *Curr.Opin.Biotechnol.* **13**: 321-328.

Pflieger, D., J. P. Le Caer, C. Lemaire, B. A. Bernard, G. Dujardin, and J. Rossier. 2002, Systematic identification of mitochondrial proteins by LC-MS/MS. *Anal.Chem.* **74**: 2400-2406.

Pieper, R. *Proteomics*, in press.

Pieper, R., Q. Su, C. L. Gatlin, S. T. Huang, N. L. Anderson, and S. Steiner. 2003, Multi-component immunoaffinity subtraction chromatography: An innovative step towards a comprehensive survey of the human plasma proteome. *Proteomics.* **3**: 422-432.

Ping, P., J. Zhang, W. M. Pierce, Jr., and R. Bolli. 2001, Functional proteomic analysis of protein kinase C epsilon signaling complexes in the normal heart and during cardioprotection. *Circ.Res.* **88**: 59-62.

Pleissner, K. P., V. Regitz-Zagrosek, C. Weise, M. Neuss, B. Krudewagen, P. Soding, K. Buchner, F. Hucho, A. Hildebrandt, and E. Fleck. 1995, Chamber-specific expression of human myocardial proteins detected by two-dimensional gel electrophoresis. *Electrophoresis* **16**: 841-850.

Pleissner, K. P., S. Sander, H. Oswald, V. Regitz-Zagrosek, and E. Fleck. 1996, The construction of the World Wide Web-accessible myocardial two-dimensional gel electrophoresis protein database "HEART-2DPAGE": a practical approach. *Electrophoresis* **17**: 1386-1392.

Pleissner, K. P., P. Soding, S. Sander, H. Oswald, M. Neuss, V. Regitz-Zagrosek, and E. Fleck. 1997, Dilated cardiomyopathy-associated proteins and their presentation in a WWW-accessible two-dimensional gel protein database. *Electrophoresis* **18**: 802-808.

Pohlner, K., I. Portig, S. Pankuweit, F. Lottspeich, and B. Maisch. 1997, Identification of mitochondrial antigens recognized by antibodies in sera of patients with idiopathic dilated cardiomyopathy by two-dimensional gel electrophoresis and protein sequencing. *Am.J.Cardiol.* **80**: 1040-1045.

Rabilloud, T. 1998, Use of thiourea to increase the solubility of membrane proteins in two-dimensional electrophoresis. *Electrophoresis* **19**: 758-760.

Rabilloud, T. 2002, Two-dimensional gel electrophoresis in proteomics: old, old fashioned, but it still climbs up the mountains. *Proteomics* **2**: 3-10.

Rabilloud, T., C. Valette, and J. J. Lawrence. 1994, Sample application by in-gel rehydration improves the resolution of two-dimensional electrophoresis with immobilized pH gradients in the first dimension. *Electrophoresis* **15**: 1552-1558.

Sanchez, J. C., V. Rouge, M. Pisteur, F. Ravier, L. Tonella, M. Moosmayer, M. R. Wilkins, and D. F. Hochstrasser. 1997, Improved and simplified in-gel sample application using reswelling of dry immobilized pH gradients. *Electrophoresis* **18**: 324-327.

Santoni, V., S. Kieffer, D. Desclaux, F. Masson, and T. Rabilloud. 2000, Membrane proteomics: use of additive main effects with multiplicative interaction model to classify plasma membrane proteins according to their solubility and electrophoretic properties. *Electrophoresis* **21**: 3329-3344.

Saurin, A. T., D. J. Pennington, N. J. Raat, D. S. Latchman, M. J. Owen, and M. S. Marber. 2002, Targeted disruption of the protein kinase C epsilon gene abolishes the infarct size reduction that follows ischaemic preconditioning of isolated buffer-perfused mouse hearts. *Cardiovasc.Res.* **55**: 672-680.

Schagger, H. and G. von Jagow. 1991, Blue native electrophoresis for isolation of membrane protein complexes in enzymatically active form. *Anal.Biochem.* **199**: 223-231.

Scheele, G. A. 1975, Two-dimensional gel analysis of soluble proteins. Charaterization of guinea pig exocrine pancreatic proteins. *J.Biol.Chem.* **250**: 5375-5385.

Scheler, C., X. P. Li, J. Salnikow, M. J. Dunn, and P. R. Jungblut. 1999, Comparison of two-dimensional electrophoresis patterns of heat shock protein Hsp27 species in normal and cardiomyopathic hearts. *Electrophoresis* **20**: 3623-3628.

Shevchenko, A., M. Wilm, O. Vorm, and M. Mann. 1996, Mass spectrometric sequencing of proteins silver-stained polyacrylamide gels. *Anal.Chem.* **68**: 850-858.

Smith, R. D., G. A. Anderson, M. S. Lipton, L. Masselon, L. Pasa-Tolic, Y. Shen, and H. R. Udseth. 2002, The use of accurate mass tags for high-throughput microbial proteomics. *OMICS.* **6**: 61-90.

Stanley, B. A., I. Neverova, H. A. Brown, and J. E. Van Eyk. 2003, Optimizing protein solubility for two-dimensional gel electrophoresis analysis of human myocardium. *Proteomics.* **3**: 815-820.

Switzer, R. C., III, C. R. Merril, and S. Shifrin. 1979, A highly sensitive silver stain for detecting proteins and peptides in polyacrylamide gels. *Anal.Biochem.* **98**: 231-237.

Taylor, S. W., E. Fahy, B. Zhang, G. M. Glenn, D. E. Warnock, S. Wiley, A. N. Murphy, S. P. Gaucher, R. A. Capaldi, B. W. Gibson, and S. S. Ghosh. 2003, Characterization of the human heart mitochondrial proteome. *Nat.Biotechnol.* **21**: 281-286.

Thiede, B., A. Otto, U. Zimny-Arndt, E. C. Muller, and P. Jungblut. 1996, Identification of human myocardial proteins separated by two-dimensional electrophoresis with matrix-assisted laser desorption/ionization mass spectrometry. *Electrophoresis* **17**: 588-599.

Toraason, M., W. Moorman, P. I. Mathias, C. Fultz, and F. Witzmann. 1997, Two-dimensional electrophoretic analysis of myocardial proteins from lead-exposed rabbits. *Electrophoresis* **18:** 2978-2982.

Unlu, M., M. E. Morgan, and J. S. Minden. 1997, Difference gel electrophoresis: a single gel method for detecting changes in protein extracts. *Electrophoresis* **18:** 2071-2077.

van, d., V, Z. Papp, N. M. Boontje, R. Zaremba, J. W. de Jong, P. M. Janssen, G. Hasenfuss, and G. J. Stienen. 2003, The effect of myosin light chain 2 dephosphorylation on Ca2+ - sensitivity of force is enhanced in failing human hearts. *Cardiovasc.Res.* **57:** 505-514.

Venter, J. C., M. D. Adams, E. W. Myers, P. W. Li, R. J. Mural, G. G. Sutton, H. O. Smith, M. Yandell, C. A. Evans, R. A. Holt, J. D. Gocayne, P. Amanatides, R. M. Ballew, D. H. Huson, J. R. Wortman, Q. Zhang, C. D. Kodira, X. H. Zheng, L. Chen, M. Skupski, G. Subramanian, P. D. Thomas, J. Zhang, G. L. Gabor Miklos, C. Nelson, S. Broder, A. G. Clark, J. Nadeau, V. A. McKusick, N. Zinder, A. J. Levine, R. J. Roberts, M. Simon, C. Slayman, M. Hunkapiller, R. Bolanos, A. Delcher, I. Dew, D. Fasulo, M. Flanigan, L. Florea, A. Halpern, S. Hannenhalli, S. Kravitz, S. Levy, C. Mobarry, K. Reinert, K. Remington, J. Abu-Threideh, E. Beasley, K. Biddick, V. Bonazzi, R. Brandon, M. Cargill, I. Chandramouliswaran, R. Charlab, K. Chaturvedi, Z. Deng, F. Di, V, P. Dunn, K. Eilbeck, C. Evangelista, A. E. Gabrielian, W. Gan, W. Ge, F. Gong, Z. Gu, P. Guan, T. J. Heiman, M. E. Higgins, R. R. Ji, Z. Ke, K. A. Ketchum, Z. Lai, Y. Lei, Z. Li, J. Li, Y. Liang, X. Lin, F. Lu, G. V. Merkulov, N. Milshina, H. M. Moore, A. K. Naik, V. A. Narayan, B. Neelam, D. Nusskern, D. B. Rusch, S. Salzberg, W. Shao, B. Shue, J. Sun, Z. Wang, A. Wang, X. Wang, J. Wang, M. Wei, R. Wides, C. Xiao, C. Yan, A. Yao, J. Ye, M. Zhan, W. Zhang, H. Zhang, Q. Zhao, L. Zheng, F. Zhong, W. Zhong, S. Zhu, S. Zhao, D. Gilbert, S. Baumhueter, G. Spier, C. Carter, A. Cravchik, T. Woodage, F. Ali, H. An, A. Awe, D. Baldwin, H. Baden, M. Barnstead, I. Barrow, K. Beeson, D. Busam, A. Carver, A. Center, M. L. Cheng, L. Curry, S. Danaher, L. Davenport, R. Desilets, S. Dietz, K. Dodson, L. Doup, S. Ferriera, N. Garg, A. Gluecksmann, B. Hart, J. Haynes, C. Haynes, C. Heiner, S. Hladun, D. Hostin, J. Houck, T. Howland, C. Ibegwam, J. Johnson, F. Kalush, L. Kline, S. Koduru, A. Love, F. Mann, D. May, S. McCawley, T. McIntosh, I. McMullen, M. Moy, L. Moy, B. Murphy, K. Nelson, C. Pfannkoch, E. Pratts, V. Puri, H. Qureshi, M. Reardon, R. Rodriguez, Y. H. Rogers, D. Romblad, B. Ruhfel, R. Scott, C. Sitter, M. Smallwood, E. Stewart, R. Strong, E. Suh, R. Thomas, N. N. Tint, S. Tse, C. Vech, G. Wang, J. Wetter, S. Williams, M. Williams, S. Windsor, E. Winn-Deen, K. Wolfe, J. Zaveri, K. Zaveri, J. F. Abril, R. Guigo, M. J. Campbell, K. V. Sjolander, B. Karlak, A. Kejariwal, H. Mi, B. Lazareva, T. Hatton, A. Narechania, K. Diemer, A. Muruganujan, N. Guo, S. Sato, V. Bafna, S. Istrail, R. Lippert, R. Schwartz, B. Walenz, S. Yooseph, D. Allen, A. Basu, J. Baxendale, L. Blick, M. Caminha, J. Carnes-Stine, P. Caulk, Y. H. Chiang, M. Coyne, C. Dahlke, A. Mays, M. Dombroski, M. Donnelly, D. Ely, S. Esparham, C. Fosler, H. Gire, S. Glanowski, K. Glasser, A. Glodek, M. Gorokhov, K. Graham, B. Gropman, M. Harris, J. Heil, S. Henderson, J. Hoover, D. Jennings, C. Jordan, J. Jordan, J. Kasha, L. Kagan, C. Kraft, A. Levitsky, M. Lewis, X. Liu, J. Lopez, D. Ma, W. Majoros, J. McDaniel, S. Murphy, M. Newman, T. Nguyen, N. Nguyen, and M. Nodell. 2001, The sequence of the human genome. *Science* **291:** 1304-1351.

Vondriska, T. M., J. B. Klein, and P. Ping. 2001a, Use of functional proteomics to investigate PKC epsilon-mediated cardioprotection: the signaling module hypothesis. *Am.J.Physiol Heart Circ.Physiol* **280:** H1434-H1441.

Vondriska, T. M., J. Zhang, C. Song, X. L. Tang, X. Cao, C. P. Baines, J. M. Pass, S. Wang, R. Bolli, and P. Ping. 2001b, Protein kinase C epsilon-Src modules direct signal transduction in nitric oxide-induced cardioprotection: complex formation as a means for cardioprotective signaling. *Circ.Res.* **88:** 1306-1313.

Washburn, M. P., D. Wolters, and J. R. Yates, III. 2001, Large-scale analysis of the yeast proteome by multidimensional protein identification technology. *Nat.Biotechnol.* **19:** 242-247.

Wasinger, V. C., S. J. Cordwell, A. Cerpa-Poljak, J. X. Yan, A. A. Gooley, M. R. Wilkins, M. W. Duncan, R. Harris, K. L. Williams, and I. Humphery-Smith. 1995, Progress with gene-product mapping of the Mollicutes: Mycoplasma genitalium. *Electrophoresis* **16:** 1090-1094.

Watson, J. D. and F. H. Crick. 1974, Molecular structure of nucleic acids: a structure for deoxyribose nucleic acid. J.D. Watson and F.H.C. Crick. Published in Nature, number 4356 April 25, 1953. *Nature* **248:** 765.

Weekes, J., K. Morrison, A. Mullen, R. Wait, P. Barton, and M. J. Dunn. 2003, Hyperubiquitination of proteins in dilated cardiomyopathy. *Proteomics.* **3:** 208-216.

Weekes, J., C. H. Wheeler, J. X. Yan, J. Weil, T. Eschenhagen, G. Scholtysik, and M. J. Dunn. 1999, Bovine dilated cardiomyopathy: proteomic analysis of an animal model of human dilated cardiomyopathy. *Electrophoresis* **20:** 898-906.

Westbrook, J. A., J. X. Yan, R. Wait, S. Y. Welson, and M. J. Dunn. 2001, Zooming-in on the proteome: very narrow-range immobilised pH gradients reveal more protein species and isoforms. *Electrophoresis* **22:** 2865-2871.

Wheeler, C. H., S. L. Berry, M. R. Wilkins, J. M. Corbett, K. Ou, A. A. Gooley, I. Humphery-Smith, K. L. Williams, and M. J. Dunn. 1996, Characterisation of proteins from two-dimensional electrophoresis gels by matrix-assisted laser desorption mass spectrometry and amino acid compositional analysis. *Electrophoresis* **17:** 580-587.

Wheeler, C. H., A. Collins, M. J. Dunn, S. J. Crisp, M. H. Yacoub, and M. L. Rose. 1995, Characterization of endothelial antigens associated with transplant-associated coronary artery disease. *J.Heart Lung Transplant.* **14:** S188-S197.

Wildgruber, R., A. Harder, C. Obermaier, G. Boguth, W. Weiss, S. J. Fey, P. M. Larsen, and A. Gorg. 2000, Towards higher resolution: two-dimensional electrophoresis of Saccharomyces cerevisiae proteins using overlapping narrow immobilized pH gradients. *Electrophoresis* **21:** 2610-2616.

Wilkins, M. R., E. Gasteiger, C. H. Wheeler, I. Lindskog, J. C. Sanchez, A. Bairoch, R. D. Appel, M. J. Dunn, and D. F. Hochstrasser. 1998, Multiple parameter cross-species protein identification using MultiIdent--a world-wide web accessible tool. *Electrophoresis* **19:** 3199-3206.

Wilkins, M. R., J. C. Sanchez, A. A. Gooley, R. D. Appel, I. Humphery-Smith, D. F. Hochstrasser, and K. L. Williams. 1996, Progress with proteome projects: why all proteins expressed by a genome should be identified and how to do it. *Biotechnol.Genet.Eng Rev.* **13:** 19-50.

Wilm, M., A. Shevchenko, T. Houthaeve, S. Breit, L. Schweigerer, T. Fotsis, and M. Mann. 1996, Femtomole sequencing of proteins from polyacrylamide gels by nano-electrospray mass spectrometry. *Nature* **379:** 466-469.

Wolters, D. A., M. P. Washburn, and J. R. Yates, III. 2001, An automated multidimensional protein identification technology for shotgun proteomics. *Anal.Chem.* **73:** 5683-5690.

Yan, J. X., A. T. Devenish, R. Wait, T. Stone, S. Lewis, and S. Fowler. 2002a, Fluorescence two-dimensional difference gel electrophoresis and mass spectrometry based proteomic analysis of Escherichia coli. *Proteomics.* **2:** 1682-1698.

Yan, J. X., R. A. Harry, C. Spibey, and M. J. Dunn. 2000, Postelectrophoretic staining of proteins separated by two-dimensional gel electrophoresis using SYPRO dyes. *Electrophoresis* **21:** 3657-3665.

Yan, J. X., Wait, R, Berkelman, T, Harry, RA, Westbrook, JA, Wheeler, CH, and Dunn, MJ. 2002b. A modified silver staining protocol for visualization of proteins compatible with matrix-assisted laser desorption/ionization and electrospray ionization-mass spectrometry. *Electrophoresis* **21:** 3666-3672.

Yan, J. X., M. R. Wilkins, K. Ou, A. A. Gooley, K. L. Williams, J. C. Sanchez, O. Golaz, C. Pasquali, and D. F. Hochstrasser. 1996, Large-scale amino-acid analysis for proteome studies. *J.Chromatogr.A* **736:** 291-302.

Yates, J. R., III. 1998, Database searching using mass spectrometry data. *Electrophoresis* **19:** 893-900.

Ytrehus, K., Y. Liu, and J. M. Downey. 1994, Preconditioning protects ischemic rabbit heart by protein kinase C activation. *Am.J.Physiol.* **266:** H1145-H1152.

Zhou, G., H. Li, D. DeCamp, S. Chen, H. Shu, Y. Gong, M. Flaig, J. W. Gillespie, N. Hu, P. R. Taylor, M. R. Emmert-Buck, L. A. Liotta, E. F. Petricoin, III, and Y. Zhao. 2002, 2D differential in-gel electrophoresis for the identification of esophageal scans cell cancer-specific protein markers. *Mol.Cell.Proteomics.* **1:** 117-124.

Zuo, X., L. Echan, P. Hembach, H. Y. Tang, K. D. Speicher, D. Santoli, and D. W. Speicher. 2001, Towards global analysis of mammalian proteomes using sample prefractionation prior to narrow pH range two-dimensional gels and using one-dimensional gels for insoluble and large proteins. *Electrophoresis* **22:** 1603-1615.

Zuo, X. and D. W. Speicher. 2000, A method for global analysis of complex proteomes using sample prefractionation by solution isoelectrofocusing prior to two-dimensional electrophoresis. *Anal.Biochem.* **284:** 266-278.

Chapter 3

Proteomics : Haematological Perspectives

JEAN-DANIEL TISSOT and PHILIPPE SCHNEIDER
*Service Régional Vaudois de Transfusion Sanguine, Lausanne, rue du Bugnon 27, CH-1005
Lausanne, Switzerland*

1.   INTRODUCTION

Blood, pure and eloquent, is the title of the book published by one of the most famous haematologist, M. M. Wintrobe (Wintrobe, 1980). This title perfectly summarizes what is blood. Since antiquity, it has been recognized as the essential component of life. Without knowledge of the circulation, ancient Egyptian religion recognized the heart as the seat of the soul (Hart, 2001). However, for centuries, blood remains a quite mysterious liquid. The key that was missing for its study was a method of scrutinizing its elements. This arrived through the efforts of two cousins, Carl Weigert and Paul Ehrlich. Weigert developed staining of tissue sections, while Ehrlich, still a medical student, applied these methods to staining heat-dried blood films. The last 50 years have seen an extraordinary increase in our knowledge of the pathophysiology of almost all the common blood diseases. This allowed the development of efficient treatments for various haematological disorders, such as pernicious anaemia, acute leukaemia or haemophilia (Weatherall, 2000).

Blood is composed of two compartments that are plasma and cells, notably red blood cells, leukocytes, and platelets. For decades, the detail microscopic evaluation of these elements, as well as of their modifications, was the basis of classical haematology. Our increasing knowledge of the molecular and cellular mechanisms offered valuable insights into the origins of blood disorders. As a result of these advances, the pattern of haematological practice has changed quite dramatically over recent years.

57

H. Hondermarck (ed.), Proteomics: Biomedical and Pharmaceutical Applications, 57–99.

From a descriptive medical discipline, haematology has evolved towards a rapidly changing science, that is situated on the cross road of genomics and proteomics.

The largely invariant genome of an individual determines its potential for protein expression. However, it neither specifies which proteins are expressed in the cells nor their level of post-translational modifications. For a long time, the complete characterization of the genome of various species has been an aim of the scientific community. Until the 70's, even a short sequence of nucleic acid was difficult to obtain. Some years later, two methods permitting the sequencing of the genomes were elaborated, one by F. Sanger and A. R. Coulson (Sanger and Coulson, 1975), the other by A. Maxam and W. Gilbert (Maxam and Gilbert, 1977).

The first genome to be sequenced was that of *Haemophilus influenzae* in 1995 (Fleischmann *et al.*, 1995). Since that time, several genomes were entirely sequenced, and in 2001 the human genome was completed (Lander *et al.*, 2001; Venter *et al.*, 2001). This allowed to estimate the number of encoded genes in humans. Because one gene does not produce only one protein, at least 30'000 genes encode for about 100'000 proteins. Furthermore, due to alternative splicing, different mRNAs are synthesized, conducting to the production of different proteins (Roberts and Smith, 2002). In addition, different post-translational modifications such as phosphorylation or glycosylation can occur in the cells (Banks *et al.*, 2000; Imam-Sghiouar *et al.*, 2002; Sarioglu *et al.*, 2000; Wilkins *et al.*, 1999; Yan *et al.*, 1999). These and other post-translational modifications are crucial for protein function. They are also of importance for protein stability as well as their cellular location. All these changes are not apparent from genomic sequence or mRNA expression data. In addition, during the life time of an individual, the synthesis of specific proteins can be activated or suppressed. Consequently, it is important to know which proteins are produced by a given cell at a given moment. To do so, the first step is to study the transcriptome, which is the pool of all transcribed mRNA molecules (Devaux *et al.*, 2001; Strausberg and Riggins, 2001). The transcriptome is translated into proteins, which characterise the status of a cell or of an organ (Kettman *et al.*, 2001; Oliver *et al.*, 2002).

The term proteome describes the expressed protein complement of a cell or a tissue at a given time (Anderson and Anderson, 1998; Banks *et al.*, 2000; Fields, 2001). The progresses made in protein sciences over the last few years have been made possible through the developments of mass spectrometry. They have recently led to the 2002 awards of the Nobel Prize in chemistry to J. B. Fenn and K. Tanaka (Cho and Normile, 2002).

The rescent advances made in bioinformatics as well as in proteomics are particularly impressive, and a number of proteomic databases dealing with plasma and/or blood cells are now available on the Internet (Table 1).

Table 1. Some Internet sites devoted to blood proteomics

Consortium for plasma science	http://www.plasmaconsortium.com/links.htm
Haematopoietic cell lines	http://www-smbh.univ-paris13.fr/lbtp/Biochemistry/Biochimie/bque.htm
Haematopoietic cells	http://proteomics.cancer.dk/2Dgallery/2Dgallery.html
Human plasma proteome project (HUPO)	http://www.hupo.org
Human platelets	http://www.bioch.ox.ac.uk/glycob/ogp/
Jurkat T-cell	http://www.mpiib-berlin.mpg.de/2D-PAGE/organisms/index.html
Myeloid development	http://bioinfo.mbb.yale.edu/expression/myelopoiesis/
Plasma proteins: clinical utility and interpretation	http://www.fbr.org/publications/plasmaprot/5553POCK.PDF
PlasmaDB Database 2D gels	http://www.lecb.ncifcrf.gov/plasmaDB/2dGels.html
Proteomics and protein structure (rat and cow serum)	http://linux.farma.unimi.it/
Swiss-2DPAGE (plasma)	http://www.expasy.org/cgi-bin/map2/def?PLASMA_HUMAN
Swiss-2DPAGE (platelets)	http://www.expasy.org/cgi-bin/map2/def?PLATELET_HUMAN
Swiss-2DPAGE (red blood cells)	http://www.expasy.org/cgi-bin/map2/def?RBC_HUMAN
The plasma proteome institute	http://www.plasmaproteome.org/

2. PLASMA/SERUM

2.1 Overview

Among body fluids, plasma/serum is probably the biological fluid that has been studied in the more deepest details. Serum is obtaine from plasma after colotting. The clot is mainly composed by fibrin. Fibrinogen, the precursor of fibrin, and several proteins involved in the blood coagulation (coagulation factors) are absent in serum. For example, prothrombin, which is present in plasma and activated by the coagulation cascade into thrombin (Seligsohn and Lubetsky, 2001), is no more detectable in serum (Fig. 1).

Plasma/serum is very complex. It contains water, electrolytes, lipids and phosopholipids, carbohydrates, free amino acids, vitamins, nucleic acids, hormones, and proteins. Key improvements have been accomplished in the study of the global protein content (reviewed in details in the classical series edited by F.W. Putnam "The Plasma Proteins. Structure, Function, and Genetic Control" (Putnam, 1975).

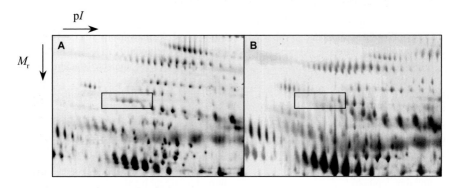

Figure 1. Details of silver stained high-resolution two-dimensional polyacrylamide gel of (A) plasma, and (B) of serum. Note the absence of the spots corresponding to prothombin in panel B. First dimension: immobilized 4 to 7 pH gradient, second dimension: 9 to 16% polyacrylamide gel electrophoresis.

Changes in plasma protein composition occur during fetal as well as during early extrauterine life (Tissot *et al.*, 1993b; Tissot *et al.*, 2000b). Not all proteins are produced by the conceptus, but many are transfered from the mother across the placenta. Some proteins remain at a fraction of their adult levels during all the intrauterine life, others are progressively produced in increasing amount when term approaches, and finally, others are present in higher concentrations during fetal life than after delivery. The plasma concentration of a given protein is governed not only by fetal synthesis and degradation, but also by exchange between the mother and the fetus through the placenta. Materno-fetal transfer of proteins involves several different mechanisms such as first-order process or active transport. The concentration of each fetal plasma protein results from a balance between opposing dynamic metabolic and physiological processes which proceed simultaneously. The relative impact of these processes continuously shifts during development, and not always in the same direction.

Plasma contains a number of well characterized proteins involved in blood coagulation (Dahlbäck, 2000), in fibrinolysis (Williams *et al.*, 2002), as well as in the control of the immune system through immunoglobulins (Igs) and the complement (Parkin and Cohen, 2001). Of interest, a single serine protease such as thrombin has been shown to be involved in many different biological functions such as conversion of fibrinogen to fibrin or activation of protein C into activated protein C. In addition, by its signalling action on different protease-activated receptors, thrombin triggers platelet activation and can also regulate the blood vessel diameter (Coughlin, 2000).

2.2   Studies of plasma/serum proteins

Arne Tiselius was the first, after having separated serum polypeptides by electrophoresis, to describe protein fractions corresponding to albumin, α-, β-, and γ-globulins. The first published diagram of human serum protein electrophoresis was published in 1939. The number of fractions slowly expanded into electrophoretic subfractions identified as α1, α2, β1, β2, γ1 and γ2. These fractions mobility characteristics and are still used to denote serum proteins such as α1-macroglobulin, α2-antiplasmin or β2-microglobulin (Tissot *et al.*, 2000c; Tissot *et al.*, 2000b). Resolution has been improved by the use of new support such as cellulose acetate, agarose, and polyacrylamide gels, which rendered electrophoretic methods very popular. Serum protein electrophoresis is still largely used in clinical laboratories. Clinical interpretation is based on the alteration of the content of one or more of the five fractions (Carlson, 2001). Over the last few years, capillary zone electrophoresis has emerged as a powerful new tool for rapid separation of various biopolymers, including proteins. Separation patterns obtained by capillary zone electrophoresis are similar to those obtained after densitometric scanning of cellulose acetate membrane electrophoresis or agarose gel electrophoresis (Bossuyt *et al.*, 1998; Oda *et al.*, 1997). Recently, dedicated automated systems for the routine analysis of human serum proteins in clinical laboratories have become commercially available. Several studies have clearly shown that the electrophoretic patterns and the clinical informations obtained by capillary zone electrophoresis are comparable with the data obtained by classical methods. In addition, the method is also suited to detect monoclonal gammopathies (Merlini *et al.*, 2001b; Tissot *et al.*, 2002b).

Another technique, high-resolution two-dimensional gel electrophoresis (2-DE), has been introduced in research laboratories. In 1975, investigators described optimised 2-D procedures in which proteins were denatured, and separated on polyacrylamide gel (Klose, 1975; O'Farrel, 1975). The first gel dimension comprised a separation by isoelectric focusing according to the peptide charge, and the second gel dimension separated proteins according to their apparent M_r (Fig. 2). Thus, peptides are separated from one another according to two independent biochemical characteristics (Tissot *et al.*, 2000a). 2-DE was shown to be particularly valuable in the study, as well as in the identification of thousands of cellular or secreted proteins, including many of those present in human plasma/serum. N. L. and N. G. Anderson, in 1977, were the first to demonstrate the usefulness of 2-DE for the study of human plasma proteins (Anderson and Anderson, 1977). Their report was a main breakthrough to the study of plasma proteins, and clearly established the first premises of plasma proteomics. Since that time, these authors have regularly reviewed the progress made on plasma proteomics over the years (Anderson and Anderson, 1991), and have recently published an important update on the human plasma proteome (Anderson and Anderson, 2002).

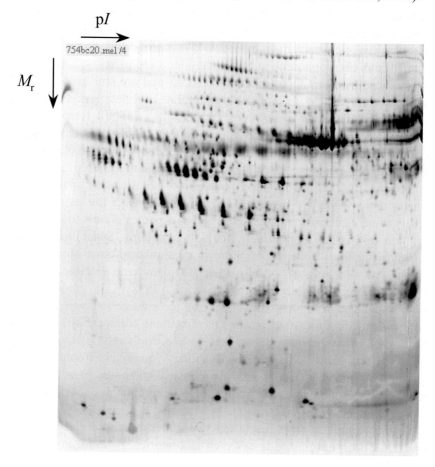

Figure 2. Silver stained high-resolution two-dimensional polyacrylamide gel of the plasma from a healthy blood donor. The first dimension comprises a separation according to the protein charge by isoelectric focusing, and the second separates proteins according to their sizes. Proteins are therefore separated from one another according to two independent biochemical properties. Polypeptides rarely appear as single spots, and most of them are resolved as multiple spots characterised by charge and size microheterogeneities (genetic polymorphism and/or post-translational modifications). The largest spot corresponds to albumin. First dimension: immobilized 4 to 7 pH gradient. Second dimension: 9 to 16% gradient polyacrylamide gel electrophoresis. Proteins can be identified by comparison with plasma protein maps available on the Internet (http://www.expasy.org/cgi-bin/map2/def?PLASMA_HUMAN).

About 300 different proteins have been now identified in human plasma (Table 2). Using on-line reversed-phase microcapillary liquid chromatography coupled with ion trap mass spectrometry, J. N. Adkins and collaborators were able to identify 490 proteins in serum (Adkins *et al.*, 2002). However, due to the high range of protein concentration (up to 12 orders of magnitude), the global analysis of plasma is all but an easy task (Corthals *et al.*, 2000). It has been estimated that up to 10'000 proteins may be present in plasma, most of which would be present in very low relative abundance. One approach to the study of these low-abundant proteins is to remove high-abundant common proteins such as albumin or immunoglobulins (Igs) (Rothenmund *et al.*, 2003; Wang *et al.*, 2003). Nowadays, efforts are done in order to identify most of the plasma proteins, by using industrial-scale proteomics (Argoud-Puy *et al.*, 2002). Using this approach, it rapidly appeared that pooling of large volumes of human plasma (up to 6 litres) is necessary in order to try to identify proteins that are present in trace amounts in plasma.

2.3   Applications of plasma/serum proteomics

Even if the technologies used in proteomics have evolved over the years (Anderson and Anderson, 1998), 2-DE is still a key and powerful tool to study genetic modifications (Anderson and Anderson, 1979; Tissot *et al.*, 1993b) as well as modifications of protein expression in plasma/serum of patients with various diseases (Hochstrasser and Tissot, 1993), or plasma treated for virus inactivation before transfusion (Tissot *et al.*, 1994a). The technique has been shown to be particularly interesting in the study of the monoclonal gammopathies (Tissot *et al.*, 1993d; Tissot *et al.*, 1993a; Tissot *et al.*, 2002b; Tissot and Spertini, 1995; Tracy *et al.*, 1982; Tracy *et al.*, 1984; Vu *et al.*, 2002) as well as in the evaluation of abnormal glycosylation of proteins (Golaz *et al.*, 1995; Gravel *et al.*, 1994; Gravel *et al.*, 1996; Henry *et al.*, 1997; Henry *et al.*, 1999; Heyne *et al.*, 1997; Zdebska *et al.*, 2001).

2.3.1   The immunoglobulinopathies

At the end of the 19th century, H. Bence Jones observed in the urine of patients an abnormal substance precipitating in the presence of nitric acid, redissolved by heating, and precipitating again by cooling [recently reviewed by R. A. Kyle (Kyle, 2000)]. This substance was later identified as Ig light chains, and its presence was linked to multiple myeloma. This disease corresponds to a neoplastic proliferation of plasma cells in the bone marrow (Bataille and Harousseau, 1997; Kyle, 1999a).

These plasma cells can produce a complete Ig constituted of two identical heavy chains of the same isotype and of two light chains of the same type, frequently in association with free light chains corresponding to an excess of their production compared to that of heavy chains (Kyle, 1994). In some cases, light chains (light chain disease) or, more rarely, heavy chain fragments (heavy chain disease) are solely produced by the malignant plasma cell clone, and in 1 to 5% of all patients with multiple myeloma, no detectable Ig-related protein is detected (nonsecretory myeloma). In about 85% of cases of myeloma without serum monoclonal gammopathy, Ig-related proteins are detected in the cytoplasm of plasma cells, and in 15%, they are completely absent (Bladé and Kyle, 1999).

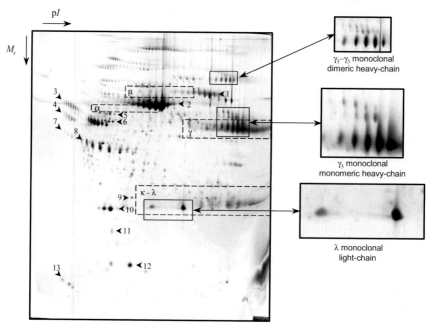

Figure 3. Silver stained high resolution *two*-dimensional polyacrylamide gel of a monoclonal IgG$_3$. 1, transferrin; 2, albumin; 3, alpha-1-antichymotrypsin; 4, , alpha-2-HS-glycoprotein; 5, antithrombin; 6, Gc-globulin; 7, lysin-rich glycoprotein; 8, haptoglobin beta-chain ; 9, serum amyloid P-component ; 10, apolipoprotein A-I; 11, retinol binding protein; 12, transthyretin; 13, apolipoprotein C-II, μ, polyclonal IgM heavy-chain area; α, polyclonal IgA heavy-chain area; γ, polyclonal IgG heavy-chain area; κ-λ, polyclonal Ig light-chain area. The monoclonal γ$_3$ heavy-chain is resolved as sets of spots characterized by charge and size microheterogeneities appearing as a monomer as well as a dimer. In this case, the monoclonal λ-light chain appears as two well defined spots, characterized by differences in their charge. First dimension: immobilized pH 3 to 10 gradient, second dimension : 9 to 16% gradient polyacrylamide gel [from (Tissot et al., 2002b) with permission of the publisher].

Table 2. Summary of proteins identified in human plasma [from (Anderson and Anderson, 2002), with permission of the authors and of the publisher].

5'-Nucleotidase	CA 72-4
Acid labile subunit of IGFBP	CA27.29/15-3 (MUC1 mucin antigens)
Acid phophatase, tartrate resistant	Calreticulin
Acid phosphatase, prostatic	Carboxypeptidase N, regulatory
Actin beta (from platelets)	Carboxypeptidase N, catalytic
Actin gamma (from platelets)	Carcinoembryonic antigen
Adenosine deaminase	Cathepsin D
Adiponectin	CD5 antigen-like protein (Spα)
Alanine aminotransferase (ALT)	Ceruloplasmin
Albumin	Cholinesterase plasma
Aldolase (muscle type)	Chorionic gonadotropin beta (hCG)
Alkaline Phosphatase (bone)	Chromogranin A
Alpha-1,3-fucosyltransferase (FUT6)	Chromogranin B (secretogranin I)
Alpha-1-acid Glycoprotein	Coagulation Factor II (prothrombin)
Alpha-1-antichymotrypsin	Coagulation Factor IX
Alpha-1-antitrypsin	Coagulation Factor V
Alpha-1B glycoprotein	Coagulation Factor VII, H
Alpha-1-microglobulin	Coagulation Factor VII, L
Alpha-2-antiplasmin	Coagulation Factor VIII
Alpha-2-HS glycoprotein	Coagulation Factor X
Alpha-2-macroglobulin	Coagulation Factor XI
Alpha-fetoprotein	Coagulation Factor XII
Amylase (pancreatic)	Coagulation Factor XIII A
Angiostatin	Coagulation Factor XIII B
Angiotensin converting enzyme (ACE)	Collagen I c-terminal propeptide
Angiotensinogen	Collagen I c-terminal telopeptide
Antithrombin III (AT3)	Collagen I n-terminal propeptide
Apolipoprotein A-I	Collagen I n-terminal telopeptide
Apolipoprotein A-II	Collagen III c-terminal propeptide
Apolipoprotein A-IV	Collagen III n-terminal propeptide
Apolipoprotein B-100	Collagen IV 7S n-terminal propeptide
Apolipoprotein B-48	Complement C1 Inhibitor
Apolipoprotein C-I	Complement C1q, A
Apolipoprotein C-II	Complement C1q, B
Apolipoprotein C-III	Complement C1q, C
Apolipoprotein C-IV	Complement C1r
Apolipoprotein D	Complement C1s
Apolipoprotein E	Complement C2
Apolipoprotein F	Complement C3A anaphylatoxin
Apolipoprotein H	Complement C3B, alpha
Apolipoprotein J (Clusterin)	Complement C3B, beta
Apolipoprotein(a)	Complement C4, gamma
Aspartate aminotransferase (AST)	Complement C4 binding protein, alpha
Beta thromboglobulin	Complement C4 binding protein, beta
Beta-2-microglobulin	Complement C5A anaphylotoxin
CA 125	Complement C1q, B
CA 19-9	Complement C1q, C

Table 2 (cont.). Summary of proteins identified in human plasma

Complement C5B, alpha	Growth hormone binding protein
Complement C5B, beta	Haptoglobin alpha-1
Complement C6	Haptoglobin alpha-2-chain
Complement C7	Haptoglobin beta chain
Complement C8, alpha	Haptoglobin beta chain, cleaved
Complement C8, beta	Haptoglobin-related gene product
Complement C8, gamma	Hemoglobin, alpha
Complement C9	Hemoglobin, beta
Complement Factor B	Hemopexin (beta-1B-glycoprotein)
Complement Factor B-Bb Fragment	Histidinerich alpha-2-glycoprotein
Complement Factor D	ICAM1, soluble
Complement Factor H	Ig kappa light chain
Complement Factor I	Ig lambda light chain
Connective tissue activating peptide III	IgA1
Corticotropin releasing hormone (CRH)	IgA2
C-reactive Protein	IgD
Creatine Kinase, B	IgE
Creatine Kinase, M	IGFBP3
CRHBP	IgG1
Cystatin C	IgG2
Elastase (neutrophil)	IgG3
Eosinophil granule major basic protein	IgG4
E-selectin, soluble	Ig J-chain
Ferritin, H	IgM
Ferritin, L	Inhibin (activin), beta A
Fibrin fragment D-dimer	Inhibin (activin), beta B
Fibrinogen extended gamma chain	Inhibin (activin), beta C
Fibrinogen, alpha	Inhibin (activin), beta E
Fibrinogen, beta	Inhibin, alpha
Fibrinogen, gamma	Insulin C-peptide
Fibronectin	Insulin, A chain
Fibulin-1	Insulin, B chain
Ficolin 1	Insulin-like growth factor IA
Ficolin 2	Insulin-like growth factor II
Ficolin 3	Inter-alpha trypsin inhibitor, H1
Follicle stimulating hormone	Inter-alpha trypsin inhibitor, H2
G6PD	Inter-alpha trypsin inhibitor, H4
Galactoglycoprotein (leukosialin)	Inter-alpha trypsin inhibitor, L
Gammaglutamyl transferase alpha	Interferon alpha
Gc-globulin	Interferon beta
GCSF	Interferon gamma
Gelsolin	Interleukin-1 beta
GHRH	Interleukin-10
Glutamate carboxypeptidase II	Interleukin-12, alpha
Glutathione S-transferase	Interleukin-12, beta
Glycoprotein hormones alpha chain	Interleukin-1 receptor antagonist
GMCSF	Interleukin-2
Growth hormone	Interleukin-4

Table 2 (cont.). Summary of proteins identified in human plasma

Interleukin-5	Prolactin
Interleukin-6	Prolyl hydroxylase, alpha
Interleukin-8	Prolyl hydroxylase, beta
IP10, small inducible cytokine B10	Prostaglandin-H2 D-isomerase
Isocitrate dehydrogenase	Prostate specific antigen
Kininogen	Protein Z
Ksp37	P-selectin, soluble
Laminin, alpha	Rantes
Laminin, beta	Renin
Laminin, gamma	Retinol Binding Protein
LDH (heart)	S100 protein
Lecithin-cholesterol acyltransferase	Secretogranin V
Leucine-rich alpha-2-glycoprotein	Serum Amyloid A
LHRH	Serum Amyloid P
Lipase	Sex Hormone Binding Globulin
Luteinizing hormone (LH), beta	Tetranectin
Mannose-binding protein	Thyroglobulin
Matrix metalloproteinase-2	Thyroid Stimulating Hormone
M-CSF	Thyrotropin-releasing hormone
Melastatin	Thyroxin Binding Globulin
MIP-1 alpha	Tissue Factor
MIP-1 beta	Tissue inhibitor of metalloproteinases 1
MSE55	Tissue inhibitor of metalloproteinases 2
Myelin basic protein	Tissue plasminogen activator
Myoglobin	Tissue plasminogen activator Inhibitor
N-acetyl-B-D-glucosaminidase, alpha	TNF-alpha
N-acetyl-B-D-glucosaminidase, beta	TNF-binding protein 1
N-acetyl-muramyl-L-alanine amidase	TNF-binding protein 2
Neuron-specific enolase	Transcobalamin
Neutrophil-activating peptide 2	Transcortin
Osteocalcin	Transferrin
Osteonectin	Transferrin (asialo, tau, beta-2-)
Pancreatic zymogen granule protein GP2	Transferrin receptor (soluble)
Paraoxonase parathyroid hormone	Transthyretin
Parathyroid hormone-related protein	Triacylglycerol lipase (pancreatic)
PASP	Troponin I (cardiac)
Pepsinogen A	Troponin I (skeletal)
Plasma hyaluronan binding protein	Troponin T (cardiac)
Plasma kallikrein	Tryptase, beta-2
Plasma serine protease inhibitor	Tyrosine hydroxylase
Plasminogen -Glu	Urokinase (High MW kidney type) A
Plasminogen -Lys	Urokinase (High MW kidney type) B
Platelet Factor 4	VCAM-1, soluble
Pre-alpha trypsin inhibitor, H3	Vitronectin
Pregnancy-associated plasma protein-A	Von Willebrand Factor
Pregnancy-associated plasma protein-A2	Zn-alpha-2-glycoprotein
Pregnancy-specific beta-1-glycoprotein 3	

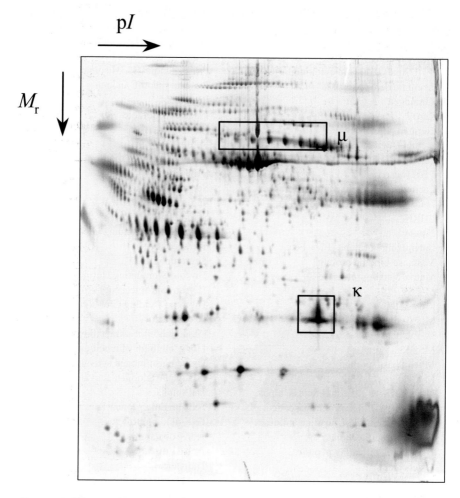

Figure 4. Silver stained high-resolution two-dimensional polyacrylamide gel of the a serum sample of a patient presenting with monoclonal IgM-κ. First dimension: immobilized pH 3 to 10 gradient, second dimension: 9 to 16% gradient polyacrylamide gel. μ, monoclonal; κ, monoclonal light chain.

Monoclonal gammopathy is related to the presence in the blood of a monoclonal Ig (Sahota *et al.*, 1999). Serum protein analysis should be done when multiple myeloma, Waldenström's macroglobulinemia, primary amyloidosis or a related B cell disorder is suspected (Attaelmannan and Levinson, 2000). The strategies to detect as well as to identify monoclonal gammopathies in the clinical laboratory have been reviewed (Keren *et al.*, 1999; Kyle, 1999b). Other, but not routinely used electrophoretic techniques, have been employed to further characterize monoclonal Igs, including isoelectric focusing and 2-DE.

The interest of 2-DE as well as its limitations in the analysis of monoclonal gammopathy, either in plasma/serum or urine samples, have been described by different groups of investigators, (Goldfarb, 1992; Harrison, 1992; Harrison *et al.*, 1993; Spertini *et al.*, 1995; Tissot *et al.*, 1992; Tissot *et al.*, 1993c; Tissot *et al.*, 1993a; Tissot *et al.*, 1993d; Tissot *et al.*, 1994b; Tissot *et al.*, 1999; Tissot and Hochstrasser, 1993; Tissot and Spertini, 1995; Tracy *et al.*, 1982; Tracy *et al.*, 1984). Notably, 2-DE has been shown to be useful in difficult cases, such as oligoclonal gammopathies, cold agglutinins or cryoglobulins [reviewed by (Tissot *et al.*, 2002a)].

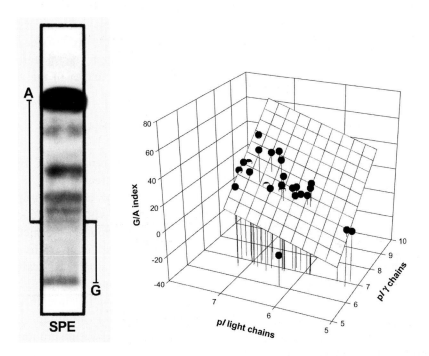

Figure 5. Serum protein electrophoresis (SPE) of a monoclonal IgG and three dimensional scatter and mesh. The latter shows a good correlation (R=0.733) between the G/A index, the pI of the light chains and that of the heavy chains (measured using 2-DE). The G/A index is obtained after SPE, and is calculated by dividing the length of the migration of the monoclonal IgG (G) from the application point with that of the migration of albumin (A). G/A index=-200.827+(19.707×pI heavy chains)+(13.053×pI light chains) [adapted from (Vu et al., 2002), with permission of the publisher].

Since proteins are separated under denaturing and reducing conditions, heavy chains are well separated from light chains. In addition, heavy chains of different isotypes are detected on non overlapping areas of the gels (Figs. 2 and 3). The map locations of IgG γ chains and of κ and λ chains were first described by N. L. Anderson and N. G. Anderson (Anderson and Anderson, 1977). Based on the analysis of monoclonal Igs of different isotypes, the position of IgA α, IgM μ, IgD δ, and IgE ε were then identified [reviewed by (Tissot and Spertini, 1995)].

Polyclonal κ and λ light chains from normal human plasma appear as a wide band in the 25'000 to 28'000 Da part of the gel map. Monoclonal light chains can be detected as sets of spots (Fig. 3) or as isolated (Fig. 4) prominent spots.

D. H. Vu *et al.*, studied the electrophoretic properties of monoclonal IgG in a series of 73 cases (Vu *et al.*, 2002). They showed that a classification of these monoclonal gammopathies, based only on their electrophoretic properties of their heavy and light chains was not possible, because of their extreme biochemical diversity. This diversity was expressed by a large macroheterogeneity expressed by the differences of the pI of each monoclonal light and heavy chains (Fig. 5), and by their microhetero-geneities, detected after 2-DE as minor variations of the pI and/or of the M_r of the corresponding spots (Fig. 3).

2.3.2   Proteomics of Igs

Interesting improvements have been achieved in the analysis of Igs separated by 2-DE using mass spectrometry. An important breakthrough was made by A. Lebaud *et al.*, in their study of a case of crystal-storing histiocytosis, which is a rare event in disorders associated with monoclonal gammopathy (Lebeau *et al.*, 2002). These investigators reported a patient presenting with IgA-κ paraproteinemia and paraproteinuria. 2-DE of liver tissue combined with immunoblotting revealed the massive storage of heavy chains of α type and light chains of κ type, each in a monoclonal pattern. Analysis of the stored κ light chain by nanoelectrospray-ionization mass spectrometry indicated that it belongs to the variable κI subgroup. In addition, the authors identified some unusual amino acid substitutions including Leu59, usually important for hydrophobic interactions within a protein, at a position where it has never been previously described in plasma cell disorders. These results suggested that conformational alterations induced by amino acid exchanges represented a crucial pathogenic factor in crystal-storing histiocytosis. As shown by this particular example, proteomics will certainly be an interesting approach to evaluate disorders associated with protein aggregation, that result of improper folding or misfolding of proteins (Merlini *et al.*, 2001a).

Proteomics was also shown to be useful to study cryoglobulins (Fig. 6). Cryoglobulins are cold-precipitable Igs associated with a number of infectious, autoimmune and neoplastic disorders (Schifferli *et al.*, 1995;

Tissot *et al.*, 1993a; Tissot *et al.*, 1994b). The precipitate is frequently composed of Igs of different isotypes, particularly of IgM and IgG, the former being the antibody, the latter being the antigen

Using Fourier transform ion cyclotron resonance (FT-ICR) mass spectrometry analysis, E. Damoc *et al.*, showed that accurate biochemical data can be obtained for both Ig heavy and light chains contained in cryoprecipitates (Damoc *et al.*, 2002). FT-ICR mass spectrometry is of all mass spectrometric techniques currently available in proteomics the method providing the highest mass determination accuracy (< 1 ppm). FT-ICR mass spectrometry permitted direct identification of the complementarity determining regions (CDR1 and CDR2) of the monoclonal IgM μ heavy chain variable region as well as that of the κ light chain (Damoc *et al.*, 2003).

In addition, using this technique, it was possible to identify more than 50% of the sequence of 2 different proteins associated with IgM that are the J-chain and CD5L (CD5 antigen-like protein, Spα). CD5L is a member of the scavenger receptor cysteine-rich superfamily of proteins, produced by immune cells that have important functions in the regulation of the immune system (Aruffo *et al.*, 1997; Gebe *et al.*, 1997; Gebe *et al.*, 2000; Tissot *et al.*, 2002a).

2.3.3   Plasma/serum protein modifications

Modifications of the expression of plasma proteins can be related to genetic variations, complex combinations of post-translational modifications, alterations directly or indirectly related to diseases, or combinations of these mechanisms. Genetic variants of proteins such as haptoglobin, Gc-globulin, apolipoprotein E, apolipoprotein A-1, transferrin, or alpha-1-antitrypsin can be identified using 2-DE (Anderson and Anderson, 1979; Rosenblum *et al.*, 1983; Tissot *et al.*, 1993b; Tissot *et al.*, 2000b).

Quantitative and qualitative changes of the protein expression during foetal and early extrauterine life have been observed. 2-DE revealed a progressively more complex protein pattern and an increase in the size of many spots (Tissot *et al.*, 1993b; Tissot *et al.*, 2000b). Three main protein pattern modifications were observed on 2-D gels of fetuses at different gestational ages : a) the progressive appearance of the protease inhibitor alpha-1-antichymotrypsin; b) the progressive increase of polyclonal IgG, which was particularly evident during the last weeks of pregnancy; c) the gradual diminution of alpha-fetoprotein which was undetectable on protein maps of term new-borns. On the other hand, before the 38th week of gestation, polyclonal IgA or IgM heavy chains, as well as the two main proteins involved in free haemoglobin transport and catabolism, that are haptoglobin and hemopexin, were undetectable.

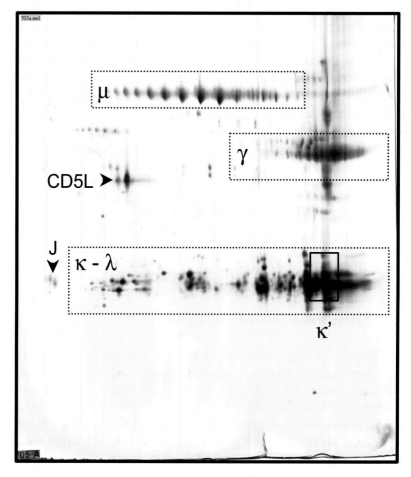

Figure 6. Silver stained high resdolution two-dimensional polyacrylamide gel of type II cryoglobulins (polyclonal IgG and monoclonal IgM). First dimension: immobilized pH 3 to 10 gradient, second dimension: 9 to 16% gradient polyacrylamide gel. μ, monoclonal IgM heavy chain; γ, polyclonal IgG heavy chains; κ-λ, polyclonal light chains; κ', monoclonal light chain hidden by polyclonal light chains, J, Ig J-chain; CD5L, CD5 antigen-like protein (Spα) [adapted from (Tissot et al., 2002b) with permission of the publisher].

A novel "foetal" polypeptide, characterised by an apparent Mr of 46'000 and a pI of 5.0 was identified. N-terminal microsequencing (25-EDPQ) and immunoblotting using specific anti-alpha-1-antitrypsin antibodies reveal that this spot most likely corresponds to a foetal specific form of alpha-1-antitrypsin (Tissot *et al.*, 2000b). The polypeptide was observed in all foetal samples and in all plasma/serum samples obtained from infants of less than 2 years of age, but was either undetectable or appears as a shaded spot in adults. This particular form of alpha-1-antitrypsin has been also identified in mouse foetal plasma (Nathoo and Finlay, 1987).

Changes of plasma/serum proteins in association with various diseases have been reviewed by different investigators over these last ten years (Anderson and Anderson, 2002; Hochstrasser and Tissot, 1993; Young and Tracy, 1995). Most of these studies reported on spot pattern modifications, as illustrated in Fig. 7, showing part of the plasma 2-D gel of a patient presenting with chronic renal insufficiency, and alterations of the pattern of retinol binding protein (Tissot *et al.*, 1991).

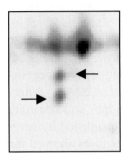

Figure 7. Enlarged zoom on retinol binding protein in renal insufficiency after two-dimensional polyacrylamide gel electrophoresis. First dimension: immobilized pH 3 to 10 gradient, second dimension: 9 to 16% gradient polyacrylamide gel. Retinol binding protein appeared as two large spots. This particular pattern modification is related to the accumulation, in renal failure, of a truncated variant of retinol binding protein (–266 Da) (arrow on the left) relative to the normal form (arrow on the right) (Kiernan et al., 2002).

Diseases-related post-translational modifications, notably those involving glycosylation have been well described by different investigators (Golaz *et al.*, 1995; Gravel *et al.*, 1994; Gravel *et al.*, 1996; Henry *et al.*, 1997; Henry *et al.*, 1999; Heyne *et al.*, 1997). In a study, undertaken to identify which plasma proteins are altered in patients with alcohol-related diseases, 2-DE and lectin blotting, followed by chemiluminescence detection, revealed abnormal microheterogeneities of haptoglobin, transferrin and alpha-1-antitrypsin (Gravel *et al.*, 1996). These data supported the hypothesis of a general mechanism of liver glycosylation alteration of plasma proteins induced by excessive alcohol consumption.

H. Henry and collaborators also showed alterations of the glycosylation of transferrin in patients with chronic alcohol abuse, producing the carbohydrate-deficient transferrin isoforms (Henry *et al.*, 1999). These alterations were very similar to those present in patients with carbohydrate-deficient glycoprotein syndrome type 1 (CDG1) (Henry *et al.*, 1997; Heyne *et al.*, 1997). In these patients, the pattern of glycoproteins was characterized by abnormal trains of isoforms with decreased mass (delta molecular weight 3'000) and all showed a cathodal shift. 2-DE and SDS-PAGE mass analysis of transferrin, alpha-1-antitrypsin, haptoglobin beta-chain, and alpha-1-acid glycoprotein after neuraminidase and N-glycosidase F treatments demonstrated that the additional trains of the isoforms found in CDG1 contain homologous species of isoforms. Some of them still showed charge differences, and all still contained glycans, except transferrin, with some unusual nonglycosylated isoforms. In addition, deficiencies in clusterin and serum amyloid P, not described so far, have been found in these CDG1 patients.

Modifications of protein patterns, and more particularly the presence of charge microheterogeneities may be due to deamidation, a relativerly commun observation after 2-DE of plasma proteins. H. Sarioglu et al., showed that several proteins, including beta-2-microglobulin and the haptoglobin chains, presented differences in p*I* due to deamidation of asparagines (Sarioglu *et al.*, 2000). After enzymatic cleavage with endopeptidases, the asparagine and deamidated asparagine containing peptides were separated and quantified by reversed-phase high performance liquid chromatography. The authors showed that the difference between the spots was a growing number of negative charges introduced in the protein by an increasing number of deamidated asparagines. As a consequence, the mass difference between two spots was exactly 1 Da. Charge shifts to a more acidic p*I* may also be due to oxidation of cystein residues to cysteic acid (Vuadens *et al.*, 2003). P. Lutter *et al.*, reported intriguing results based on the analysis of rViscumin heterodimer (recombinant mistletoe lectin) using 2-DE and mass spectrometry (Lutter *et al.*, 2001). In this study, a series of spots, with different p*I* values, were observed, independently from the experimental conditions (urea concentration, heat treatment or the use of cysteine alkylating agents). Comparative studies of the most important spots using matrix assisted laser desorption/ionization-mass spectrometry, liquid chromatography-electrospray ionization mass spectrometry and liquid chromatography-electrospray ionization mass spectrometry-tandem mass spectrometry after tryptic in-gel digestion resulted in a sequence coverage of more than 90% for the two chains of rViscumin. No molecular differences related to common chemical, post-translational modifications or nonenzymatic deamidation were found to be the cause of the different charge values of the separated spots. Therefore, in order to gain insight in the biochemical cause of the p*I* heterogeneity, the protein spots were extracted from the 2-D gels and separated again by 2-DE. Each of the single spots tested in such experiments split up in the same heterogeneous pattern concerning the p*I* values, suggesting that the observed charge variants of the spots were the result of conformational protein variants, existing in an equilibrium during sample preparation and/or isoelectric focusing and were not caused from microheterogeneity in the primary structure of the protein. This important observation clearly indicates that modifications of the sopt patterns must be explored in details using modern proteomic tools in order to explore the biochemical background related to the electrophoretic microheterogeneities observed after 2-DE. Nowadays, it has been shown that mass spectrometry can be particularly useful to explore the biochemical origins of the modifications of plasma protein expression, as exemplified by Kiernan *et al.*, showing changes of retinol binding protein in patients with renal insufficiency (Kiernan *et al.*, 2002) or by J. M. Skehel *et al.*, in phenotyping apolipoprotein E (Skehel *et al.*, 2000).

Finally, it must be mentioned that significant proteomic studies have been published dealing with the modifications of serum protein expression that may occur in healthy as well as in animals presenting with various experimental diseases (Eberini *et al.*, 1999; Ebrini *et al.*, 2000; Gianazza *et al.*, 2002; Haynes *et al.*, 1998; Miller *et al.*, 1998; Miller *et al.*, 1999; Wait *et al.*, 2001; Wait *et al.*, 2002).

3. RED BLOOD CELLS

3.1 Overview

The production of erythroid cells is a dynamic and exquisitely regulated process. Haematopoitic stem cells, upon complex transcriptional control by various cytokines, notably erythropoietin, differentiate in the bone marrow into mature erythrocytes. These latter then circulate in the blood vessels. The primary function of red blood cells is to transport oxygen efficiently from lungs to tissues. They also transport carbon dioxide, and, as recently described, have a physiological role in the regulation of vascular tone through their interactions with nitric oxide (Pawloski and Stamler, 2002). Red blood cells also have roles in blood rheology, in platelet independent bleeding haemostasis (Blajchman *et al.*, 1994) or in the transport of immune complexes (Schifferli *et al.*, 1986).

Approximately 3×10^9 new erythrocytes are produced /kg/day in an adult. It is well established that erythropoietin is responsible both for maintaining normal erythropoiesis and for increasing red cell production in response to hypoxia.

Red blood cells are constituted by a plasma membrane and cytoplasm. The red blood cell membrane proteins, as well as the proteins of the

cytoskeleton have been particularly well characterized over these last ten
years (Avent and Reid, 2000; Cartron *et al.*, 1998; Cartron, 2000; Cartron
and Colin, 2001; Cartron and Salmon, 1983; Yazdanbakhsh *et al.*, 2000).

Enzymes of red cell metabolism have been isolated, and the structure and
the function of haemoglobin have been elucidated. Numerous diseases
related to the presence of abnormal haemoglobins, including the thalassemia
syndromes (Rund and Rachmilewitz, 2001; Schrier, 2002) have been
described. Finally, progresses have been accomplished in the understanding
of the cascades involved in the intermediate metabolism of the red blood
cells and its disorders (Beutler, 1989; Beutler, 1994; Beutler, 2001a; Beutler,
2001b) as well as those of the heme biosynthesis including porphyria
(Phillips *et al.*, 2001; Sassa, 2000).

3.2   Proteomic studies of red blood cells

2-D maps of red blood cells as well as of proteins associated with
erythrocyte ghosts have been published from the early years of 2-DE
(Bhakdi *et al.*, 1975; Conrad and Penniston, 1976; Copeland *et al.*, 1982;
Dockham *et al.*, 1986; Edwards *et al.*, 1979; Golaz *et al.*, 1993; Harell and
Morrison, 1979; Low *et al.*, 2002; Olivieri *et al.*, 2001; Rabilloud *et al.*,
1999; Rosenblum *et al.*, 1982; Rosenblum *et al.*, 1984). Several
intracytoplasmic proteins have been identified such as acetylcholinesterase,
aldehyde dehydrogenase 1A1, Alzheimer's disease amyloid A4 protein,
antioxidant protein 2, carbonic anydrase, catalase, flavin reductase, fructose
biphosphate aldolase, glucose-6-phosphate 1-dehydrogenase, glutathione
peroxidase, glyceraldehyde 3-phosphate dehydrogenase, glyoxalase,
haemoglobin alpha and beta chains, hypoxantine phosphoribosyl transferase,
isozymes R/L, lactate dehydrogenase, peroxiredoxin 2, porphobilinogen
synthase, proteasome beta chain, protein disulfide isomerase, pyruvate
kinase, ribose-phosphate pyrophosphokinase III, superoxide dismutase,
thioredoxin, or ubiquitin, 6 phosphogluconic deshydrogenase.

By contrast, few proteins of the cytoskeleton or of the erythrocyte
membrane were detected with 2-DE. The few proteins identified were actin,
band 3 and spectrin. T. Rabilloud and collaborators unambiguously
identified band 3 from normal erythrocyte membrane. They also observed
several protein spots from erythrocyte ghosts infected by *Plasmodium
falciparum* that were not present from those non infected (Rabilloud *et al.*,
1999). When working with red blood cell membrane proteins separated by 1-
DE and analysed by mass spectrometry, T. Y. Low and collaborators were
able to identify 44 polypeptides, of which only 19 were also found on 2-D
gels (Low *et al.*, 2002). Some of the proteins isolated by 1-DE were
polypeptides with high hydrophobicity. Only one protein associated with a

blood group system (Kell blood group protein) was found using this proteomic approach. These data clearly show the actual limits of 2-DE to study membrane proteins, and more particularly the blood group associated proteins. However, the use of novel zwitterionic detergents to solubilize proteins as those developed by C. Tastet and collaborators (Tastet *et al.*, 2003) or S. Luche and collaborators (Luche *et al.*, 2003), will certainly allow the identification of new red blood cell membrane proteins.

4.   PLATELETS

4.1   Overview

Platelets, also known as thrombocytes, are the smallest blood particles that circulate in blood. They derive from megakaryocytes and have critical roles in haemostasis and in the control of blood coagulation. Megakaryocytes development is a process in which a wide variety of signals is working to direct a highly specific response. Megakaryopoiesis and subsequent thrombopoiesis occur through complex biologic steps (Matsumura and Kanakura, 2002). Megakaryocyte precursors develope from haematopoietic stem cells. Then, they differentiate into mature polyploid megakaryocytes, and finally release platelets. Although a number of growth factors can augment megakaryopoiesis in vitro, thrombopoietin is the physiologic and the most potent regulator of megakaryopoiesis in vitro and in vivo (Kaushansky and Drachman, 2002). Thrombopoietin induces the growth of megakaryocyte precursors through activation of multiple signaling cascades. About two third of platelets circulate, and one thrird reside in the spleen. Platelets lack a nucleus and therefore have little ability to alter their biochemical composition or strucure.

Circulating platelets do not adhere to normal endothelium or to each other. However, they do adhere to subendothelium that is exposed when the endothelial lining of the vessel is broken. Platelet adhesion requires endothelial cell secretion of the protein von Willebrand factor, which is found in the vessel wall and in plasma (de Groot, 2002). Von Willebrand factor binds during platelet adhesion to a glycoprotein receptor of the platelet surface membrane glycoprotein Ib. Collagen and the first thrombin formed at the injury site activate platelets. These reactions activate phospholipase C, an enzyme that hydrolyzes inositol phospholipids. Products of this reaction activate protein kinase C and increase the calcium concentration of platelet cytosol, resulting in a series of overlapping events. Platelets also release factors to augment vasoconstriction (serotonin, thromboxane A_2) and initiate vessel wall repair (platelet-derived growth factor), and they provide surface membrane sites and components for formation of enzyme/cofactor complexes in blood coagulation reactions.

The clinical consequences of alterations in platelets are well known, ranging from severe thromboembolic episodes to hemorrhage resulting from thrombocytopenia. In the latter situation, platelet transfusion can be life-saving. It was observed in 1911 that transfusion of fresh whole blood boosted the platelet count in some cases of thrombocytopenia and controlled bleeding. Nowedays, platelets are often collected from blood donors by apheresis (Simon, 1994). This approach allows the supply of a therapeutically beneficial component, and platelets collected by apheresis have not been shown to be haemostatically different from platelets separated from whole blood donations.

4.2   Platelet proteomics

Platelets have been studied by 2-DE since more than 15 years (Hanash *et al.*, 1986b; Sarraj-Reguieg *et al.*, 1993), and several platelets 2-D maps have been published (Gravel *et al.*, 1995; Immler *et al.*, 1998; Marcus *et al.*, 2000). 2-DE has been shown to be an interesting tool to study the protein changes occurring during storage of platelet concentrates that are used in transfusion medicine (Sarraj-Reguieg *et al.*, 1993; Snyder *et al.*, 1987). However, the most important publication on this topics is that of E. E. O'Neil, who reported on the largest available analysis of the human platelet proteome (O'Neill *et al.*, 2002). These investigators, using large-scale proteome analysis, were able to dress a comprehensive protein expression data for platelets. Among the protein spots characterized, 123 proteins were identified (28 cytoskeletal proteins, 4 metabolic enzymes, 5 proteins with extracellular functions, 4 mitochondrial proteins, 24 proteins involved in protein processing, 31 proteins of the signaling process, 16 proteins of vesicules, 5 membrane receptors, and 7 miscellaneous proteins). This work is certainly a step towards the complete analysis of the human platelet proteins.

An interesting proteomic approach to understand the signalling events following platelet activation was recently reported (Maguire *et al.*, 2002). The investigators studied phosphotyrosine proteomes separated by 2-DE. In resting platelets, a small number of phosphorylated proteins were observed. In thrombin activated human platelets, however, there was a large increase in the number of tyrosine phosphorylated signalling proteins. Ten of these proteins were positively identified either by Western blotting or by matrix-assisted laser desorption/ionisation-time of flight mass spectrometry. They included focal adhesion kinase (FAK), tyrosine-protein kinase Syk, activin type I receptor (ALK-4), purinoreceptor P2X6 and mitogen-acivated protein kinase (MAPK). These results clearly showed the usefulness of proteomics to study the proteins involveld in the signalling of platelet activation.

5. LEUKOCYTES

5.1 Overview

As all blood cells, leucocytes derived from a common precursor, the pluripotent stem cell (Lapidot and Petit, 2002; Verfaillie, 2002; Zubair *et al.*, 2002). The cell has the capability of self-renewal and of differentiating into different cell lineages. A common myeloid progenitor gives rise to both granulocytes and monocytes. The early stages of granulopoiesis are mediated by several well characterized transcription factors (Friedman, 2002b). The early stages of granulopoiesis are mediated by the C/EBPalpha, PU.1, RAR, CBF, and c-Myb, and the later stages require C/EBPepsilon, PU.1, and CDP. Monocyte development requires PU.1 and interferon consensus sequence binding protein and can be induced by Maf-B, c-Jun, or Egr-1. Cytokine receptor signals modulate transcription factor activities but do not determine cell fates. Several mechanisms orchestrate the myeloid developmental program, including cooperative gene regulation, protein:protein interactions, regulation of factor levels, and induction of cell cycle arrest (Friedman, 2002a).

Granulocytes comprise neutrophils, eosinophils, and basophils. Neutrophils and monocytes (macrophages) constitue two important elements of the host defense system. Basophils and mast cells contribute to mucosal immunity, whereas eosinophils, basophils and mast cells aid in response to allergens and parasitic infections.

There are evidences that the lymphoid lineages, consisting of B, T and natural killer cells, are generated from a common lymphoid progenitor (Kondo *et al.*, 1997). The growth and the development of lymphoid cells occur in multiple anatomic locations, where different factors influence these processes. Evidence now exists that natural killer cells may be more related to T cells than to B cells. B and T cells are of great importance either in the control or as effectors of the immune response (Delves and Roitt, 2000a; Delves and Roitt, 2000b; von Andrian and Mackay, 2000). B cells can differentiate into mature plasmocytes that produce antibodies, and are therefore implicated in the humoral immune response. They express CD19, a B cell restricted 95 kDa protein present throughout the B cell lineage until plasma cell differentiation.

B and T cells are involved in the control as well as effectors of the immune response. B cells can differentiate into mature plasmocytes that produce antibodies, and are therefore implicated in the humoral immune response. T lymphocytes are the major actors of the cellular immune response (Akashi *et al.*, 1999; Hardy and Hayakawa, 2001).

T lymphocytes are separated in two main subsets, that can be differentiated according to the expression of different molecular markers, CD4 and CD8. The CD4 and CD8 proteins are cell surface molecules related to the lineage functions. They are intimate participants in the recognition of peptide-major histocompatibility complex ligands by T cell receptors. The T suppressor-inducer lymphocytes express predominantly the CD4 molecule, whereas the T cytotoxic-suppressor lymphocytes show on their surfaces the CD8 molecule. Activation of naive CD4 positive helper T (Th) cells through the T cell receptor causes these cells to proliferate and differentiate into effector T helper cells. Two subsets of effector Th cells have been defined on the basis of their distinct cytokine secretion patterns and their immunomodulatory effects (Murphy and Reiner, 2002). Th1 cells produce primarily interferon-gamma and tumor necrosis factor, which are required for cell-mediated inflammatory reactions, whereas Th2 cells secrete interleukin (IL)-4, IL-5, IL-10 and IL-13, which mediate B cell activation and antibody production. In general, an efficient clearance of intracellular pathogens is based on innate cell activation, while antibody responses are best suited for extracellular infections. The decision of naive CD4^{+} T cells to become Th1 and Th2 has important consequences in the success of an immune response and the progression of diseases. For example, a Th1 response against *Leishmania major* results in the resolution of disease, while a Th2-type response allows the progression of disease. A predominant Th1 response has been observed in several autoimmune diseases, such as rheumatoid arthritis, experimental autoimmune encephalomyelitis and insulin-dependent diabetes mellitus (Diehl and Rincon, 2002).

B cell differentiation is a highly regulated process with pathways and steps that have been well delineated (Hardy and Hayakawa, 2001; Rolink *et al.*, 2001). This development depends on intrinsic as well as extrinsic factors. These latter factors mainly originate from interactions of differentiating progenitor B cells with supportive stromal cells. In particular, chemokines, as well as their receptors, play important roles in lymphopoiesis and in lymphoid organ development (Ansel and Cyster, 2001; Parkin and Cohen, 2001). It has been estimated that lymphocytes have the potential to produce about 10^{15} different antibodies in humans. This extraordinary feat is achieved from the use of fewer than 400 genes, with genetic components lying on 3 chromosomes. The *IGH* cluster is located on chromosome 14; the *IGK* cluster is present on chromosome 2, whereas the *IGL* cluster is located on chromosome 22 (Delves and Roitt, 2000a). Part of the extremely large antibody diversity is generated through a mechanism called V(D)J recombination (Grawunder and Harfst, 2001). The variable part of the Ig light and heavy chains are respectively encoded by two (variable L, V_L ; joining L, J_L) or three (variable H, V_H ; diversity, D ; joining H, J_H) discrete minigenes among many. These minigenes, under the strict influence of two recombination activator genes (RAG-1, and RAG-2), must be rearranged before they can generate functional genes (Tonegawa, 1983).

Naive B lymphocytes develop continuously from pluripotential progenitors in the adult bone marrow and subsequently migrate to peripheral lymphoid organs where they encounter antigens. The initial activation of B cells and their fate following antigen contact are intimately linked to their stimulation by cytokines and to the ligation of co-stimulation cell surface receptors. Among those latter, CD40 appears critical for both B cell proliferation and survival as well as for the preferential formation of memory cells rather than plasma cells (Duchosal, 1997). The intracellular pathways leading to plasma cell formation are complexe and their identification is still ongoing. Expression of B lymphocyte-induced maturation protein-1 (Blimp-1) and of transcription factor XBP-1 appear critical for commitment to plasma cells (Angelin-Duclos *et al.*, 2000).

5.2 Proteomics of stem cells and cell differentiation

The study of human stem cells is of importance, due to the potential therapeutic roles. Proteomic techniques were used to approach the protein profile associated with the early-stage differentiation of embryonic stem cells (Guo *et al.*, 2001). They are totipotent stem cells, which can differentiate into various kinds of cell types including heamatopoietic or neuronal cells. They are widely used as a model system for investigating mechanisms of differentiation events during early development. The protein profile of parent embryonic stem cells were comprised with neural-committed cells, induced by all-trans retinoic acid for four days. Several proteins were selected from 2-DE gels, and were identified by peptide mass fingerprinting. Nine proteins were known to being involved in the process of neural differentiation and/or neural survival. Of those, alpha-3/alpha-7 tubulin and vimentin were down-regulated, while cytokeratin 8, cytokeratin 18, G1/S-special cyclin D2, follistatin-related protein, NEL protein, platelet-activating factor acetylhydrolase IB alpha-subunit, and thioredoxin peroxidase 2 were upregulated during differentiation. These results indicated that the molecular mechanisms of differentiation of embryonic stem cells in vitro might be identified, and that proteomic analysis is an effective strategy to comprehensively unravel the regulatory network of differentiation.

Myeloid development in a murine cell line was analyzed by 2-DE with mass spectrometry combined with oligonucleotide chip hybridization for a comprehensive and quantitative study of the temporal patterns of protein and mRNA expression (Lian *et al.*, 2002). This global analysis detected 123 known proteins and 29 "new" proteins out of 220 protein spots identified. Bioinformatic analysis of these proteins revealed clusters with functional importance to myeloid differentiation. Combined oligonucleotide microarray and proteomic approaches have been used to study genes associated with dendritic cell differentiation (Le Naour *et al.*, 2001). Dendritic cells are antigen-presenting cells that play a role in initiating primary immune responses. They derive from human CD14$^+$ blood monocytes. Analysis of gene expression changes at the RNA level using oligonucleotide microarrays complementary to 6300 human genes showed that approximately 40% of the genes were expressed in dentritic cells. About 4% of the genes were found to be regulated during dendritic differentiation or maturation. Protein analysis of the same cell populations was done, and about 4% of the protein spots separated by 2-DE exhibited quantitative changes during differentiation and maturation. The differentially expressed proteins were identified by mass spectrometry. They represented proteins with calcium binding, fatty acid binding, or chaperone activities as well as proteins involved in cell motility.

A "biomic approach" that integrated three independent methods, DNA microarray, proteomics and bioinformatics, was used to study the differentiation of human myeloid leukemia cell line HL-60 into macrophages (Juan *et al.*, 2002). Analysis of gene expression changes at the RNA level using cDNA against an array of 6033 human genes showed that 98.6% of the genes were expressed in the HL-60 cells, and that 10.5% of these genes were regulated during cell differentiation. Among these genes, some encoded for secreted proteins, other were involved in cell adhesion, signaling transduction, and metabolism. Protein analysis using 2-DE showed a total of 682 distinct protein spots, and 19.9% exhibited quantitative changes during differentiation. These differentially expressed proteins were identified by mass spectrometry.

5.3   Proteomics of monocytes, granulocytes and lymphocytes

One of the first reports dealing with the global study of proteins of monocytes, granulocytes and lymphocytes was that published by M. A. Gemmell and N. L. Anderson (Gemmell and Anderson, 1982). Although most proteins in each cell type appeared to be common to all three, there were several specific marker proteins that distinguish one cell type from another.

There are accumulating evidences that the lymphoid lineages, consisting of B, T and natural killer cells, are generated from a common lymphoid progenitor by successive differentiation steps that are tightly controlled by various growth factors or cytokines as well as by the microenvironment.

Since the pioneer 2-DE investigations of lymphocyte populations (Gemmell and Anderson, 1982; Giometti *et al.*, 1982; Hanash *et al.*, 1986a; Hochstrasser *et al.*, 1986; Kuick *et al.*, 1988; Lester *et al.*, 1982; Nyman *et al.*, 2001b; Pasquariello *et al.*, 1993; Willard-Gallo *et al.*, 1984; Willard-Gallo, 1984; Willard-Gallo *et al.*, 1988; Willard, 1982a; Willard *et al.*, 1982; Willard, 1982b; Willard and Anderson, 1981), a number of proteomic studies have been published (Caron *et al.*, 2002; Lefkovits *et al.*, 2000; Nyman *et al.*, 2001a). Some important achievement in this filed have been those of S. M. Hanash's group (Hanash *et al.*, 1993). These investigators have undertaken an effort aimed at developing a database of lymphoid proteins detectable by 2-DE. The database contains 2-DE patterns and derived information pertaining to: (i) polypeptide constituents of non-stimulated and stimulated mature T cells and immature thymocytes; (ii) cultured T cells and cell lines that have been manipulated by transfection with a variety of constructs or by treatment with specific agents; (iii) single cell-derived T and B cell clones; (iv) cells obtained from patients with lymphoproliferative disorders and leukemia; and (v) a variety of other relevant cell populations (Hanash and Teichroew, 1998).

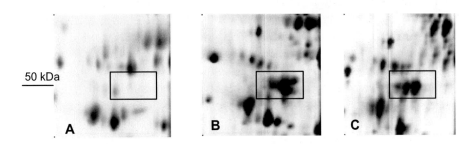

Figure 8. Details of high-resolution silver stained two-dimensional polyacrylamide gels of the blood lymphocytes obtained from healthy donors. A; B lymphocytes (CD19[+]), B; T lymphocytes (CD8[+]), C; T lymphocytes (CD4[+]). The spots corresponding to vimentin. tubulin, desmin or cytokeratin are in the frame. First dimension: immobilized 3 to 10 pH gradient, second dimension: 9 to 16% polyacrylamide gel electrophoresis [adapted from (Vuadens et al., 2002), with permission of the publisher].

Proteomic studies of purified populations of human lymphocytes showed that proteins such vimentin, tubulin, desmin or cytokeratin were more expressed in T lymphocytes when compared to B lymphocytes (Fig. 8) (Vuadens *et al.*, 2002). In addition, an apparently CD8 over-expressed protein was identified and named Swiprosin 1 (SwissProt Q96C19) (Fig. 9).

This protein, constituted by 240 amino acids, contains 2 potential EF-hand calcium-binding domains and is a member of a new family of proteins.

Signalling by immunoreceptors of T cells was recently investigated using proteomics (Bini *et al.*, 2002). Plasma membrane microdomains, referred to lipid rafts, were analysed following T cell antigen receptor triggering. This study showed that T cell antigen receptor engagement promoted the temporally regulated recruitment of proteins participating in the T cell antigen receptor signalling cascade. Interactions of proteins with CD4 was studied by O. K Bernhard and collaborators (Bernhard *et al.*, 2003). In this study, protein interactions were studied using mass spectrometry of the whole protein complex purified by affinity chromatography, in order to identify directly the proteins present in the CD4 receptor complex. Using appropriate control experiments, the authors demonstrated the specific nature of the CD4-p56lck interaction.

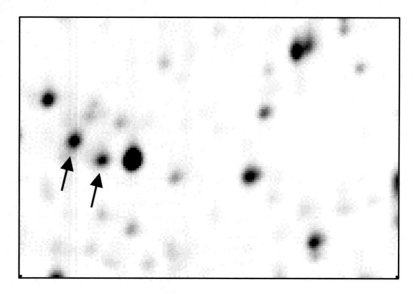

Figure 9. Details of high-resolution silver stained two-dimensional polyacrylamide gel of T lymphocytes (CD8[+]) showing the spots corresponding to Swiprosin 1 (SwissProt Q96C19) are highlighted by arrows. First dimension: immobilized 3 to 10 pH gradient, second dimension: 9 to 16% polyacrylamide gel electrophoresis [adapted from (Vuadens et al., 2002), with permission of the publisher].

By evaluating T cells exposed to oxidants, M. Fratelli and collaborators showed formation of mixed disulfide bounds between glutathione and the cysteines of some proteins (glutathionylation). These findings suggested that protein glutathionylation might be a common mechanism for the global regulation of protein functions (Fratelli *et al.*, 2002).

Remarkable proteomic studies on lymphoblastoid proteins, including phosphoproteins, were published (Caron *et al.*, 2002; Feuillard *et al.*, 2000; Fouillit *et al.*, 2000; Imam-Sghiouar *et al.*, 2002; Joubert-Caron *et al.*, 2000; Poirier *et al.*, 2001a; Poirier *et al.*, 2001b; Poirier *et al.*, 2001c; Toda *et al.*, 2000). More than 60 different proteins were identified, and several phosphorylated peptides were characterized using two different strategies (Toda *et al.*, 2000). The first one was the use of affinity separation of proteins phosphorylated on tyrosine after immunoprecipitation, the second one was metal chelate affinity chromatography on ferric ions.

5.4   Leukaemia and lymphoma

Some of the current technologies for proteome profiling and the application of proteomics to the analysis of leukaemia have been recently reviewed (Hanash *et al.*, 2002; Singh *et al.*, 2002). In 1982 and 1983, the first studies reporting on the analysis of proteins of leukaemic cells separated by 2-DE were published (Anderson *et al.*, 1983; Hanash *et al.*, 1982; Takeda *et al.*, 1983), and in 1988, a polypeptide was identified in association with malignant leukaemia (Hanash *et al.*, 1988). When analysing leukaemic cells of infants and older children with acute lymphoblastic leukaemia by 2-DE, these investigators were able to show the occurrence of a cytosolic polypeptide (designated L3) that was limited to non-T acute leukaemia. This cellular polypeptide distinguished between acute lymphoblastic leukaemia in infants and in older children (Hanash *et al.*, 1989). In addition, it was shown that the occurrence of particular polypeptide markers detected by 2-DE was related to a myeloid origin of the blasts of two children who presented with undifferentiated leukaemia at the time of diagnosis (Hanash and Baier, 1986).

2-DE was used to identify cellular proteins in T and B acute lymphoblastic leukaemia cell lines (Mohammad *et al.*, 1994b). Two B cell and two T cell lines as well as a normal Epstein-Barr virus-transformed line of B cell origin were studied. Despite the great similarities in the patterns of the B and T cell derived from 2-DE gels, three proteins appeared to be unique to B cells, whereas eight proteins were unique to T cell lines. It is not known if these proteins are markers for B or T acute lymphoblastic leukaemia, respectively, or if they are unique to the cell lines that were studied.

The protein expression in hairy cell leukaemia was compared to that of B cell chronic lymphocytic leukaemia (Talebian-Ziai *et al.*, 1992). Five distinct polypeptides were detected in hairy cell leukaemia, but not in cells of most chronic lymphocytic leukaemia and normal B lymphocytes. Two out of these 5 polypeptides appeared to be specific of hairy cell leukaemia. In addition,

2-DE showed distinct patterns in 3 splenic lymphoma with villous lymphocytes as well as in 2 cases of hairy cell leukaemia variants.

F. K. Saunder and collaborators evaluated protein synthesis in leukaemic cells from 10 chronic lymphocytic leukaemia patients using 2-DE (Saunders *et al.*, 1993). Only minor differences between each of the chronic lymphocytic leukaemia cell samples were observed. However, comparison between the chronic lymphocytic leukaemia cells and normal B lymphocytes demonstrated marked differences in protein synthesis between the leukaemic and non-malignant cells. In a series of 15 patients presenting with different stages of B cell chronic lymphocytic leukaemia, 2-DE revealed modifications of the protein expression that correlated with the staging of the disease, and more particularly with the more advanced stage (stage C) (Saunders *et al.*, 1994). This type of analysis showed that proteomics could prove invaluable for identifying proteins whose expression could intimately be associated with the evolution of chronic lymphocytic leukaemia.

The protein expression patterns obtained by 2-DE in B cell chronic lymphocytic leukaemia was also studied by T. Voss and collaborators in 24 patient samples (Voss *et al.*, 2001). This analysis allowed the identification of proteins that clearly discriminated between different patient groups (with defined chromosomal characteristics or according to clinical parameters). More particularly, this study showed that B cell chronic lymphocytic leukaemia patients with shorter survival exhibit changed levels of redox enzymes such as heat shock protein 27 and protein disulfide isomerase.

F. Poirier and collaborators reported they data evaluating the proteome of a Burkitt lymphoma cell line (DG 75), as well as its modification following treatment with 5'-azycytidine (Poirier *et al.*, 2001c). Their results brought new insights into lymphoma cell gene regulations following treatment with 5'-azycytidine.

R. M. Mohammad and collaborators assessed proteins in human non-Hodgkin's B lymphoma (Mohammad *et al.*, 1994a). Studies using 2-DE of five cell lines representing four different non-Hodgkin's B lymphoma maturational stages and a normal Epstein-Barr virus-transformed line of B cell origin revealed that some proteins were present in malignant non-Hodgkin's B lymphoma, but were absent in the normal Epstein-Barr virus-cell line. However, whether or not these proteins were potential malignant markers able to distinguish between different non-Hodgkin's B lymphoma maturational stages needs further exploration.

In a study of normal lymph nodes and lymphoma samples, using 2-DE in association with immuno-detection, the expression of leukocyte differentiation and tumour markers such as CD3 and CD5, as well as cell cycle regulatory molecules such as cyclin dependent kinases, notably CDK6, was evaluated (Antonucci *et al.*, 2002). A much increased level of expression of CD5 in mantle cell lymphoma, up to ten times higher than in the control was detected. In addition, CD5 in tumour tissues appeared to be microheterogeneous as compared to normal samples.

6.   CONCLUDING REMARKS

Proteomics is still in its young age, even if the first steps, notably 2-DE, were developed 27 years ago. Certainly, in the next near future, proteomics will change the face of haematology as well. It is a rapidly developing science, depending on many basic developments. New technologies such as protein chips, allowing a rapid and efficient isolation of multiple families of proteins, certainly will modify the approaches that will be used in the future. More and more sophisticated mass spectrometers will be available, at a lower cost. Studies of membrane proteins as well as of protein-protein interactions will be made possible. The rapidly growing development of bioinformatics will also transform the handling of the multitude of data that can accumulate in a set of proteomic experiments. But despite these fantastic changes of the experimental approaches, it still will be necessary to ask the good questions in order to try to give the best answers with the aim to treat patients more efficiently and more adequately.

ACKNOWLEDGEMENTS

The authors are indebted to all the collaborators who actively participated over the years to research activities at the Service régional vaudois de transfusion sanguine (SRTS VD). They also want to acknowledge the financial support of the Fondation Cetrasa, in Lausanne.

REFERENCES

Adkins, J.N., Varnum, S.M., Auberry, K.J., Moore, R.J., Angell, N.H., Smith, R.D., Springer, D.L., and Pounds, J.G., 2002, Toward a human blood serum proteome: analysis by multidimensional separation coupled with mass spectrometry. *Mol. Cell Proteomics* **1:** 947-955.

Akashi, K., Traver, D., Kondo, M., and Weissman, I.L., 1999, Lymphoid development from hematopoietic stem cells. *Int. J. Hematol.* **69:** 217-226.

Anderson, L. and Anderson, N.G., 1977, High resolution two-dimensional electrophoresis of human plasma proteins. *Proc. Natl. Acad. Sci. U. S. A.* **74:** 5421-5425.

Anderson, N.L. and Anderson, N.G., 1979, Microheterogeneity of serum transferrin, haptoglobin and alpha 2 HS glycoprotein examined by high resolution two-dimensional electrophoresis. *Biochem. Biophys. Res. Commun.* **88:** 258-265.

Anderson, N.L. and Anderson, N.G., 1991, A two-dimensional gel database of human plasma proteins. *Electrophoresis* **12:** 883-906.

Anderson, N.L. and Anderson, N.G., 1998, Proteome and proteomics: new technologies, new concepts, and new words. *Electrophoresis* **19:** 1853-1861.

Anderson, N.L. and Anderson, N.G., 2002, The human plasma proteome: History, character, and diagnostic prospects. *Mol. Cell. Proteomics* **1:** 845-867.

Anderson, N.L., Wiltsie, J.C., Li, C.Y., Willard-Gallo, K.E., Tracy, R.P., Young, D.S., Powers, M.T., and Anderson, N.G., 1983, Analysis of human leukemic cells by use of high-resolution two- dimensional electrophoresis. I: results of a pilot study. *Clin. Chem.* **29:** 762-767.

Angelin-Duclos, C., Cattoretti, G., Lin, K.I., and Calame, K., 2000, Commitment of B lymphocytes to a plasma cell fate is associated with Blimp-1 expression in vivo. *J. Immunol.* **165:** 5462-5471.

Ansel, K.M. and Cyster, J.G., 2001, Chemokines in lymphopoiesis and lymphoid organ development. *Curr. Opin. Immunol.* **13:** 172-179.

Antonucci, F., Chilosi, M., Santacatterina, M., Herbert, B., and Righetti, P.G., 2002, Proteomics and immunomapping of reactive lymph-node and lymphoma. *Electrophoresis* **23:** 356-362.

Argoud-Puy, G., Baussant, T., Böhm, G., Botti, P., Bougueleret, L., Colinge, J., Cusin, I., Gaertner, H., Heller, M., Jimenez, S., *et al.*, 2002, Industrial-scale proteomics analysis of human plasma. In *Applied Proteomics* (P. Palagi, M. Quadroni, J. Rossier, J. C. Sanchez, and R. Stöcklin, eds.), FontisMedia, Lausanne, pp. 17-18.

Aruffo, A., Bowen, M.A., Patel, D.D., Haynes, B.F., Starling, G.C., Gebe, J.A., and Bajorath, J., 1997, CD6-ligand interactions: a paradigm for SRCR domain function? *Immunol. Today* **18:** 498-504.

Attaelmannan, M. and Levinson, S.S., 2000, Understanding and identifying monoclonal gammopathies. *Clin. Chem.* **46:** 1230-1238.

Avent, N.D. and Reid, M.E., 2000, The Rh blood group system: a review. *Blood* **95:** 375-387

Banks, R.E., Dunn, M.J., Hochstrasser, D.F., Sanchez, J.C., Blackstock, W., Pappin, D.J., and Selby, P.J., 2000, Proteomics: new perspectives, new biomedical opportunities. *Lancet* **356:** 1749-1756.

Bataille, R. and Harousseau, J.L., 1997, Multiple myeloma. *N. Engl. J. Med.* **336:** 1657-1664.

Bernhard, O.K., Burgess, J.A., Hochgrebe, T., Sheil, M.M., and Cunningham, A.L., 2003, Mass spectrometry analysis of CD4-associating proteins using affinity chromatography and affinity tag-mediated purification of tryptic peptides. *Proteomics.* **3:** 139-146.

Beutler, E., 1989, Glucose-6-phosphate dehydrogenase: new perspectives. *Blood* **73:** 1397-1401.

Beutler, E., 1994, G6PD deficiency. *Blood* **84:** 3613-3636.

Beutler, E., 2001a, Discrepancies between genotype and phenotype in hematology: an important frontier. *Blood* **98:** 2597-2602.

Beutler, E., 2001b, DNA-based diagnosis of red cell enzymopathies: how we threw out the baby with the bathwater. *Blood* **97:** 3325.

Bhakdi, S., Knufermann, H., and Wallach, D.F., 1975, Two-dimensional separation of erythrocyte membrane proteins. *Biochim. Biophys. Acta* **394:** 550-557.

Bini, L., Pacini, S., Liberatore, S., Valensin, S., Pellegrini, M., Raggiaschi, R., Pallini, V., and Baldari, C.T., 2002, Extensive temporally regulated reorganization of the lipid raft proteome following T cell antigen receptor triggering. *Biochem. J.* **369:** 301-309.

Bladé, J. and Kyle, R.A., 1999, Nonsecretory myeloma, immunoglobulin D myeloma, and plasma cell leukemia. *Hematol. Oncol. Clin. North Am.* **13:** 1259-1272.

Blajchman, M.A., Bordin, J.O., Bardossy, L., and Heddle, N.M., 1994, The contribution of the haematocrit to thrombocytopenic bleeding in experimental animals. *Br. J. Haematol.* **86:** 347-350.

Bossuyt, X., Schiettekatte, G., Bogaerts, A., and Blanckaert, N., 1998, Serum protein electrophoresis by CZE 2000 clinical capillary electrophoresis system. *Clin. Chem.* **44:** 749-759.

Carlson, J., 2001, From paper electrophoresis to computer-supported interpretation of capillary electrophoresis--clinical plasma protein analysis in Malmo, Sweden. *Clin. Chem. Lab. Med.* **39:** 1020-1024.

Caron, M., Imam-Sghiouar, N., Poirier, F., Le Caer, J.P., Labas, V., and Joubert-Caron, R., 2002, Proteomic map and database of lymphoblastoid proteins. *J. Chromatogr. B Analyt. Technol. Biomed. Life Sci.* **771:** 197-209.

Cartron, J.P., 2000, Molecular basis of red cell protein antigen deficiencies. *Vox Sang.* **78:** 7-23.

Cartron, J.P., Bailly, P., Le Van, K.C., Cherif-Zahar, B., Matassi, G., Bertrand, O., and Colin, Y., 1998, Insights into the structure and function of membrane polypeptides carrying blood group antigens. *Vox Sang.* **74 Suppl 2:** 29-64.

Cartron, J.P. and Colin, Y., 2001, Structural and functional diversity of blood group antigens. *Transfus. Clin. Biol.* **8:** 163-199.

Cartron, J.P. and Salmon, C., 1983, Red cell membrane diseases and blood group abnormalities. *Rev. Fr. Transfus. Immunohematol.* **26:** 599-623.

Cho, A. and Normile, D., 2002, Nobel Prize in Chemistry. Mastering macromolecules. *Science* **298:** 527-528.

Conrad, M.J. and Penniston, J.T., 1976, Resolution of erythrocyte membrane proteins by two-dimensional electrophoresis. *J. Biol. Chem.* **251:** 253-255.

Copeland, B.R., Todd, S.A., and Furlong, C.E., 1982, High resolution two-dimensional gel electrophoresis of human erythrocyte membrane proteins. *Am. J. Hum. Genet.* **34:** 15-31.

Corthals, G.L., Wasinger, V.C., Hochstrasser, D.F., and Sanchez, J.C., 2000, The dynamic range of protein expression: A challenge for proteomic research. *Electrophoresis* **21:** 1104-1115.

Coughlin, S.R., 2000, Thrombin signalling and protease-activated receptors. *Nature* **407:** 258-264.

Dahlbäck, B., 2000, Blood coagulation. *Lancet* **355:** 1627-1632.

Damoc, E., Buddrus, S., Becker, S., Crettaz, D., Tissot, J.D., and Przybylski, M., 2002, High resolution proteome analysis of cryoglobulins using FT-ICR mass spectrometry. In *Applied Proteomics* (P. M. Palagi, M. Quadroni, J. Rossier, J. C. Sanchez, and R. Stöcklin, eds.), FontisMedia, Lausanne, pp. 154-157.

Damoc, E., Youhnovski, N., Crettaz, D., Tissot, J.D., and Przybylski, M,.(2003, High resolution proteome analysis of cryoglobulins using Fourier transform-ion cyclotron resonance mass spectrometry. *Proteomics*, in press.

de Groot, P.G., 2002, The role of von Willebrand factor in platelet function. *Semin. Thromb. Hemost.* **28:** 133-138.

Delves, P.J. and Roitt, I.M., 2000a, The immune system. First of two parts. *N. Engl. J. Med.* **343:** 37-49.

Delves, P.J. and Roitt, I.M., 2000b, The immune system. Second of two parts. *N. Engl. J. Med.* **343:** 108-117.

Devaux, F., Marc, P., and Jacq, C., 2001, Transcriptomes, transcription activators and microarrays. *FEBS Lett.* **498:** 140-144.

Diehl, S. and Rincon, M., 2002, The two faces of IL-6 on Th1/Th2 differentiation. *Mol. Immunol.* **39:** 531-536.

Dockham, P.A., Steinfeld, R.C., Stryker, C.J., Jones, S.W., and Vidaver, G.A., 1986, An isoelectric focusing procedure for erythrocyte membrane proteins and its use for two-dimensional electrophoresis. *Anal. Biochem.* **153**: 102-115.

Duchosal, M.A., 1997, B-cell development and differentiation. *Semin. Hematol.* **34**: 2-12.

Eberini, I., Miller, I., Zancan, V., Bolego, C., Puglisi, L., Gemeiner, M., and Gianazza, E., 1999, Proteins of rat serum IV. Time-course of acute-phase protein expression and its modulation by indomethacine. *Electrophoresis* **20**: 846-853.

Ebrini, I., Agnello, D., Miller, I., Villa, P., Fratelli, M., Ghezzi, P., Gemeiner, M., Chan, J., Aebersold, R., and Gianazza, E., 2000, Proteins of rat serum V: adjuvant arthritis and its modulation by nonsteroidal anti-inflammatory drugs. *Electrophoresis* **21**: 2170-2179.

Edwards, J.J., Anderson, N.G., Nance, S.L., and Anderson, N.L., 1979, Red cell proteins. I. Two-dimensional mapping of human erythrocyte lysate proteins. *Blood* **53**: 1121-1132.

Feuillard, J., Schuhmacher, M., Kohanna, S., Asso-Bonnet, M., Ledeur, F., Joubert-Caron, R., Bissieres, P., Polack, A., Bornkamm, G.W., and Raphael, M., 2000, Inducible loss of NF-kappaB activity is associated with apoptosis and Bcl-2 down-regulation in Epstein-Barr virus-transformed B lymphocytes. *Blood* **95**: 2068-2075.

Fields, S., 2001, Proteomics - Proteomics in genomeland. *Science* **291**: 1221-1223.

Fleischmann, R.D., Adams, M.D., White, O., Clayton, R.A., Kirkness, E.F., Kerlavage, A.R., Bult, C.J., Tomb, J.F., Dougherty, B.A., Merrick, J.M., *et al.*, 1995, Whole-genome random sequencing and assembly of Haemophilus influenzae Rd. *Science* **269**: 496-512.

Fouillit, M., Joubert-Caron, R., Poirier, F., Bourin, P., Monostori, E., Levi-Strauss, M., Raphael, M., Bladier, D., and Caron, M., 2000, Regulation of CD45-induced signaling by galectin-1 in Burkitt lymphoma B cells. *Glycobiology* **10**: 413-419.

Fratelli, M., Demol, H., Puype, M., Casagrande, S., Eberini, I., Salmona, M., Bonetto, V., Mengozzi, M., Duffieux, F., Miclet, E., Bachi, A., Vandekerckhove, J., Gianazza, E., and Ghezzi, P., 2002, Identification by redox proteomics of glutathionylated proteins in oxidatively stressed human T lymphocytes. *Proc. Natl. Acad. Sci. U. S. A* **99**: 3505-3510.

Friedman, A.D., 2002a, Transcriptional regulation of granulocyte and monocyte development. *Oncogene* **21**: 3377-3390.

Friedman, A.D., 2002b, Transcriptional regulation of myelopoiesis. *Int. J. Hematol.* **75**: 466-472.

Gebe, J.A., Kiener, P.A., Ring, H.Z., Li, X., Francke, U., and Aruffo, A., 1997, Molecular cloning, mapping to human chromosome 1 q21-q23, and cell binding characteristics of Spalpha, a new member of the scavenger receptor cysteine-rich (SRCR) family of proteins. *J. Biol. Chem.* **272**: 6151-6158.

Gebe, J.A., Llewellyn, M., Hoggatt, H., and Aruffo, A., 2000, Molecular cloning, genomic organization and cell-binding characteristics of mouse Spalpha. *Immunology* **99**: 78-86.

Gemmell, M.A. and Anderson, N.L., 1982, Lymphocyte, monocyte, and granulocyte proteins compared by use of two- dimensional electrophoresis. *Clin. Chem.* **28**: 1062-1066.

Gianazza, E., Eberini, I., Villa, P., Fratelli, M., Pinna, C., Wait, R., Gemeiner, M., and Miller, I., 2002, Monitoring the effects of drug treatment in rat models of disease by serum protein analysis. *J. Chromatogr. B Analyt. Technol. Biomed. Life Sci.* **771**: 107-130.

Giometti, C.S., Willard, K.E., and Anderson, N.L., 1982, Cytoskeletal proteins from human skin fibroblasts, peripheral blood leukocytes, and a lymphoblastoid cell line compared by two-dimensional gel electrophoresis. *Clin. Chem.* **28**: 955-961.

Golaz, O., Gravel, P., Walzer, C., Turler, H., Balant, L., and Hochstrasser, D.F., 1995, Rapid detection of the main human plasma glycoproteins by two- dimensional polyacrylamide gel electrophoresis lectin affinoblotting. *Electrophoresis* **16**: 1187-1189.

Golaz, O., Hughes, G.J., Frutiger, S., Paquet, N., Bairoch, A., Pasquali, C., Sanchez, J.-C., Tissot, J.D., Appel, R.D., Walzer, C., Balant, L., and Hochstrasser, D.F., 1993, Plasma and red blood cell protein maps: Update 1993. *Electrophoresis* **14:** 1223-1231.

Goldfarb, M.F., 1992, Two-dimensional electrophoretic analysis of immunoglobulin patterns in monoclonal gammopathies. *Electrophoresis* **13:** 440-444.

Gravel, P., Golaz, O., Walzer, C., Hochstrasser, D.F., Turler, H., and Balant, L.P., 1994, Analysis of glycoproteins separated by two-dimensional gel electrophoresis using lectin blotting revealed by chemiluminescence. *Anal. Biochem.* **221:** 66-71.

Gravel, P., Sanchez, J.C., Walzer, C., Golaz, O., Hochstrasser, D.F., Balant, L.P., Hughes, G.J., Garcia-Sevilla, J., and Guimon, J., 1995, Human blood platelet protein map established by two-dimensional polyacrylamide gel electrophoresis. *Electrophoresis* **16:** 1152-1159.

Gravel, P., Walzer, C., Aubry, C., Balant, L.P., Yersin, B., Hochstrasser, D.F., and Guimon, J., 1996, New alterations of serum glycoproteins in alcoholic and cirrhotic patients revealed by high resolution two-dimensional gel electrophoresis. *Biochem. Biophys. Res. Commun.* **220:** 78-85.

Grawunder, U. and Harfst, E., 2001, How to make ends meet in V(D)J recombination. *Curr. Opin. Immunol.* **13:** 186-194.

Guo, X., Ying, W., Wan, J., Hu, Z., Qian, X., Zhang, H., and He, F., 2001, Proteomic characterization of early-stage differentiation of mouse embryonic stem cells into neural cells induced by all-trans retinoic acid in vitro. *Electrophoresis* **22:** 3067-3075.

Hanash, S.M. and Baier, L.J., 1986, Two-dimensional gel electrophoresis of cellular proteins reveals myeloid origin of blasts in two children with otherwise undifferentiated leukemia. *Cancer* **57:** 1539-1543.

Hanash, S.M., Baier, L.J., Welch, D., Kuick, R., and Galteau, M., 1986a, Genetic variants detected among 106 lymphocyte polypeptides observed in two-dimensional gels. *Am. J. Hum. Genet.* **39:** 317-328.

Hanash, S.M., Kuick, R., Strahler, J., Richardson, B., Reaman, G., Stoolman, L., Hanson, C., Nichols, D., and Tueche, H.J., 1989, Identification of a cellular polypeptide that distinguishes between acute lymphoblastic leukemia in infants and in older children. *Blood* **73:** 527-532.

Hanash, S.M., Madoz-Gurpide, J., and Misek, D.E., 2002, Identification of novel targets for cancer therapy using expression proteomics. *Leukemia* **16:** 478-485.

Hanash, S.M., Neel, J.V., Baier, L.J., Rosenblum, B.B., Niezgoda, W., and Markel, D., 1986b, Genetic analysis of thirty-three platelet polypeptides detected in two-dimensional polyacrylamide gels. *Am. J. Hum. Genet.* **38:** 352-360.

Hanash, S.M., Strahler, J.R., Chan, Y., Kuick, R., Teichroew, D., Neel, J.V., Hailat, N., Keim, D.R., Gratiot-Deans, J., Ungar, D., and ., 1993, Data base analysis of protein expression patterns during T-cell ontogeny and activation. *Proc. Natl. Acad. Sci. U. S. A.* **90:** 3314-3318.

Hanash, S.M., Strahler, J.R., Kuick, R., Chu, E.H., and Nichols, D., 1988, Identification of a polypeptide associated with the malignant phenotype in acute leukemia. *J. Biol. Chem.* **263:** 12813-12815.

Hanash, S.M. and Teichroew, D., 1998, Mining the human proteome: Experience with the human lymphoid protein database. *Electrophoresis* **19:** 2004-2009.

Hanash, S.M., Tubergen, D.G., Heyn, R.M., Neel, J.V., Sandy, L., Stevens, G.S., Rosenblum, B.B., and Krzesicki, R.F., 1982, Two-dimensional gel electrophoresis of cell proteins in childhood leukemia, with silver staining: a preliminary report. *Clin. Chem.* **28:** 1026-1030.

Hardy, R.R. and Hayakawa, K., 2001, B cell development pathways. *Annu. Rev. Immunol.* **19:** 595-621.

Harell, D. and Morrison, M., 1979, Two-dimensional separation of erythrocyte membrane proteins. *Arch. Biochem. Biophys.* **193**: 158-168.

Harrison, H.H., 1992, Patient-specific microheterogeneity patterns of monoclonal immunoglobulin light chains as revealed by high resolution, two-dimensional electrophoresis. *Clin. Biochem.* **25**: 235-243.

Harrison, H.H., Miller, K.L., Abu-Alfa, A., and Podlasek, S.J., 1993, Immunoglobulin clonality analysis: Resolution of ambiguities in immunofixation electrophoresis results by high-resolution, two- dimensional electrophoretic analysis of paraprotein bands eluted from agarose gels. *Am. J. Clin. Pathol.* **100**: 550-560.

Hart, G.D., 2001, Descriptions of blood and blood disorders before the advent of laboratory studies. *Br. J. Haematol.* **115**: 719-728.

Haynes, P., Miller, I., Aebersold, R., Gemeiner, M., Eberini, I., Lovati, M.R., Manzoni, C., Vignati, M., and Gianazza, E., 1998, Proteins of rat serum: I. Establishing a reference two-dimensional electrophoresis map by immunodetection and microbore high performance liquid chromatography-electrospray mass spectrometry. *Electrophoresis* **19**: 1484-1492.

Henry, H., Froehlich, F., Perret, R., Tissot, J.D., Eilers-Messerli, B., Lavanchy, D., Dionisi-Vici, C., Gonvers, J.J., and Bachmann, C., 1999, Microheterogeneity of serum glycoproteins in patients with chronic alcohol abuse compared with carbohydrate-deficient glycoprotein syndrome type I. *Clin. Chem.* **45**: 1408-1413.

Henry, H., Tissot, J.D., Messerli, B., Markert, M., Muntau, A., Skladal, D., Sperl, W., Jaeken, J., Weidinger, S., Heyne, K., and Bachmann, C., 1997, Microheterogeneity of serum glycoproteins and their liver precursors in patients with carbohydrate-deficient glycoprotein syndrome type I: Apparent deficiencies in clusterin and serum amyloid P. *J. Lab. Clin. Med.* **129**: 412-421.

Heyne, K., Henry, H., Messerli, B., Bachmann, C., Stephani, U., Tissot, J.D., and Weidinger, S., 1997, Apolipoprotein J deficiency in type I and IV of carbohydrate-deficient glycoprotein syndrome (glycanosis CDG). *Eur. J. Pediatr.* **156**: 247-248.

Hochstrasser, D., Roux, P., Dayer, J.M., Cruchaud, A., Appel, R., Funk, M., Pellegrini, C., and Muller, A.F., 1986, [Protein mapping of normal circulating cells: lymphocytes and monocytes]. *Schweiz. Med. Wochenschr.* **116**: 1773-1775.

Hochstrasser, D.F. and Tissot, J.D., 1993, Clinical applications of high-resolution two dimensional polyacrylamide gel electrophoresis. *Adv. Electrophoresis* **6**: 268-375.

Imam-Sghiouar, N., Laude-Lemaire, I., Labas, V., Pflieger, D., Le Caer, J.P., Caron, M., Nabias, D.K., and Joubert-Caron, R., 2002, Subproteomics analysis of phosphorylated proteins: application to the study of B-lymphoblasts from a patient with Scott syndrome. *Proteomics.* **2**: 828-838.

Immler, D., Gremm, D., Kirsch, D., Spengler, B., Presek, P., and Meyer, H.E., 1998, Identification of phosphorylated proteins from thrombin-activated human platelets isolated by two-dimensional gel electrophoresis by electrospray ionization tandem mass spectrometry (ESI-MS/MS) and liquid chromatography electrospray ionization mass spectrometry (LC-ESI-MS). *Electrophoresis* **19**: 1015-1023.

Joubert-Caron, R., Le Caer, J.P., Montandon, F., Poirier, F., Pontet, M., Imam, N., Feuillard, J., Bladier, D., Rossier, J., and Caron, M., 2000, Protein analysis by mass spectrometry and sequence database searching: a proteomic approach to identify human lymphoblastoid cell line proteins [In Process Citation]. *Electrophoresis* **21**: 2566-2575.

Juan, H.F., Lin, J.Y., Chang, W.H., Wu, C.Y., Pan, T.L., Tseng, M.J., Khoo, K.H., and Chen, S.T., 2002, Biomic study of human myeloid leukemia cells differentiation to macrophages using DNA array, proteomic, and bioinformatic analytical methods. *Electrophoresis* **23**: 2490-2504.

Kaushansky, K. and Drachman, J.G., 2002, The molecular and cellular biology of thrombopoietin: the primary regulator of platelet production. *Oncogene* **21:** 3359-3367.

Keren, D.F., Alexanian, R., Goeken, J.A., Gorevic, P.D., Kyle, R.A., and Tomar, R.H., 1999, Guidelines for clinical and laboratory evaluation patients with monoclonal gammopathies. *Arch. Pathol. Lab. Med.* **123:** 106-107.

Kettman, J.R., Frey, J.R., and Lefkovits, I., 2001, Proteome, transcriptome and genome: top down or bottom up analysis? *Biomol. Eng.* **18:** 207-212.

Kiernan, U.A., Tubbs, K.A., Nedelkov, D., Niederkofler, E.E., and Nelson, R.W., 2002, Comparative phenotypic analyses of human plasma and urinary retinol binding protein using mass spectrometric immunoassay. *Biochem. Biophys. Res. Commun.* **297:** 401-405.

Klose, J., 1975, Protein mapping by combined isoelectric focusing and electrophoresis of mouse tissues. A novel approach to testing for induced point mutations in mammals. *Humangenetik.* **26:** 231-243.

Kondo, M., Weissman, I.L., and Akashi, K., 1997, Identification of clonogenic common lymphoid progenitors in mouse bone marrow. *Cell* **91:** 661-672.

Kuick, R.D., Hanash, S.M., and Strahler, J.R., 1988, Somatic mutation studies in human lymphoid cells: the detectability of quantitative genetic variants in two-dimensional gels. *Electrophoresis* **9:** 192-198.

Kyle, R.A., 1994, The monoclonal gammopathies. *Clin. Chem.* **40:** 2154-2161.

Kyle, R.A., 1999a, Clinical aspects of multiple myeloma and related disorders including amyloidosis. *Pathol. Biol. (Paris)* **47:** 148-157.

Kyle, R.A., 1999b, Sequence of testing for monoclonal gammopathies. [Review] [17 refs]. *Arch. Pathol. Lab. Med.* **123:** 114-118.

Kyle, R.A., 2000, Multiple myeloma: An odyssey of discovery. *Br. J. Haematol.* **111:** 1035-1044.

Lander, E.S., Linton, L.M., Birren, B., Nusbaum, C., Zody, M.C., Baldwin, J., Devon, K., Dewar, K., Doyle, M., FitzHugh, W., *et al.*, 2001, Initial sequencing and analysis of the human genome. *Nature* **409:** 860-921.

Lapidot, T. and Petit, I., 2002, Current understanding of stem cell mobilization: the roles of chemokines, proteolytic enzymes, adhesion molecules, cytokines, and stromal cells. *Exp. Hematol.* **30:** 973-981.

Le Naour, F., Hohenkirk, L., Grolleau, A., Misek, D.E., Lescure, P., Geiger, J.D., Hanash, S., and Beretta, L., 2001, Profiling changes in gene expression during differentiation and maturation of monocyte-derived dendritic cells using both oligonucleotide microarrays and proteomics. *J. Biol. Chem.* **276:** 17920-17931.

Lebeau, A., Zeindl-Eberhart, E., Muller, E.C., Muller-Hocker, J., Jungblut, P.R., Emmerich, B., and Lohrs, U., 2002, Generalized crystal-storing histiocytosis associated with monoclonal gammopathy: molecular analysis of a disorder with rapid clinical course and review of the literature. *Blood* **100:** 1817-1827.

Lefkovits, I., Kettman, J.R., and Frey, J.R., 2000, Global analysis of gene expression in cells of the immune system I. Analytical limitations in obtaining sequence information on polypeptides in two-dimensional gel spots. *Electrophoresis* **21:** 2688-2693.

Lester, E.P., Lemkin, P., and Lipkin, L., 1982, A two-dimensional gel analysis of autologous T and B lymphoblastoid cell lines. *Clin. Chem.* **28:** 828-839.

Lian, Z., Kluger, Y., Greenbaum, D.S., Tuck, D., Gerstein, M., Berliner, N., Weissman, S.M., and Newburger, P.E., 2002, Genomic and proteomic analysis of the myeloid differentiation program: global analysis of gene expression during induced differentiation in the MPRO cell line. *Blood* **100:** 3209-3220.

Low, T.Y., Seow, T.K., and Chung, M.C., 2002, Separation of human erythrocyte membrane associated proteins with one- dimensional and two-dimensional gel electrophoresis followed by identification with matrix-assisted laser desorption/ionization-time of flight mass spectrometry. *Proteomics.* **2:** 1229-1239.

Luche, S., Santoni, V., and Rabilloud, T., 2003, Evaluation of nonionic and zwiterionic detergents as membrane protein solubilizers in two-dimensional electrophoresis. *Proteomics* **3:** 249-253.

Lutter, P., Meyer, H.E., Langer, M., Witthohn, K., Dormeyer, W., Sickmann, A., and Bluggel, M., 2001, Investigation of charge variants of rViscumin by two-dimensional gel electrophoresis and mass spectrometry. *Electrophoresis* **22:** 2888-2897.

Maguire, P.B., Wynne, K.J., Harney, D.F., O'Donoghue, N.M., Stephens, G., and Fitzgerald, D.J., 2002, Identification of the phosphotyrosine proteome from thrombin activated platelets. *Proteomics.* **2:** 642-648.

Marcus, K., Immler, D., Sternberger, J., and Meyer, H.E., 2000, Identification of platelet proteins separated by two-dimensional gel electrophoresis and analyzed by matrix assisted laser desorption/ioniztion-time of flight-mass spectrometry and detection of tyrosine-phosphorylated proteins. *Electrophoresis* **21:** 2622-2636.

Matsumura, I. and Kanakura, Y., 2002, Molecular control of megakaryopoiesis and thrombopoiesis. *Int. J. Hematol.* **75:** 473-483.

Maxam, A.M. and Gilbert, W., 1977, A new method for sequencing DNA. *Proc. Natl. Acad. Sci. U. S. A* **74:** 560-564.

Merlini, G., Bellotti, V., Andreola, A., Palladini, G., Obici, L., Casarini, S., and Perfetti, V., 2001a, Protein aggregation. *Clin. Chem. Lab. Med.* **39:** 1065-1075.

Merlini, G., Marciano, S., Gasparro, C., Zorzoli, I., Bosoni, T., and Moratti, R., 2001b, The Pavia approach to clinical protein analysis. *Clin. Chem. Lab. Med.* **39:** 1025-1028.

Miller, I., Haynes, P., Eberini, I., Gemeiner, M., Aebersold, R., and Gianazza, E., 1999, Proteins of rat serum: III. Gender-related differences in protein concentration under baseline conditions and upon experimental inflammation as evaluated by two-dimensional electrophoresis. *Electrophoresis* **20:** 836-845.

Miller, I., Haynes, P., Gemeiner, M., Aebersold, R., Manzoni, C., Lovati, M.R., Vignati, M., Eberini, I., and Gianazza, E., 1998, Proteins of rat serum: II. Influence of some biological parameters of the two-dimensional electrophoresis pattern. *Electrophoresis* **19:** 1493-1500.

Mohammad, R.M., Maki, A., Vistisen, K., and Al-Katib, A., 1994a, Protein studies of human non-Hodgkin's B-lymphoma: Appraisal by two-dimensional gel electrophoresis. *Electrophoresis* **15:** 1566-1572.

Mohammad, R.M., Vistisen, K., and al Katib, A., 1994b, Protein study of T and B acute lymphoblastic leukemia cell lines. *Electrophoresis* **15:** 1218-1224.

Murphy, K.M. and Reiner, S.L., 2002, The lineage decisions of helper T cells. *Nat. Rev. Immunol.* **2:** 933-944.

Nathoo, S.A. and Finlay, T.H., 1987, Fetal-specific forms of alpha 1-protease inhibitors in mouse plasma. *Pediatr. Res.* **22:** 1-5.

Nyman, T.A., Rosengren, A., Syyrakki, S., Pellinen, T.P., Rautajoki, K., and Lahesmaa, R., 2001a, A proteome database of human primary T helper cells. *Electrophoresis* **22:** 4375-4382.

Nyman, T.A., Rosengren, A., Syyrakki, S., Pellinen, T.P., Rautajoki, K., and Lahesmaa, R., 2001b, A proteome database of human primary T helper cells. *Electrophoresis* **22:** 4375-4382.

O'Farrel, P.H., 1975, High resolution two-dimensional electrophoresis. *J. Biol. Chem.* **250:** 5375-5385.

O'Neill, E.E., Brock, C.J., von Kriegsheim, A.F., Pearce, A.C., Dwek, R.A., Watson, S.P., and Hebestreit, H.F., 2002, Towards complete analysis of the platelet proteome. *Proteomics.* **2:** 288-305.

Oda, R.P., Clark, R., Katzmann, J.A., and Landers, J.P., 1997, Capillary electrophoresis as a clinical tool for the analysis of protein in serum and other body fluids. *Electrophoresis* **18:** 1715-1723.

Oliver, D.J., Nikolau, B., and Wurtele, E.S., 2002, Functional genomics: high-throughput mRNA, protein, and metabolite analyses. *Metab Eng* **4:** 98-106.

Olivieri, E., Herbert, B., and Righetti, P.G., 2001, The effect of protease inhibitors on the two-dimensional electrophoresis pattern of red blood cell membranes. *Electrophoresis* **22:** 560-565.

Parkin, J. and Cohen, B., 2001, An overview of the immune system. *Lancet* **357:** 1777-1789.

Pasquariello, A., Ferri, C., Moriconi, L., La Civita, L., Longombardo, G., Lombardini, F., Greco, F., and Zignego, A.L., 1993, Cryoglobulinemic membranoproliferative glomerulonephritis associated with hepatitis C virus. *Am. J. Nephrol.* **13:** 300-304.

Pawloski, J.R. and Stamler, J.S., 2002, Nitric oxide in RBCs. *Transfusion* **42:** 1603-1609.

Phillips, J.D., Parker, T.L., Schubert, H.L., Whitby, F.G., Hill, C.P., and Kushner, J.P., 2001, Functional consequences of naturally occurring mutations in human uroporphyrinogen decarboxylase. *Blood* **98:** 3179-3185.

Poirier, F., Bourin, P., Bladier, D., Joubert-Caron, R., and Caron, M., 2001a, Effect of 5-azacytidine and galectin-1 on growth and differentiation of the human B lymphoma cell line bl36. *Cancer Cell Int.* **1:** 2.

Poirier, F., Imam, N., Pontet, M., Joubert-Caron, R., and Caron, M., 2001b, The BPP (protein biochemistry and proteomics) two-dimensional electrophoresis database. *J. Chromatogr. B Biomed. Sci. Appl.* **753:** 23-28.

Poirier, F., Pontet, M., Labas, V., Le Caër, J.P., Sghiouar-Imam, N., Raphaël, M., Caron, M., and Joubert-Caron, R., 2001c, Two-dimensional database of a Burkitt lymphoma cell line (DG 75) proteins: Protein pattern changes following treatment with 5'-azycytidine. *Electrophoresis* **22:** 1867-1877.

Putnam,F.W. 1975 *The plasma proteins. Structure, Function and Genetic Control (Volumes 1, 2, 3 and 4),* Academic Press, New York.

Rabilloud, T., Blisnick, T., Heller, M., Luche, S., Aebersold, R., Lunardi, J., and Braun-Breton, C., 1999, Analysis of membrane proteins by two-dimensional electrophoresis: Comparison of the proteins extracted from normal or *Plasmodium falciparum* - infected erythrocyte ghosts. *Electrophoresis* **20:** 3603-3610.

Roberts, G.C. and Smith, C.W., 2002, Alternative splicing: combinatorial output from the genome. *Curr. Opin. Chem. Biol.* **6:** 375-383.

Rolink, A.G., Schaniel, C., Andersson, J., and Melchers, F., 2001, Selection events operating at various stages in B cell development. *Curr. Opin. Immunol.* **13:** 202-207.

Rosenblum, B.B., Hanash, S.M., Yew, N., and Neel, J.V., 1982, Two-dimensional electrophoretic analysis of erythrocyte membranes. *Clin. Chem.* **28:** 925-931.

Rosenblum, B.B., Neel, J.V., and Hanash, S.M., 1983, Two-dimensional electrophoresis of plasma polypeptides reveals "high" heterozygosity indices. *Proc. Natl. Acad. Sci. U. S. A.* **80:** 5002-5006.

Rosenblum, B.B., Neel, J.V., Hanash, S.M., Joseph, J.L., and Yew, N., 1984, Identification of genetic variants in erythrocyte lysate by two- dimensional gel electrophoresis. *Am. J. Hum. Genet.* **36:** 601-612.

Rothenmund, D., Locke, V.L., Liew, A., Thomas, T.M., Wasinger, V., and Rylatt, D.B., 2003, Depletion of highly abundant protein albumin from human plasma using the Gradiflow. *Proteomics* **3:** 279-287.

Rund, D. and Rachmilewitz, E., 2001, Pathophysiology of alpha- and beta-thalassemia: therapeutic implications. *Semin. Hematol.* **38**: 343-349.

Sahota, S.S., Garand, R., Mahroof, R., Smith, A., Juge-Morineau, N., Stevenson, F.K., and Bataille, R., 1999, V_H gene analysis of IgM-secreting myeloma indicates an origin from a memory cell undergoing isotype switch events. *Blood* **94**: 1070-1076.

Sanger, F. and Coulson, A.R., 1975, A rapid method for determining sequences in DNA by primed synthesis with DNA polymerase. *J. Mol. Biol.* **94**: 441-448.

Sarioglu, H., Lottspeich, F., Walk, T., Jung, G., and Eckerskorn, C., 2000, Deamidation as a widespread phenomenon in two-dimensional polyacrylamide gel electrophoresis of human blood plasma proteins. *Electrophoresis* **21**: 2209-2218.

Sarraj-Reguieg, A., Tissot, J.D., Hochstrasser, D.F., Von Fliedner, V., Bachmann, F., and Schneider, P., 1993, Effect of prestorage leukocyte reduction on proteins of platelets obtained by apheresis. *Vox Sang.* **65**: 279-285.

Sassa, S., 2000, Hematologic aspects of the porphyrias. *Int. J. Hematol.* **71**: 1-17.

Saunders, F.K., Sharrard, R.M., Winfield, D.A., Lawry, J., Goepel, J.R., Hancock, B.W., and Goyns, M.H., 1993, 2D-gel analysis of proteins in chronic lymphocytic leukemia cells and normal B-lymphocytes. *Leuk. Res.* **17**: 223-230.

Saunders, F.K., Winfield, D.A., Goepel, J.R., Hancock, B.W., Sharrard, R.M., and Goyns, M.H., 1994, 2D-gel analysis of protein synthesis profiles of different stages of chronic lymphocytic leukaemia. *Leuk. Lymphoma* **14**: 319-322.

Schifferli, J.A., French, L.A., and Tissot, J.D., 1995, Hepatitis C virus infection, cryoglobulinemia, and glomerulonephritis. In *Advances in Nephrology* (J.L. Funck-Brentano, J.F. Bach, H. Kreis and J.P. Grünfeld, eds.), Year Book Medical Publishers, Chicago, pp. 107-129.

Schifferli, J.A., Ng, Y.C., and Peters, D.K., 1986, The role of complement and its receptor in the elimination of immune complexes. *N. Engl. J. Med.* **315**: 488-495.

Schrier, S.L., 2002, Pathophysiology of thalassemia. *Curr. Opin. Hematol.* **9**: 123-126.

Seligsohn, U. and Lubetsky, A., 2001, Genetic susceptibility to venous thrombosis. *N. Engl. J. Med.* **344**: 1222-1231.

Simon, T.L., 1994, The collection of platelets by apheresis procedures. *Transfus. Med. Rev.* **8**: 132-145.

Singh, S.M., Zada, A.A., Hiddemann, W., Tenen, D.G., Reddy, V.A., and Behre, G., 2002, Proteomic analysis of transcription factor interactions in myeloid stem cell development and leukaemia. *Expert. Opin. Ther. Targets.* **6**: 491-495.

Skehel, J.M., Schneider, K., Murphy, N., Graham, A., Benson, G.M., Cutler, P., and Camilleri, P., 2000, Phenotyping apolipoprotein E*3-leiden transgenic mice by two-dimensional polyacrylamide gel electrophoresis and mass spectrometric identification. *Electrophoresis* **21**: 2540-2545.

Snyder, E.L., Dunn, B.E., Giometti, C.S., Napychank, P.A., Tandon, N.N., Ferri, P.M., and Hofmann, J.P., 1987, Protein changes occurring during storage of platelet concentrates. A two-dimensional gel electrophoretic analysis. *Transfusion* **27**: 335-341.

Spertini, F., Tissot, J.D., Dufour, N., Francillon, C., and Frei, P.C., 1995, Role of two-dimensiomal electrophoretic analysis in the diagnosis and characterization of IgD monoclonal gammopathy. *Allergy* **50**: 664-670.

Strausberg, R.L. and Riggins, G.J., 2001, Navigating the human transcriptome. *Proc. Natl. Acad. Sci. U. S. A* **98**: 11837-11838.

Takeda, A., Waldron, J.A., Jr., Ruddle, N.H., and Cone, R.E., 1983, Analysis of normal and neoplastic lymphocyte surface-labeled proteins by two-dimensional polyacrylamide gel electrophoresis. *Exp. Cell Res.* **148**: 83-93.

Talebian-Ziai, S., Azgui, Z., Valensi, F., Sigaux, F., and Charron, D., 1992, Specific polypeptide markers in chronic B cell malignancies detected by two-dimensional gel electrophoresis. *Electrophoresis* **13**: 388-393.

Tastet, C., Charmont, S., Chevallet, M., Luche, S., and Rabilloud, T., 2003, Structure-efficiency relationships of zwitterionic detergents as protein solubilizers in two-dimensional electrophoresis. *Proteomics.* **3**: 111-121.

Tissot,J.D., Duchosal,M.A., and Schneider,P., 2000a, Two-dimensional polyacrylamide gel electrophoresis. In *Encyclopedia of Separation Science* (I. Wilson, T.R. Adlar, C.F. Poole and M. Cook, eds.), Academic Press, London, pp. 1364-1371.

Tissot, J.D., Helg, C., Chapuis, B., Zubler, R.H., Jeannet, M., Hochstrasser, D.F., Hohlfeld, P., and Schneider, P., 1992, Clonal imbalances of serum immunoglobulins after allogeneic bone marrow transplantation. *Bone Marrow Transplant.* **10**: 347-353.

Tissot, J.D. and Hochstrasser, D.F., 1993, Analysis of plasma/serum immunoglobulins by two-dimensional polyacrylamide gel electrophoresis. *Clin. Immunol. Newsletter* **13**: 97-101.

Tissot, J.D., Hochstrasser, D.F., Schneider, B., Morgenthaler, J.J., and Schneider, P., 1994a, No evidence for protein modifications in fresh frozen plasma after photochemical treatment: an analysis by high-resolution two- dimensional electrophoresis. *Br. J. Haematol.* **86**: 143-146.

Tissot, J.D., Hochstrasser, D.F., Spertini, F., Schifferli, J.A., and Schneider, P., 1993a, Pattern variations of polyclonal and monoclonal immunoglobulins of different isotypes analyzed by high-resolution two-dimensional electrophoresis. *Electrophoresis* **14**: 227-234.

Tissot, J.D., Hohlfeld, P., Forestier, F., Tolsa, J.F., Calame, A., Plouvier, E., Bossart, H., and Schneider, P., 1993b, Plasma/serum protein patterns in human fetuses and infants: a study by high-resolution two-dimensional polyacrylamide gel electrophoresis. *Appl. Theor. Electroph.* **3**: 183-190.

Tissot, J.D., Hohlfeld, P., Hochstrasser, D.F., Tolsa, J.F., Calame, A., and Schneider, P., 1993c, Clonal imbalances of plasma/serum immunoglobulin production in infants. *Electrophoresis* **14**: 245-247.

Tissot,J.D., Hohlfeld,P., Layer,A., Forestier,F., Schneider,P., and Henry,H., 2000b, Clinical applications. Gel electrophoresis. In *Encyclopedia of Separation Science* (I. Wilson, T.R. Adlar, C.F. Poole and M. Cook, eds.), Academic Press, London, pp. 2468-2475.

Tissot, J.D., Invernizzi, F., Schifferli, J.A., Spertini, F., and Schneider, P., 1999, Two-dimensional electrophoretic analysis of cryoproteins: A report of 335 samples. *Electrophoresis* **20**: 606-613.

Tissot,J.D., Layer,A., Schneider,P., and Henry,H., 2000c, Clinical applications. Electrophoresis. In *Encyclopedia of Separation Science* (I. Wilson, T.R. Adlar, C. F. Poole and M. Cook, eds.), Academic press, London, pp. 2461-2467.

Tissot, J.D., Sanchez, J.C., Vuadens, F., Scherl, A., Schifferli, J.A., Hochstrasser, D.F., Schneider, P., and Duchosal, M.A., 2002a, IgM are associated to Spalpha (CD5 antigen-like). *Electrophoresis* **23**: 1203-1206.

Tissot, J.D., Schifferli, J.A., Hochstrasser, D.F., Pasquali, C., Spertini, F., Clément, F., Frutiger, S., Paquet, N., Hughes, G.J., and Schneider, P., 1994b, Two-dimensional polyacrylamide gel electrophoresis analysis of cryoglobulins and identification of an IgM-associated peptide. *J. Immunol. Methods* **173**: 63-75.

Tissot, J.D., Schneider, P., Hohlfeld, P., Spertini, F., Hochstrasser, D.F., and Duchosal, M.A., 1993d, Two-dimensional electrophoresis as an aid in the analysis of the clonality of immunoglobulins. *Electrophoresis* **14**: 1366-1371.

Tissot, J.D., Schneider, P., James, R.W., Daigneault, R., and Hochstrasser, D.F., 1991, High-resolution two-dimensional protein electrophoresis of pathological plasma/serum. *Appl. Theor. Electroph.* **75:** 7-12.

Tissot, J.D. and Spertini, F., 1995, Analysis of immunoglobulins by two-dimensional gel electrophoresis. *J. Chromatogr. A* **698:** 225-250.

Tissot, J.D., Vu, D.H., Schneider, P., Vuadens, F., Crettaz, D., and Duchosal, M.A., 2002b, The immunoglobulinopathies: From physiopathology to diagnosis. *Proteomics* **2:** 813-824

Toda, T., Sugimoto, M., Omori, A., Matsuzaki, T., Furuichi, Y., and Kimura, N., 2000, Proteomic analysis of Epstein-Barr virus-transformed human B- lymphoblastoid cell lines before and after immortalization. *Electrophoresis* **21:** 1814-1822.

Tonegawa, S., 1983, Somatic generation of antibody diversity. *Nature* **302:** 575-581.

Tracy, R.P., Currie, R.M., Kyle, R.A., and Young, D.S., 1982, Two-dimensional gel electrophoresis of serum specimens from patients with monoclonal gammopathies. *Clin. Chem.* **28:** 900-907.

Tracy, R.P., Kyle, R.A., and Young, D.S., 1984, Two-dimensional electrophoresis as an aid in the analysis of monoclonal gammopathies. *Hum. Pathol.* **15:** 122-129.

Venter, J.C., Adams, M.D., Myers, E.W., Li, P.W., Mural, R.J., Sutton, G.G., Smith, H.O., Yandell, M., Evans, C.A., Holt, R.A., *et al.*, 2001, The sequence of the human genome. *Science* **291:** 1304-1351.

Verfaillie, C.M., 2002, Adult stem cells: assessing the case for pluripotency. *Trends Cell Biol.* **12:** 502-508.

von Andrian, U.H. and Mackay, C.R., 2000, T-cell function and migration. Two sides of the same coin. *N. Engl. J. Med.* **343:** 1020-1034.

Voss, T., Ahorn, H., Haberl, P., Dohner, H., and Wilgenbus, K., 2001, Correlation of clinical data with proteomics profiles in 24 patients with B-cell chronic lymphocytic leukemia. *Int. J. Cancer* **91:** 180-186.

Vu, D.H., Schneider, P., and Tissot, J.D., 2002, Electrophoretic characteristics of monoclonal immunoglobulin G of different subclasses. *J. Chromatogr. B Analyt. Technol. Biomed. Life Sci.* **771:** 355-368.

Vuadens, F., Crettaz, D., Scelatta, C., Servis, C., Quadroni, M., Bienvenut, W.V., Schneider, P., Hohlfeld, P., Applegate, L.A., and Tissot, J.D,. 2003, Plasticity of protein expression during culture of fetal skin cells. *Electrophoresis*, in press.

Vuadens, F., Gasparini, D., Deon, C., Sanchez, J.C., Hochstrasser, D.F., Schneider, P., and Tissot, J.D., 2002, Identification of specific proteins in different lymphocyte populations by proteomic tools. *Proteomics* **2:** 105-111.

Wait, R., Gianazza, E., Eberini, I., Sironi, L., Dunn, M.J., Gemeiner, M., and Miller, I., 2001, Proteins of rat serum, urine, and cerebrospinal fluid: VI. Further protein identifications and interstrain comparison. *Electrophoresis* **22:** 3043-3052.

Wait, R., Miller, I., Eberini, I., Cairoli, F., Veronesi, C., Battocchio, M., Gemeiner, M., and Gianazza, E., 2002, Strategies for proteomics with incompletely characterized genomes: the proteome of Bos taurus serum. *Electrophoresis* **23:** 3418-3427.

Wang, Y.Y., Cheng, P., and Chan, D.W., 2003, A simple affinity spin tube method for removing high-abundant proteins or enriching low-abundant biomarkers for serum proteomic analysis. *Proteomics* **3:** 243-248.

Weatherall, D.J., 2000, Haematology in the new millennium. *Br. J. Haematol.* **108:** 1-5.

Wilkins, M.R., Gasteiger, E., Gooley, A.A., Herbert, B.R., Molloy, M.P., Binz, P.A., Ou, K.L., Sanchez, J.C., Bairoch, A., Williams, K.L., and Hochstrasser, D.F., 1999, High-throughput mass spectrometric discovery of protein post-translational modifications. *J. Mol. Biol.* **289:** 645-657.

Willard-Gallo, K.E., 1984, Analysis of normal subset-specific and disease-specific human leukocyte proteins by cell sorting and two-dimensional electrophoresis. *Ann. N. Y. Acad. Sci.* **428:** 201-222.

Willard-Gallo, K.E., Houck, D.W., and Loken, M.R., 1988, Analysis of human lymphocyte protein expression. I. Identification of subpopulation markers by two-dimensional polyacrylamide gel electrophoresis. *Eur. J. Immunol.* **18:** 1453-1461.

Willard-Gallo, K.E., Humblet, Y., and Symann, M., 1984, Leukocyte membrane proteins in chronic lymphocytic leukemia, as studied by two-dimensional gel electrophoresis. *Clin. Chem.* **30:** 2069-2077.

Willard, K.E., 1982a, Two-dimensional analysis of human lymphocyte proteins II. Regulation of lymphocyte proteins by a molecule present in normal human urine. *Clin. Chem.* **28:** 1074-1083.

Willard, K.E., 1982b, Two-dimensional analysis of human lymphocyte proteins. III. Preliminary report on a marker for the early detection and diagnosis of infectious mononucleosis. *Clin. Chem.* **28:** 1031-1035.

Willard, K.E. and Anderson, N.G., 1981, Two-dimensional analysis of human lymphocyte proteins: I. An assay for lymphocyte effectors. *Clin. Chem.* **27:** 1327-1334.

Willard, K.E., Thorsrud, A.K., Munthe, E., and Jellum, E., 1982, Two-dimensional electrophoretic analysis of human leukocyte proteins from patients with rheumatoid arthritis. *Clin. Chem.* **28:** 1067-1073.

Williams, M.D., Chalmers, E.A., and Gibson, B.E.S., 2002, The investigation and management of neonatal haemostasis and thrombosis. *Br. J. Haematol.* **119:** 295-309.

Wintrobe, M.M., 1980., *Blood, pure and eloquent.* McGraw-Hill, New York.

Yan, J.X., Sanchez, J.C., Binz, P.A., Williams, K.L., and Hochstrasser, D.F., 1999, Method for identification and quantitative analysis of protein lysine methylation using matrix-assisted laser desorption ionization time-of-flight mass spectrometry and amino acid analysis. *Electrophoresis* **20:** 749-754.

Yazdanbakhsh, K., Lomas-Francis, C., and Reid, M.E., 2000, Blood groups and diseases associated with inherited abnormalities of the red blood cell membrane. *Transf. Med. Rev.* **14:** 364-374.

Young, D.S. and Tracy, R.P., 1995, Clinical applications of two-dimensional electrophoresis. *J. Chromatogr. A* **698:** 163-179.

Zdebska, E., Musielak, M., Jaeken, J., and Koscielak, J., 2001, Band 3 glycoprotein and glycophorin A from erythrocytes of children with congenital disorder of glycosylation type-Ia are underglycosylated. *Proteomics* **1:** 269-274.

Zubair, A.C., Silberstein, L., and Ritz, J., 2002, Adult hematopoietic stem cell plasticity. *Transfusion* **42:** 1096-1101.

Chapter 4

Proteomics in the Neurosciences

MARK O. COLLINS*, HOLGER HUSI* and SETH G. N. GRANT[†]
*Division of Neuroscience, University of Edinburgh, 1 George Square, Edinburgh, EH8 9JZ
UK †Wellcome Trust Sanger Institute, Hinxton, CAMBS, CB10 1SA UK

1.   INTRODUCTION

The field of proteomics aims to provide a comprehensive view of the characteristics and function of every protein present within a given biological system. The genome, which encodes the entire complement of genes in an organism, has been seen as the fundamentally complex component of cells. However it has emerged in recent years that its complexity is overshadowed by that of the transcriptome, which describes the full complement of mRNA transcripts transcribed from the genome of the cell. This in turn gives rise to yet another level of complexity see in that of the cellular proteome. This proteome is derived from the variety of alternative splicing events that occur at the level of transcription and the use of alternative start and stop sites and frame-shifting which occurs at the level of translation. This complexity is compounded by the 300 or more different protein modifications, which have been, reported (Aebersold and Goodlett 2001) of which phosphorylation and glycosylation are prototypic.

Comprehensive proteome analysis also seeks to elucidate protein sub-cellular distribution and protein-protein interactions, which are crucial in determining the integration of large numbers of proteins into interaction networks, and functional molecular complexes that ultimately perform a myriad of cellular functions. Finally, the proteome is highly dynamic and temporal and thus describes the complement of proteins expressed in a cell at a single time point.

H. Hondermarck (ed.), Proteomics: Biomedical and Pharmaceutical Applications, 101–121.
© 2004 *Kluwer Academic Publishers. Printed in the Netherlands.*

Proteomic analysis in any system encompasses many areas and integration of this data into biologically meaningful knowledge is the ultimate goal. The progress and future objectives of each area in relation to the field of neuroscience is discussed and the future of bioinformatic integration of this data is suggested (Figure 1).

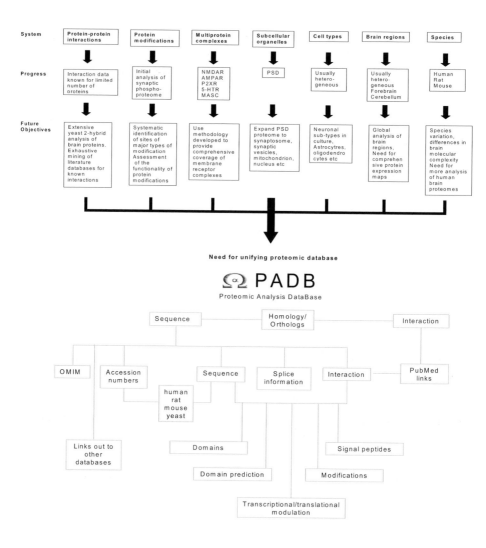

Figure 1. (a) Overview of the progress and objectives of proteomics and (b) the structure of PADB, a proteomic analysis database.

2. GLOBAL NEUROPROTEOMICS: APPROACHES & OBJECTIVES

2.1 Two-dimensional electrophoresis

The biochemical approach of two-dimensional electrophoresis which has become the classical proteomic approach to whole proteome analysis has the capacity to display a large number of proteins expressed the studied system under given physiological conditions. Construction of global expression maps for defined proteomes is the most widely used application of proteomics and when used in combination with mass spectrometry (MS) techniques can be a powerful approach. There have been a number of studies focused on global neuroproteomics from whole brain analysis to the analysis of synaptic components. Two-dimensional maps have been constructed for whole human (Langen, Berndt et al. 1999) mouse (Gauss, Kalkum et al. 1999) and rat brains (Fountoulakis, Schuller et al. 1999) to analyse the entire brain proteome. These studies yielded more than 200 discrete proteins with significant over lap between datasets. The mouse brain proteome study is the largest to date with 8767 resolved spots and over 300 proteins identified by a combination of proteomics and genetic methods, (Figure 2).

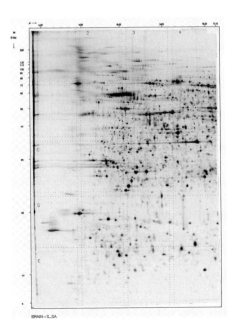

Figure 2. Protein standard map of the mouse brain supernatant fraction obtained by large-gel 2-D electro-phoresis followed by densitometry and computer-assisted spot detect-ion (image from Gauss, C., Kalkum, M., Lowe, M., Lehrach, H. & Klose, J. Analysis of the mouse proteome. (I) Brain proteins: separation by two-dimensional electrophoresis and identification by mass spectrometry and genetic variation. Electro-phoresis 20, 575-600. (1999)

A number of brain region proteomes have been studied to investigate differences in protein expression. Preliminary analysis of the mouse cerebellar proteome has identified 30 proteins (Beranova-Giorgianni, Giorgianni et al. 2002) and analysis of the porcine cerebellum led to identification of 56 spots (Friso and Wikstrom 1999). A developmental proteomic study of the rat cerebellum yielded resolution of over 3000 spots and identification of 67 of these (Taoka, Wakamiya et al. 2000). Most proteins showed an increase in abundance as the cerebellum matured, however, 42 spots appeared to be exclusively expressed in the immature cerebellum. Some of the latter were identified by MS and included proteins with defined roles in nervous system development.

Although, the approach of using 2D in conjunction with MS techniques is a classical proteomic approach, it is evident that it falls short as a sole platform for global proteomics. The range of biochemically different proteins, co-migration of proteins and difficulty of resolution of low abundance molecules has hampered the use of this technology to comprehensively analyse highly complex mixtures of proteins present in brain preparations. Problems associated with 2D resolution of membrane proteins have been overcome to some extent (Santoni, Molloy et al. 2000), however gel-free proteomic solutions such as multidimensional protein identification technology (MudPIT) (Wu, MacCoss et al. 2003) allow greatly increased coverage of the membrane proteome. However, 2D-electrophoresis will remain a useful tool in proteomics, due to sophisticated image analysis, software and fluorescent dyes, which now permit more sensitive and accurate comparative proteome analysis.

2.2   Subcellular Proteomics

Another approach which has proved to be very successful in the systematic identification of proteins, has been used to identify the molecular constituents of the post-synaptic proteome (PSP) (Husi 2003). The post-synaptic density is an amorphous structure usually 400 nm long and 40 nm wide and is located beneath the synaptic membrane, and is visible under the electron microscope as tight complexes of post-synaptic junctional proteins (Figure 3). It is well established that the regulation of synaptic proteins that occur in the PSP plays an important role in synaptic plasticity, a mechanism which is believed to underlie the synaptic mechanism of learning and memory (Kennedy 1993).

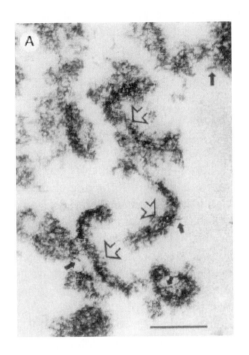

Figure 3. PSDs appear as slightly curved bars (open arrows indicate the cytoplasmic side). Adhering postsynaptic membranes are absent. Although little substructure is apparent, occaisional rod-like elements (4.5 x 28nm, small arrows) can be seen. (Image from Ziff, E.B. Enlightening the postsynaptic density. Neuron 19, 1163-1174 (1997))

The PSP was purified from mouse forebrain by density centrifugation as described previously (Carlin, Grab et al. 1980) with minor modifications. Protein samples were separated by SDS-PAGE, stained by Coomassie blue, and the entire gel lane was excised into 42 individual protein bands, reduced, alkylated and digested with trypsin . The resultant peptides were identified by online LC-MS/MS analysis and corresponded to 620 unique proteins. This approach has far exceeded past studies, one of which only identified 26 molecules in the post-synaptic density (Walikonis, Jensen et al. 2000). Another study, which employed a 2D electrophoresis approach, resolved 1700 spots. MS identified a total of 90 spots, containing 47 different protein species. Unsurprisingly, within the dataset reported by Husi et al, were many previously described synaptic proteins including ion channels, scaffold proteins, signaling enzymes and cytoskeletal proteins. The majority of the proteins were not known to be postsynaptic proteins. Although this proteomic strategy attempted to use a relatively unbiased strategy, it cannot be completely comprehensive as low abundance proteins may escape detection, and the selection of whole forebrain as starting material will overlook molecules found in small subsets of neurons. A further limitation imposed by the method of biochemical isolation is that some proteins may be contaminants not normally found at synapses. Despite this concern they observed a significant number of known synaptic proteins and a lack of many nuclear and other proteins that would not be expected. A systematic study using electron microscopy would be necessary to confirm or exclude each protein's synaptic localization.

This postsynaptic protein profile is useful in several ways. First, it appears that many of these proteins appear to be excellent candidates for various aspects of cell biology at the postsynaptic terminal. For example, 24 components of the protein synthetic machinery (elongation factors, ribosomal proteins) were identified; proteins encoded by synaptically localized mRNAs; cytoskeletal components and modulators of the cytoskeleton; motor proteins involved with ion channel trafficking channels (e.g. KIF proteins, NSF, AMPA); ubiquitination mechanisms and so on. Since many of these proteins have well characterized functions in non-neuronal settings, their identification may facilitate tests of their function at synapses. A second insight relates to the overall organization of the postsynaptic terminal into sets of macromolecular complexes. In addition to components of NMDA receptor complexes and AMPA complexes, there are ribosomal and other complexes, consistent with the view that the postsynaptic terminal is organized into macromolecular machines. Third, a number of proteins are involved with neurological diseases and this list may prove useful in directing human genetic studies of synaptic pathology. In addition to its use in understanding the biology of synapses, this dataset may prove useful in guiding human genetic studies of the nervous system.

The approach of biochemical purification combined with high-resolution mass spectrometry used for the definition of the PSP could be easily applied to many other sub-cellular structures. Indeed, subcellular structures, such as axo-glial junctions and nodes of Ranvier would be prime candidates for such an approach. Also, refined or new biochemical approaches to isolate structures such as dendrites, axons and growth cones would facilitate definition of other neuronal proteomes.

2.3   Neuro-phosphoproteomics

Another global neuroproteomic study carried out by Collins et al, (unpublished results) focused on the neurophosphoproteome, i.e. the collection of phosphoproteins at the synapse. Protein phosphorylation is an essential regulator of protein function and signaling mechanisms in any biological system. It has been estimated that one third of all proteins are phosphorylated in the mammalian cell and if it is assumed that one third of all proteins are expressed in the brain, then a conservative estimate of the number of phosphoproteins expressed in the brain would be in the order of

2000-3000. This is underpinned by the fact that up to 5% of the expressed genome may code for kinases.

Protein phosphorylation has been shown to be intimately involved in the assembly and function of the post-synaptic density (Yamauchi 2002), synaptic vesicle turnover (Turner, Burgoyne et al. 1999), LTP (Hedou and Mansuy 2003) and synaptic plasticity/memory (Schulman 1995). It is therefore obvious that study of protein phosphorylation on a proteomic platform will yield a wealth of data relevant to the molecular basis of brain function.

Identification of phosphoproteins and characterisation of phosphorylation sites has proved difficult for many reasons, most notably due to the low stochiometry and transient nature of protein phosphorylation. This has been especially true for global analysis of phosphorylation and there was a need for specific enrichment techniques to be developed to tackle these issues.

A large-scale phosphoprotein isolation strategy was developed using a pre-existing methodology for phosphopeptide purification. Phosphoproteins were purified by large-scale metal affinity chromatography (Andersson and Porath 1986) (Muszynska, Andersson et al. 1986) from denatured synaptosomes which contain the pre- and post-synaptic terminals, including the postsynaptic density. Purified phosphoproteins were digested in solution and analysed by DRA-LC-MS-MS. This approach has yielded the largest mammalian proteomic dataset to date, identifying 236 potential phosphoproteins from many protein classes (channels, transporters/transporting ATPases, scaffolding and adaptor proteins, kinases, phosphatases, small G-proteins and modulators, other signaling molecules, metabolic enzymes, cell adhesion and cytoskeletal proteins, nuclear/transcriptional, ribosomal and novel proteins. 28 phosphopeptides were detected and characterised and contained 37 phosphorylated residues. Of these 236 putative phosphoproteins it was shown by extensive literature mining that 81 of these had been shown previously to be phosphorylated, thus bringing confidence to the purification strategy used.

This dataset has considerable overlap with the post-synaptic density proteome, with one hundred common proteins (Husi 2003), fifty-four of which, are known to be phosphorylated. Also, thirty-one proteins contained in the MAGUK-associated signaling complex (MASC) (Husi 2003) are found in this phosphoproteomic dataset with twenty-four of these proteins known to be phosphorylated. This phosphoproteomic approach is being optimised in a number of ways, from using a double IMAC strategy, i.e. sequential purification of phosphoproteins and then phosphopeptides, to using more refined MS techniques such as precursor-ion scanning. These changes should yield much better coverage and depth to the synaptic phosphoproteome. Integration of this data on dynamic regulation of proteins into phosphorylation networks (Figure 4), and in also into the PSP network will add an extra dimension and understanding to the organisation and dynamism of the post-synaptic density.

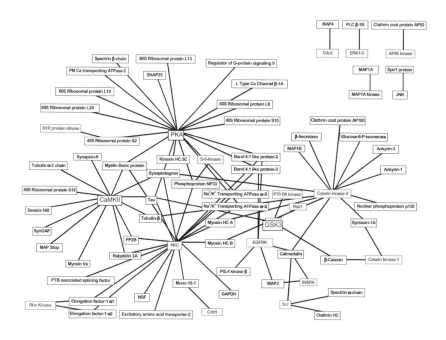

Figure 4. Literature based kinase substrate interaction map of proteins identified in the phosphoproteomic dataset by Collins et al.

The next major challenge for phosphoproteomics will be identification of cognate kinases and phosphatases for the potential wealth of phosphorylation sites which will arise from large-scale phosphorylation site studies. Also, other functional aspects of protein phosphorylation will need to be addressed in a high throughput fashion, namely the effects of phosphorylation on protein-protein interactions and protein stability. The development of high-density peptide and protein arrays (Lesaicherre, Uttamchandani et al. 2002) should facilitate the study of some of these aspects of the functional consequences of protein phosphorylation in a large-scale manner.

3.   MULTIPROTEIN COMPLEXES: A PROTEOMIC APPROACH

3.1   NMDA receptor-adhesion protein signaling complex

Protein-protein interactions are the basis on which the cellular and structure and function are built and interaction partners are an immediate lead into biological function which can be exploited for therapeutic purposes. Proteomics has been shown to make a crucial contribution to the study of protein-protein interactions (Neubauer, Gottschalk et al. 1997). Proteomic approaches to tackle multiprotein complexes usually involve purification of the entire complex by a variety of affinity methods, and protein identification by western blotting or mass spectrometry based approaches. The first proteomic analysis of multiprotein complexes relevant to the brain was the purification and identification of the molecular constituents of the NMDA receptor-adhesion protein signaling complexes (NRC) (Husi and Grant 2001) (Husi, Ward et al. 2000) and this will be described as a prototypic approach to multiprotein complex proteomics.

3.1.1   Methodology

Similar to most neurotransmitter receptors, the NMDAR has a low abundance in neurones compared with the overall protein content. Therefore, it is important to generate highly enriched samples containing the target molecule. This can be achieved by a variety of methods, generally involving affinity chromatography steps. These can range from ligand affinity (drugs, substrates or co-factors) to the more widely used approach of immunoprecipitation, involving specific antibodies to the molecule of interest. A major drawback in using antibody-based approaches for proteomic analysis is the introduction of the IgG itself, which interferes with the analysis by overshadowing proteins of interest that co-migrate during the gel-separation. This same problem arises with protein-affinity based methods, in which a recombinant protein carries one or several affinity-tags (e.g. poly-histidines), or is expressed as a GST-fusion protein. This problem can be eliminated using either the tandem affinity purification (TAP) method (Rigaut, Shevchenko et al. 1999) or with the use of non-protein based methods, which exploit substrate or ligand affinities, such as peptide-based isolation procedures (Husi and Grant 2001). All of these ligands, groups or proteins are then coupled to a resin either irreversibly (usually in the case of small molecules), or through a secondary affinity linkage specific for the bait-molecule. Neuronal extracts are then incubated with the resins bearing the bait and bound proteins separated from each other as well as from the matrix by conventional gel analysis.

Identification of isolated proteins can be readily achieved using western blotting or mass spectrometry. Western blotting (also known as immunoblotting) utilizes specific antibodies to detect a protein that has been transferred from a gel onto a membrane. A major advantage of western blotting is its sensitivity and ability to detect amounts of protein beyond the range of current sequencing-based or mass spectrometry methodologies. It is obvious that this approach is biased by the assumption that a given molecule might be present, and cannot be used to identify unknown or unsuspected proteins. However, there is a large amount of neurobiological evidence concerning the putative involvement of molecules that influence many target proteins and such proteins pose ideal targets for western blotting approaches. Additionally, changes in protein levels, as a result of modulation or modification in the receptor's environment, can easily be visualized using specific antibodies against the proteins under investigation. A disadvantage of this method is its labour intensive nature and thus difficulty in performing on a large scale.

The main approaches to protein identification are peptide-mass fingerprinting using matrix-assisted laser desorption/ionization (MALDI) (Refs [(Henzel, Billeci et al. 1993), (Shevchenko, Wilm et al. 1996)) and electrospray ionization using a tandem mass spectrometer (Shevchenko, Chernushevich et al. 1997) or a combination of both (Shevchenko, Loboda et al. 2000). Recent developments led to new methodologies using direct liquid chromatography/tandem mass spectrometry (LC/MS-MS) techniques, which allow fast and reliable mapping of very low amounts of peptides in an almost completely automated fashion (Shevchenko, Chernushevich et al. 2000). Furthermore, using this method, it is also possible to determine if a protein or peptide carries any post-translational modifications, such as glycosylation or phosphorylation (Pandey, Podtelejnikov et al. 2000).

The sample is usually separated on a gel and the bands of interest excised and treated with trypsin in order to generate fragments of proteins that can subsequently be extracted from the gel slice. There are several reasons for analyzing peptides rather than proteins. The peptide's mass can be measured at much greater precision than the mass of intact proteins. Peptides give rise to a library of molecular masses that are directly derived from the proteins, which can be identified using computational techniques (Berndt, Hobohm et al. 1999).

Using the approaches described above the NRC was shown to comprise 77 proteins organized into receptor, adaptor, signaling, cytoskeletal and novel proteins, of which 30 are implicated from binding studies and another 19 participate in NMDAR signaling (Figure 5) (Husi, Ward et al. 2000).

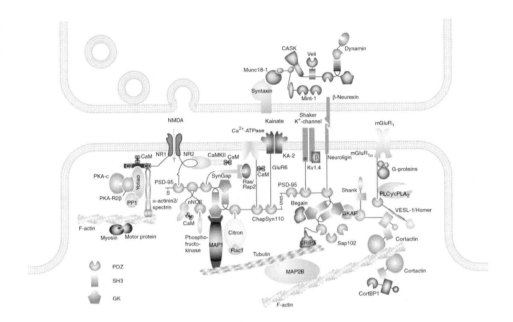

Figure 5. Schematic diagram of synaptic multiprotein complexes. Post-synaptic complexes of proteins associated with the NMDA receptor and PSD-95, found at excitatory mammalian synapses, are shown. Individual proteins are illustrated with arbitrary shapes and known interactions indicated. Proteins shown in colour are those found in a proteomic screen, whereas those shown in grey are inferred from bioinformatic studies. The specific protein–protein interactions are predicted, based on published reports from yeast two-hybrid studies. Membrane proteins (such as receptors, channels and adhesion molecules) are attached to a network of intracellular scaffold, signaling and cytoskeletal proteins, as indicated. (See also Colour Plate Section, page 385)

NMDAR and metabotropic glutamate receptor subtypes were linked to cadherins and L1 cell-adhesion molecules in complexes lacking AMPA receptors. These neurotransmitter-adhesion receptor complexes were bound to kinases, phosphatases, GTPase-activating proteins and Ras with effectors including MAPK pathway components. A striking feature of the composition is that there appears to be 'modules' or sets of signaling proteins that are known to comprise key components of signal transduction pathways that can be distinctly regulated. For example, all of the molecules necessary to induce the phosphorylation of mitogen-activated protein kinase (MAPK) following NMDAR stimulation are present in the NRC (including

CaMKII, SynGAP, Ras, MEK and ERK) (Figure 5). There modules may allow the NMDAR and mGluR to integrate signals within the complex and then couple to down stream cellular effector mechanisms, such as trafficking of AMPA receptors and cytoskeletal changes that mediate structural and physiological plasticity. Furthermore, at least 18 NRC constituents are regulated by synaptic activity indicating that the composition of the complex is dynamic. In the hippocampus, these activity-dependent genes are known to undergo specific temporal changes following the induction of plasticity.

Genetic or pharmacological interference with 15 NRC proteins impairs learning and with 22 proteins alters synaptic plasticity in rodents. Mutations in three human genes (NF1, Rsk-2, L1) are associated with learning impairments, indicating the NRC also participates in human cognition.

This NRC dataset was further expanded using a combination of peptide affinity chromatography and mass spectrometry identification of new components of the complex (Husi H et al 2003). Purification of MASCs (MAGUK-associated signaling complexes) was achieved by peptide affinity chromatography using a hexapeptide corresponding to the C-terminus of the NMDA-R2B subunit. This approach extended the NRC dataset to comprise 184 proteins which constitute the MASC(Husi 2003). Also, α-amino-3-hyroxy-5-methyklisoxazole-4-proprionic acid (AMPA) complexes were purified using immuno-affinity chromatography, resulting in the identification of ten unique proteins. These two receptor complex datasets, together with the PSD provides for the first time a general composition of the postsynaptic proteome, which totaled 698 distinct proteins. The main overlap was seen between MASC and the PSD, with 108 common proteins.

3.2   Other Membrane Receptor Complexes

Since the characterisation of the NRC, two other receptor complexes have been reported. P2X receptors are ATP-gated ion channels in the plasma membrane, and activation of the P2X7 receptor also leads to rapid cytoskeletal re-arrangements such as membrane blebbing. 11 proteins in human embryonic kidney cells that interact with the rat P2X7 receptor were identified by affinity purification followed by mass spectroscopy and immunoblotting (Kim, Jiang et al. 2001). Also using a proteomic approach based on peptide affinity chromatography followed by mass spectrometry and immunoblotting, another study identified 15 proteins that interact with the C- terminal tail of the 5-hydroxytryptamine 2C (5-HT (2C)) receptor, a GPCR (Becamel, Alonso et al. 2002). These proteins include several synaptic multidomain proteins containing one or several PDZ domains (PSD-95 and the proteins of the tripartite complex Veli3-CASK-Mint1), proteins of the actin/spectrin cytoskeleton and signaling proteins. Co-immunoprecipitation experiments showed that 5-HT (2C) receptors interact with PSD-95 and the Veli3-CASK- Mint1 complex in vivo. Electron microscopy also indicated a synaptic enrichment of Veli3 and 5-HT (2C) receptors and their co-localisation in microvilli of choroidal cells. These results indicate that the 5-HT (2C) receptor is associated with protein networks that are important for its synaptic localization and its coupling to the signaling machinery (Becamel, Alonso et al. 2002).

It is now emerging that multiprotein signaling complexes or "signaling machines", of which the NRC is prototypic, are responsible for orchestrating the complex signaling events at ionotrophic glutamate receptors (Husi, Ward et al. 2000), ATP receptors (Kim, Jiang et al. 2001) and G-protein coupled receptors (Becamel, Alonso et al. 2002). They seem to share a common organisation of receptor, protein scaffold, (in which the signaling molecules are localised) and membrane to cytoskeletal interactions. Therefore, study of individual receptor multiprotein complexes such as the NRC will most likely provide a mechanistic basis that will be very useful and applicable for the systematic identification and study of membrane receptor complexes associated with the estimated 100 other receptor types believed to exist in the brain.

4. DISEASE PROTEOMICS OF THE BRAIN

4.1 Alzheimer's Disease as a case study

The main approach of proteomic analysis of brain pathology is to use comparative 2D-page separation of constituent proteins in diseased versus normal samples. This is usually followed by gel image analysis to identify alterations in the intensity of specific spots compared to the normal state and subsequent excision of the abnormal spots and identification by mass spectrometry or peptide sequencing. The main types of samples, which have been analysed by these means, are human biopsies or postmortem samples and mouse models of human disease.

There have been a number of studies focused on Alzheimer's disease (AD), an increasingly prevalent neurodegenerative disease. Comparative proteome analysis was performed on post-mortem brain tissue samples from patients with AD and compared with profiles from non-demented control brain tissue (Schonberger, Edgar et al. 2001). Proteins were resolved by 2D-PAGE and identified by NH_2-terminal sequencing. Thirty-seven proteins differed significantly between AD and normal tissue and could be grouped into several functional categories. A number of proteins identified in this study, such as α–crystallin, superoxide dismutase, glyceraldehyde-3-phosphate dehydrogenase, and dihydropyrimidinase-related protein had been implicated previously in the pathogenesis of AD.

This work has been complemented by studies of mouse models of AD, namely mice transgenically overexpressing glycogen synthase kinase-3β (GSK-3β) and microtubule-associated protein tau, both of which have been implicated in the pathogenesis of AD. A proteomic approach was used to identify cellular abnormalities in mice over expressing human tau with the aim of understanding the role of tau in the pathogenesis of AD. Proteins were resolved by 2D-PAGE and 34 proteins whose expression levels in wild-type and tau transgenic mice differed at least 1.5 fold were identified by ESI-MS (Tilleman, Van den Haute et al. 2002). A similar study by the same group was performed on GSK-3β overexpressing mice (Tilleman, Stevens et al. 2002). GSK-3β is a serine-threonine protein kinase capable of phosphorylating tau and hyperphosphorylated tau is a principle component of neurofibrillary tangles, which are characteristic of AD brains. It is regulated by serine (Inhibitory) and tyrosine (stimulatory) phosphorylation and also by protein complex formation and its intracellular localisation. GSK-3β is a key regulator of regulating neuronal plasticity and gene expression controlling the function of many metabolic, signaling and structural proteins. It also regulates cell survival, as it facilitates a variety of proapoptotic mechanisms (Harwood 2001). It has been linked to all of the primary abnormalities associated with AD and was shown to interact with components of the plaque producing amyloid system (Grimes and Jope 2001).

A comparison between GSK-3β transgenic mice and wild type revealed 51 proteins whose expression differed significantly. Twenty-five of these proteins have been identified to be present in the postsynaptic proteome, 14 of which are components of NMDAR-PSD-95 complexes that constitute MASCs.

There was a significant decrease in the relative abundance of cytoskeletal and energy metabolism proteins and a significant increase in proteins involved in signal transduction and oxidative stress in the GSK-3β transgenic mice. A comparison of the profile of proteins with altered expression in the tau and GSK-3β transgenic mice showed that there were a number of proteins with similar changes consistent with the mutual involvement of these two proteins in the process of neurodegeneration.

Complementary large-scale approaches to the study of AD and other brain diseases have been developed, notably cDNA microarrays. These arrays can detect thousands of mRNA transcripts present in the biological sample and can be used in a similar way as 2D-PAGE for detecting alterations in diseased versus normal tissues. The use of microarrays is now routine and due to the facile methodology can be used as a first step in detecting alterations in transcript levels in various pathologies. However, the correlation between mRNA levels and protein levels can be poor, so microarray data should be interpreted with caution and confirmed by looking at protein levels by methods previously discussed. A recent example of the use of cDNA microarrays for the study of AD was reported by Colangelo et al. in which they profiled 12633 genes in human post-mortem hippocampal CA1 in normal versus AD patients (Colangelo, Schurr et al. 2002). They found decreases in transcript levels of 19 genes encoding metal ion-sensitive factors, transcription and neurotrophic factors, and signal-transduction elements involved in synaptic plasticity and organisation which correlated well with previously reported down-regulation of such function categories in AD. Also, they found increases in transcript levels of 19 genes involved in proapoptotic activities and inflammation, also consistent with AD pathology. Interpretation and integration of such data with that obtained from proteomic experiments is presently difficult as specific changes in transcripts and protein levels are not seen in many consistent sets of proteins/genes. Also, expression levels (mRNA/protein) seen in normal versus diseased experiments can be due to secondary effects and not central to the pathogenesis. However, consolidation of mRNA/protein expression data for a particular disease in a database, may uncover common disturbances in pathways relevant to the actual pathology and not due to secondary effects.

4.2 Networks and psychiatric disease

Complex diseases are ideally suited for study by proteomic approaches. Their complexity of causation requires a global approach and an example of one such study relates to the work described earlier on MASC.

The identification of MASC and its network properties provides a unique opportunity to explore human cognition and its disorders. Although it is generally accepted that cognitive mechanisms are conserved between mice and humans, it is unclear how much the rodent molecular studies map onto human psychiatric conditions. The possibility that MASC proteins may be involved with human psychiatric and neurological disorders was investigated and it was found that 46 MASC proteins implicated in mental illness in the literature (Figure 6). Although all mental disorders were searched 26 were found in schizophrenia, 18 in mental retardation, 7 in Bipolar disorder and 6 in depressive illness. This apparent bias toward schizophrenia and mental retardation could be biologically relevant since they both have a major cognitive component to their primary symptoms unlike the affective disorders.

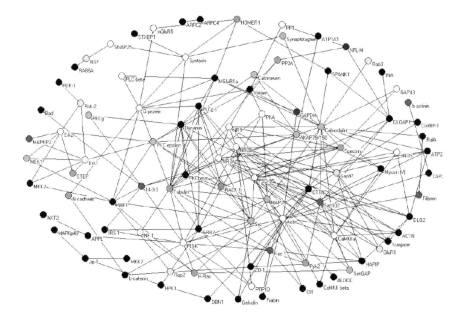

Figure 6. Network representation of MASC proteins with phenotypic annotation. 97 MASC proteins with known direct protein interactions to other MASC proteins are plotted. The NMDA receptor subunits (NR1, NR2A, NR2B) are centrally located. The association of each protein with plasticity, rodent behaviour or human psychiatric disorders is shown in the colour of the node. Key: Black – no known association; Red – Psychiatric Disorder; Green – Plasticity; Blue – Rodent Behaviour; Yellow – Psychiatric Disorders and Plasticity; Cyan – Plasticity and Rodent Behaviour; White – all three phenotypes. Orange for Psychiatric Disorders and Rodent Behaviour alone had no hits.
(See also Colour Plate Section, page 386)

The network provides a common connection or association between these disease molecules. Network simulations also show that disruption of combinations of proteins produce more severe effects on the network. This suggests that combinations of proteins (or mutant alleles) underpin the polygenic nature of the disorder.

4.3 Future perspectives

The number of comparative 2D/MS studies of disease and disease models has increased over the last number of years. This has led to the accumulation of lists of proteins whose expression levels have been shown to be altered in the particular system, but with little integration of these lists to provide a global view. This is emerging problem in the study of complex diseases, not only at the level of proteomics but also at the level of study of individual protein function in relation to disease. The pathogenesis of neurodegenerative diseases such as AD is likely to consist of a number of initial pathways converging to form the distinctive pathology of amyloid plaques and NFT's (neurofibrillary tangles) characteristic of AD. Proteins implicated in this pathogenesis may also be characteristic of the stage of progression of the disease and may be transiently altered. The components and formation of amyloid plaques and NFT's are reasonably well understood but the many pathways leading to this end point are still elusive. It is clear that there is a need to integrate the large amount of data concerning proteins involved in the pathogenesis of AD into a coherent global view of the disease.

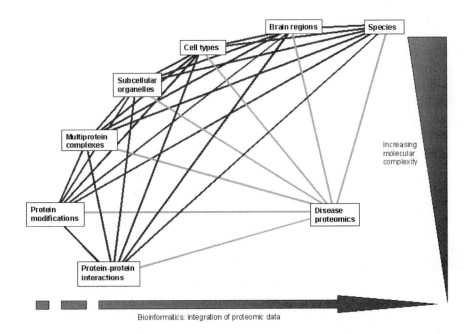

Figure 7. *Overview of levels of proteomic complexity*

It may be quite likely that an understanding of the fundamental components and mechanisms of the pathways leading to AD pathology will be gained from a holistic view that can be obtained by using a proteomic strategy combined with bioinformatics and network analysis (Figure 7).

5.   BIOINFORMATICS: INTEGRATION OF PROTEOMIC DATA

It is evident that there is a growing need for bioinformatic tools and databases to deal with the wealth of data being generated by proteomic analyses. Increased image resolution and reproducibility in 2D-electrophoresis is allowing 2D maps to be generated for particular systems (http://us.expasy.org/ch2d/). The samples and the protocols used to generate these maps are being standardised to allow easy cross-referencing between gels. Comprehensive protein identification by MS allows assignment of protein spots on these master maps yielding a characteristic spot position for each protein species. This kind of approach, in which a proteome is defined on a 2D map, will allow rapid comparative analysis. This will be particularly useful in disease proteomics, as it will bypass the necessity for routine MS.

Integrative proteomic databases such as PADB (http://www.PPID.org) (Husi 2002) will be essential for the analysis of proteomic data. The ability to submit MS datasets to such databases and retrieve comprehensive protein information, including protein-protein interactions will greatly reduce the current bottleneck of manual analysis of large numbers of proteins. As discussed for the MASC and PSP, such assembly of proteins into networks has been facilitated by this tailored proteomic database and such insights into global basic biology would not be possible in its absence. Bioinformatic analysis of proteomes is in its infancy, it can be seen in figure 1 that the future objectives of the many proteomes that exist are ambitious, but necessary.

Proteomic data must be dealt with in a holistic way if any meaningful biological questions are to be answered. Basic information on protein-protein interactions, protein modifications and multiprotein complexes must be obtained for relevant subcellular structures in various cell types and brain regions and integrated with previously published functional information. Within this framework of biological knowledge, complex biological problems and pathologies can be addressed in a less biased and comprehensive way than ever possible. Integration of diverse proteomic data to provide the ultimate global view is a formidable long-term task, which will only be achieved by standardisation of data, very powerful bioinformatic tools and comprehensive databases.

ACKNOWLEDGEMENTS

We would like to thank:- J.D. Armstrong, J.Choudhary and W.Blackstock for discussions; J.V. Turner for editorial assistance; The Wellcome Trust, Gatsby Foundation and BBSRC for support.

REFERENCES

Aebersold, R. and D. R. Goodlett, 2001, Mass spectrometry in proteomics. *Chem Rev* **101:** 269-95.

Andersson, L. and J. Porath, 1986, Isolation of phosphoproteins by immobilized metal (Fe3+) affinity chromatography. *Anal Biochem* **154:** 250-4.

Becamel, C., G. Alonso, et al. (2002). "Synaptic multiprotein complexes associated with 5-HT(2C) receptors: a proteomic approach." *EMBO J.* **21:** 2332-42.

Beranova-Giorgianni, S., F. Giorgianni, et al., 2002, Analysis of the proteome in the human pituitary. *Proteomics* **2:** 534-42.

Berndt, P., U. Hobohm, et al., 1999, Reliable automatic protein identification from matrix-assisted laser desorption/ionization mass spectrometric peptide fingerprints. *Electrophoresis* **20:** 3521-6.

Carlin, R. K., D. J. Grab, et al., 1980, Isolation and characterization of postsynaptic densities from various brain regions: enrichment of different types of postsynaptic densities. *J. Cell Biol.* **86:** 831-45.

Colangelo, V., J. Schurr, et al., 2002, Gene expression profiling of 12633 genes in Alzheimer hippocampal CA1: transcription and neurotrophic factor down-regulation and up-regulation of apoptotic and pro-inflammatory signaling. *J Neurosci Res* **70:** 462-73.

Fountoulakis, M., E. Schuller, et al., 1999, Rat brain proteins: two-dimensional protein database and variations in the expression level. *Electrophoresis* **20:** 3572-9.

Friso, G. and L. Wikstrom, 1999, Analysis of proteins from membrane-enriched cerebellar preparations by two-dimensional gel electrophoresis and mass spectrometry. *Electrophoresis* **20:** 917-27.

Gauss, C., M. Kalkum, et al., 1999, Analysis of the mouse proteome. (I) Brain proteins: separation by two- dimensional electrophoresis and identification by mass spectrometry and genetic variation. *Electrophoresis* **20:** 575-600.

Grimes, C. A. and R. S. Jope, 2001, The multifaceted roles of glycogen synthase kinase 3beta in cellular signaling. *Prog. Neurobiol* **65:** 391-426.

Harwood, A. J., 2001, Regulation of GSK-3: a cellular multiprocessor. *Cell* **105:** 821-4.

Hedou, G. and I. M. Mansuy, 2003, Inducible molecular switches for the study of long-term potentiation. *Philos Trans R Soc Lond B Biol Sci* **358:** 797-804.

Henzel, W. J., T. M. Billeci, et al., 1993, Identifying proteins from two-dimensional gels by molecular mass searching of peptide fragments in protein sequence databases. *Proc Natl Acad Sci U S A* **90:** 5011-5.

Husi, H., and Grant, S. G., 2002, Construction of a Protein-Protein Interaction Database (PPID) for Synaptic Biology. In Neuroscience Databases: A practical Guide (Boston/Dordrecht/London, Kluwer Academic Publishers): pp. 51-62.

Husi, H., Choudhary, J., Yu, L., Cumiskey, M., Blackstock, W., O'Dell, T.J., Visscher, P.M., Armstrong, J.D. & Grant, S.G.N., 2003, Synapse proteomes show scale-free network properties underlying plasticity, cognition and mental illness. Submitted.

Husi, H. and S. G. Grant, 2001, Isolation of 2000-kDa complexes of N-methyl-D-aspartate receptor and postsynaptic density 95 from mouse brain. *J. Neurochem.* **77**: 281-91.

Husi, H., M. A. Ward, et al., 2000, Proteomic analysis of NMDA receptor-adhesion protein signaling complexes. *Nat. Neurosci.* **3**: 661-9.

Kennedy, M. B., 1993, The postsynaptic density. *Curr. Opin. Neurobiol.* **3**: 732-7.

Kim, M., L. H. Jiang, et al., 2001, Proteomic and functional evidence for a P2X7 receptor signalling complex. *EMBO J.* **20**: 6347-58.

Langen, H., P. Berndt, et al., 1999, Two-dimensional map of human brain proteins. *Electrophoresis* **20**: 907-16.

Lesaicherre, M. L., M. Uttamchandani, et al., 2002, Antibody-based fluorescence detection of kinase activity on a peptide array. *Bioorg Med Chem Lett* **12**: 2085-8.

Muszynska, G., L. Andersson, et al., 1986, Selective adsorption of phosphoproteins on gel-immobilized ferric chelate. *Biochemistry* **25**: 6850-3.

Neubauer, G., A. Gottschalk, et al., 1997, Identification of the proteins of the yeast U1 small nuclear ribonucleoprotein complex by mass spectrometry. *Proc. Natl. Acad. Sci. USA* **94**: 385-90.

Pandey, A., A. V. Podtelejnikov, et al., 2000, Analysis of receptor signaling pathways by mass spectrometry: identification of vav-2 as a substrate of the epidermal and platelet-derived growth factor receptors. *Proc. Natl. Acad. Sci. USA* **97**: 179-84.

Rigaut, G., A. Shevchenko, et al., 1999, A generic protein purification method for protein complex characterization and proteome exploration. *Nat. Biotechnol.* **17**: 1030-2.

Santoni, V., M. Molloy, et al., 2000, Membrane proteins and proteomics: un amour impossible? *Electrophoresis* **21**: 1054-70.

Schonberger, S. J., P. F. Edgar, et al., 2001, Proteomic analysis of the brain in Alzheimer's disease: molecular phenotype of a complex disease process. *Proteomics* **1**: 1519-28.

Schulman, H., 1995, Protein phosphorylation in neuronal plasticity and gene expression. *Curr. Opin. Neurobiol.* **5**: 375-81.

Shevchenko, A., I. Chernushevich, et al., 1997, Rapid 'de novo' peptide sequencing by a combination of nanoelectrospray, isotopic labeling and a quadrupole/time-of-flight mass spectrometer. *Rapid. Commun. Mass Spectrom.* **11**: 1015-24.

Shevchenko, A., I. Chernushevich, et al., 2000, De Novo peptide sequencing by nanoelectrospray tandem mass spectrometry using triple quadrupole and quadrupole/time-of-flight instruments. *Methods Mol. Biol.* **146**: 1-16.

Shevchenko, A., A. Loboda, et al., 2000, MALDI quadrupole time-of-flight mass spectrometry: a powerful tool for proteomic research. *Anal. Chem.* **72**: 2132-41.

Shevchenko, A., M. Wilm, et al., 1996, A strategy for identifying gel-separated proteins in sequence databases by MS alone. *Biochem. Soc. Trans.* **24**: 893-6.

Taoka, M., A. Wakamiya, et al., 2000, Protein profiling of rat cerebella during development. *Electrophoresis* **21**: 1872-9.

Tilleman, K., I. Stevens, et al., 2002, Differential expression of brain proteins in glycogen synthase kinase-3 transgenic mice: a proteomics point of view. *Proteomics* **2**: 94-104.

Tilleman, K., C. Van den Haute, et al., 2002, Proteomics analysis of the neurodegeneration in the brain of tau transgenic mice. *Proteomics* **2**: 656-65.

Turner, K. M., R. D. Burgoyne, et al., 1999, Protein phosphorylation and the regulation of synaptic membrane traffic. *Trends Neurosci.* **22:** 459-64.

Walikonis, R. S., O. N. Jensen, et al., 2000, Identification of proteins in the postsynaptic density fraction by mass spectrometry. *J. Neurosci.* **20:** 4069-80.

Wu, C. C., M. J. MacCoss, et al., 2003, A method for the comprehensive proteomic analysis of membrane proteins. *Nature Biotech.* **21:** 532-8.

Yamauchi, T., 2002, Molecular constituents and phosphorylation-dependent regulation of the post-synaptic density. *Mass Spectrom. Rev.* **21:** 266-86.

Chapter 5

Ocular Proteomics

MICHEL FAUPEL, ERIC BERTRAND and JAN VAN OOSTRUM
Novartis Pharma Research, Functional Genomics, WJS 88.701, CH4002 Basel Switzerland

1.   INTRODUCTION

In these early years of the post-genomic era where proteomics is developing into a mature field, some people consider that biology will soon become an easy thing to do. To a certain extent, this is true, the new approaches have already enjoyed a number of successes and they will have a major impact on the way research is conducted. However, considering the previous achievements and short-comings in vision and ocular research, a number of challenges should still be expected. The purpose of this section is to describe the past dynamics of vision research as well as the current frame of proteomic studies in ophthalmics, so as to make a few predictions concerning ocular proteomics and its supporting technologies.

2.   THE RETINA AND VISUAL FUNCTION

Vision researcher have been concerned with post-translational modifications and differential expression methods years before the term proteomics existed. Thanks to their dedication we now understand the basic mechanisms of photo-transduction. Already in the eighties, investigators performed differential studies using 1D or 2D electrophoresis, localised differentially expressed proteins on their gels and were left wondering how to find out their identity. We will mention several of them, who managed, through their remarkable persistence, to identify their protein of interest by a variety of methods.

H. Hondermarck (ed.), Proteomics: Biomedical and Pharmaceutical Applications, 123–137.
© 2004 *Kluwer Academic Publishers. Printed in the Netherlands.*

We will start with the work of Susan Semple-Rowland who studied an autosomal recessive mutation responsible for retinal degeneration (rd) in chickens. In 1989, she published the results of a differential analysis performed with 2D gels and reported several alleles specific of the rd strain (Semple-Rowland and Ulshafer, 1989; Fig.1). Unfortunately, she was not able to identify these alleles as the required mass spectrometry techniques were not available at that time, also even today the genome of the chicken is not sequenced. Meanwhile, it was discovered in 1996 that several mutations in the guanylate cyclase 1 (GC1) gene cause Leber congenital amaurosis type 1 (LCA1), subjects affected by this autosomal recessive disease are born blind (Perrault et al, 1996). Semple-Rowland noticed a striking similarity between the rd and LCA1 phenotypes and decided to investigate the chicken GC1 gene. Finally, in 1998, she proved that a null mutation of this gene was responsible for the retinal degeneration (Semple-Rowland et al, 1998). One of the differentially expressed alleles had a pI of 6.0 and a MW of 110 kD which would be consistent with the characteristics GC1, however the identity of this allele has not been established.

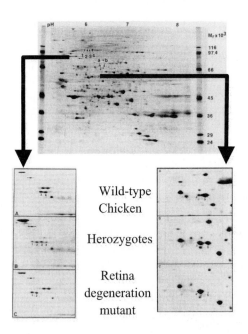

Figure 1. Differential analysis of wild-type and rd (retina degeneration) chicken using 2-D gels. Differentially expressed alleles are shown in the lower part, the position of the spots on a 2-D gel is indicated in the upper part. Adapted from Semple-Rowland and Ulshafer (1989).

The second story also begins with a successful differential expression study, this time with 1D gels. Barbara-Anne Battelle investigates the modulation of the photo-responses by efferent input in the eyes of the limulus which is the experimental preparation of choice for this kind of question. In 1989, she found that a 122 kD was phosphorylated after a stimulation of the photoreceptors with octopamine or cAMP (Edwards et al, 1989). The same substrate is phosphorylated in response to endogenous circadian efferent input, electrical stimulation of the afferent and to light (for review see Battelle, 2002). The substrate has been cloned, sequenced and identified in 1998, it turned out to be a class III myosin, homologous to the ninaC gene product of drosophila (Battelle et al, 1998). This myosin seems to be a convergence point for circadian and light signal, it might mediate some of the structural and functional changes occurring in the photoreceptor during the transition from day to night (Battelle, 2002).

The two former examples suggest that without a proper genomic database for the organism being studied the identification process is not trivial. In a recent paper Angelika Schmitt and Uwe Wolfrum demonstrated that a clever combination of immunological methods and molecular biology can circumvent this problem (Schmitt and Wolfrum, 2001). They managed to identify several component of the connecting cilium, a thin structure that links the inner and the outer segment of the photoreceptor. The connecting cilium is important for maintaining the organization and polarity of the photoreceptors, it presumably mediates the transport of newly synthesized molecules in the inner segment to the outer photosensitive segment. Some evidences also suggest that a dysfunction of the cilium might be involved in the retinal degeneration observed in Usher's syndrome. Schmitt and Wolfrum raised an antiserum against a purified axoneme fraction of bovine photoreceptors. Afterwards they performed an immuno-screening with the antiserum on a rat cDNA expression library, sequenced the positive clones and, for several gene products, confirmed the localization in the connecting cilium by immuno-cytochemistry. A proteomic study would nicely complement these results and provide new information on post-translational modifications and allelic variants. Such an approach might soon prove feasible as the team of Hiroyuki Matsumoto is cataloguing bovine photoreceptor proteins (Matsumoto and Komori, 2000).

We will close the chapter with the earlier work of Matsumoto and his group on the visual system of the drosophila which proved pioneering in many respects. In 1982, Matsumoto and his colleagues localized on 2D gels 3 classes of retinal proteins that are differentially expressed after exposure of the eye to light (Matsumoto et al, 1982). These 3 classes have been designed

Michel Faupel et al.

as 39 kD, 49 kD and 80 kD according to their apparent molecular weight, all of them undergo light-induced phosphorylation which account for the changes observed on 2D gels (Matsumoto and Pak, 1984). In 1990, the 39 kD and 49 kD phosphoproteins were identified as arrestin by several groups including Matsumoto's team (Hyde et al, 1990; Levine et al, 1990; Smith et al, 1990; Yamada et al, 1990). In 1994, they used HPLC-electrospray mass spectrometry to localize the phosphorylation site of the 49 kD arrestin by CaMK (Matsumoto et al, 1994). The 80 kD phosphoprotein was finally identified as the InaD gene product which belong to the PDZ family and serve as a scaffold for the formation of a signalling complex containing rhodopsin, calmodulin, PKC, PI-PLC and DGqα (Matsumoto et al, 1999; Fig. 2). This group also took advantage of the rich resources of drosophila's genetics and analysed the effect of several mutations on light-induced phosphorylation (Matsumoto et al, 1994).

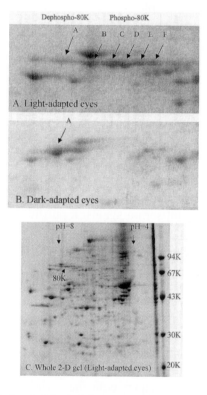

Figure 2. Differential analysis of light and dark adapted drosophila retinas. Expression pattern of a 80 kD protein for light (A) and dark (B) adapted retinas on 2-D gels. Position of the 80 kD protein on a 2-D gel (C). With kind permission of Matsumoto (1999).

3.   APPLICATION TO OPHTHALMICS

Studies at the protein level have been instrumental in elucidating visual function; will proteomics help us to understand the aetiology of ocular diseases? A first reason to be optimistic is that, so far, the results of basic research have been very efficiently translated into medically-relevant knowledge. Mutations of many genes involved in photo-transduction have been shown to provoke retinal degeneration, this is the case for rhodopsin, guanylate cyclase 1, phosphodiesterase and the cGMP channel. Furthermore, I'll discuss thoroughly two factors that are critical to the success of proteomics in the ophthalmic area: the availability of animal models to investigate ocular pathologies and the role of post-translational modifications in these diseases.

One lesson we might draw from the case of the drosophila, is that proteomics works best on organisms whose genome has been sequenced and for which interesting mutants are available. The mouse meets both conditions, powerful tools to generate mutants are established and a number of ocular phenotypes have been produced. In the 'reverse genetics' approach a gene of interest is modified by gene targeting (knock-out), insertional mutagenesis (genetrap) or transgenesis (knock-in) (Nolan, 2000). In many cases, knock-out mice are used to investigate the function of a gene, and the phenotype cannot be predicted or there might be no change in phenotype. Nonetheless, interesting ocular phenotypes are obtained every year, a very recent example is the inactivation of the BETA2/NeuroD1 transcription factor which causes photoreceptor degeneration (Pennesi et al, 2003). Also, a gene that causes disease in human can be targeted in mice with a reasonable chance of obtaining a similar phenotype, a mouse model for Ocular albinism type 1 has been generated in that way (Incerti et al., 2000). It is even possible to replace wild type alleles with a pathogenic variant either by homologous recombination or by the combining a knock-in with a knock-out: Expression of the G90D rhodopsin variant in mice causes night blindness without photoreceptor degeneration, a phenotype resembling very much human congenital blindness which is caused by the same allele (Sieving PA, 2001).

The 'forward genetics' approach is phenotype-driven: mutagen agents are used to create a large number of mutants which are selected according to their phenotype (Nolan, 2000). In this case, the gene that has been mutated is not known, but a line of mice is established and can be studied. For practical reasons, the alkylating agent ENU (N-ethyl-N-nitrosurea) is widely used in phenotypic screens, it induces point mutations in male germ cell at high frequency (Nolan, 2000). The simplest screening procedure is to look for dominant mutations, in that case ENU treated males are mated with wild type females and the first generation progeny is examined for abnormal phenotype (Nolan, 2000). The phenotype can be assessed with any ophthalmic procedure or vision test, however as hundreds or even thousands of animals have to be screened, the examination cannot be too labor-intensive. Dominant screens have produced models for cataract, retinal degeneration, corneal opacity and a number of other impairments (Ehling, 1982; Thaung, 2002; Fig. 3). An intrinsic limitation of 'forward genetic' is that once a phenotype of interest is obtained it requires an important effort to clone the mutated gene. Also, recessive mutations are more relevant to human diseases than dominant ones, but ENU recessive screens are much more demanding as the mating scheme involves three generations (Nolan, 2000). Nevertheless, the 'forward genetics' approach is generating unique phenotypes that are extremely valuable to study.

| Normal Retina | Enlarged and escavated optic disc | Scattered white spots |

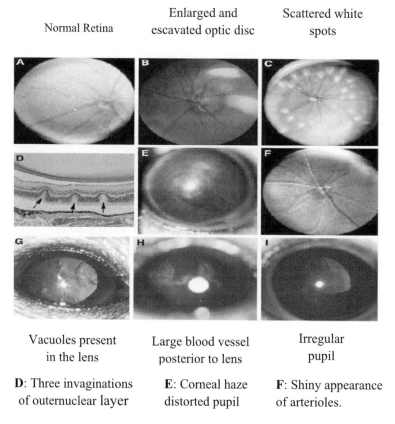

| Vacuoles present in the lens | Large blood vessel posterior to lens | Irregular pupil |

D: Three invaginations   **E**: Corneal haze   **F**: Shiny appearance
of outernuclear layer   distorted pupil   of arterioles.

Figure 3. Ocular phenotypes obtained by a dominant ENU screen. Adapted fromThaung et al (2002).

(See also Colour Plate Section, page 387)

Many ocular pathologies, either in humans or animal models, are caused by changes in specific eye tissues (for review see Steely and Clark, 2000). The refractive properties of the cornea and the lens are altered respectively in corneal dystrophies and cataracts. The main risk factor for open angle glaucoma is elevated intraocular pressure which occurs when the outflow of aqueous humour through the trabecular meshwork is obstructed, a possible way to lower the pressure would be to reduce the production of aqueous humor by the ciliary body. The various tissues we mentioned have proven appropriate for proteomic investigations (Steely and Clark, 2000), moreover several aspects of eye function should be addressed at the protein level. For instance, there are very few cell nuclei in the stroma of the cornea and none at all in the nucleus (central part) of the lens or in the aqueous humor, making gene-based approaches unpractical. Also, the extensive post-translational modifications that take place in many eye tissues are likely to have important physiological functions and might be involved in the development of ocular diseases (Steely and Clark, 2000). The transparency of the cornea depends on a very specific organisation of collagen fibrils, the post-translational modifications of collagen and its interactions with proteoglycans certainly play a central role in this organisation (Ottani et al, 2002). Proteolytic processing of extracellular matrix components occurs in the cornea and the trabecular meshwork which contain a number of matrix metalloproteinases (MMP), these enzymes have also been found in virtually all areas of the eye (Sivak and Fini, 2002). The lens is the tissue that has been the most studied using 2D gels, about 2000 spots are observed, mostly α, β, γ crystallins and their modified forms (Jungblut et al, 1998). Post-translational modifications like oxydation or glycation of the crystallins lead to protein aggregation and eventually to lens opacification, that is cataract (Steely and Clark, 2000). The expression pattern of crystallin is modified in several models of cataract and an atypical form of αB crystallin has been found in the lens of some patients (Jimenez-Asensio et al, 1999; Fig. 4).

4. NEW PROTEOMICS TECHNOLOGIES

Studies of the lens and the retina at the protein level have provided essential informations on the mechanisms of ocular disease, it will certainly be fruitful to apply the same approach to other eye tissues where a range of post-translational modifications is likely to occur. As animal models are available for many ocular pathologies it is desirable to compare the relevant control and diseased tissues by generating proteomics maps that are as complete as possible. The current state of the art maps are based on 2D gels and they encompass soluble proteins within a molecular weight range of 10-

80 kD. The recent developments in protein fractionation; separation and analysis technologies, will allow us to cover much larger portion of the proteome, principally for low abundance proteins.

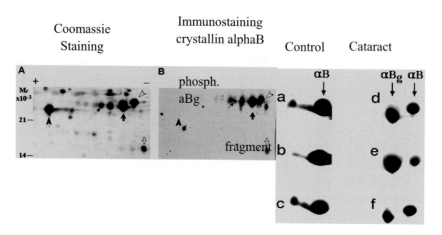

Figure 4. Post-translational modifications of crytallins involved in cataract. Coomassie (A) and immunoblot (B) staining of crystallin alphaB on 2-D gel. Differential analysis (C) of cortical lens samples from controls (a, b. c) and cataract patients (d, e, f) using 2-D immunoblot staining. Adapted from Jimenez-Asensio et al).

4.1   New pre-fractionation approaches

Now, good sample pre-fractionation strategies appear to be some of the most promising approaches for enhancing detection of low abundance proteins and increasing the total number of proteins that can be detected from complex proteomes. An optimal pre-fractionation method would have the capacity to separate complex protein mixtures into a small number of well-resolved fractions. Faupel and Schindler, developed a device and method for sample pre-fractionation: "MOST" for Multi-dimensional Off-gel Separation Technology (Faupel, M., and Schindler, P., 2002, patent pending), based on the segmented isoelectric focusing principle described by Faupel, M., and Righetti, P.G., 1990 .

The MOST technology, is achieved using micro preparative isoelectric membrane technology based on the segmented immobilised pH Gradient (IPG) technology by covalently linking acrylamido buffers to a polyacrylamide gel, fixing the buffering pH at any desired value between pH 3 to 10. Once an electric field has been applied across an array of small volume chambers (each containing up to 400μl sample), only the proteins with isoelectric points (pI's) at or above the pH of the gel membranes will pass through to the next chamber; all other proteins will remain in solution. The focusing is contained within an apparatus providing a tandem of cavities arranged in titration well plate format (Fig. 5). These are positioned to allow for a number of focusing modes in multi-dimensional separation. The cover to this apparatus consequently provided electrodes positioned respectively to accommodate these focusing modes. The objective of the MOST is to fractionate protein samples within a narrow pH range i.e. < 0.1pH unit.

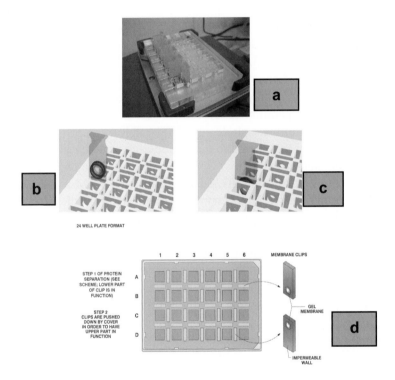

Figure 5. Multi-dimensional Off-gel Separation Technology: The apparatus (a), Scheme of membrane holder functions (b, c), and scheme of a 24 well plate format with in or out membrane function (d).

4.2   Differential expression with and without 2-D gels

Two-dimensional polyacrylamide gel electrophoresis (2DGE) is today, the only proven method for simultaneously separating thousands of proteins and quantitatively comparing changes of protein profiles in cells, tissues or whole organisms. Recently Hoving, S., et al, 2002, and Hoving, S., et al, 2000, developed new immobilized pH gradient gel strips (IPG's) for the first dimension of 2DGE, (Fig.6) This way, they improved protein resolution, loading capacity and reproducibility.

A number of mass spectrometry-based methods for the simultaneous identification and quantification of individual proteins within complex mixtures have been reported by Creasser, C.S, et al 2000, Aebersold, R., et al. 2000,Gysi, S.P., et al. 2002, Ranish, J.A., et al. 2003. Promising new technologies such as isotope coded affinity tags (ICAT), that may complement, or by-pass, 2-D gel technology will also provide a means for global expression analysis as reported by Reifenberger, J., et al. 2002, Zhang, R. and Regnier, F.E, 2002, Zhang, R., et al. 2002, Koyama, T., et al 2002.

The expression proteomics approach is highly adaptable for analysis and categorisation of drug influence and disease state as well as the influence of biological stimuli, identification of body fluids and tissue biopsies, examining PTMs and discovery of new disease markers.

Chemicals and heavy-isotope-labelling are commonly used in the early stages of sample preparation (except in differential fluorescence labelling protocols). The goal of these strategies is to be able to identify every protein expressed in a cell or tissue, and to determine each protein's abundance, state of modification, and possible involvement in multi-protein complexes.

Characterization of post-translational modifications (PTMs) is an essential task in proteomics. PTMs represent the major reason for the variety of protein isoforms and they can influence protein structure and function. Upon MALDI (matrix-assisted laser desorption / ionization), most post-translationally modified peptides form a fraction of labile molecular ions (which lose PTM-specific residues after acceleration). Compared to fully accelerated ions these fragment ions are defocused and show in reflector mass spectra reduced resolution. A short time Fourier transform using a Hanning window function now uses this difference in resolution to detect the metastable fragments. Its application over the whole mass range yields frequency distributions and amplitudes as a function of mass, where an increased low frequency proportion is highly indicative for metastable fragments. Applications on the detection of metastable losses originating from carboxamidomethylated cysteines, oxidized methionines, phosphorylated and glycosylated amino acid residues where reported by Wirth et al. 2002. The metastable loss of mercaptoacetamide detected with this procedure represents a new feature and its integration in search algorithms will improve the specificity of MALDI peptide mass fingerprinting.

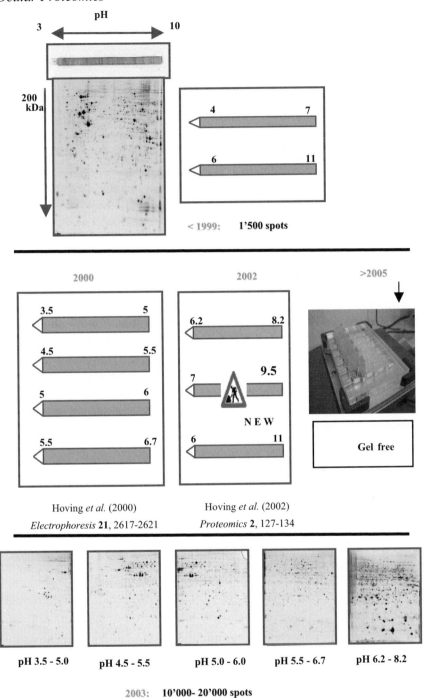

Figure 6. Progress of 2-D gels range and resolutions. Status before 1999, Introduction of ultra zoom gels and results using ultra zoom gels showing very high resolution in the pH 3.5 to pH 8.2 range.

4.3   Improving proteomic maps

Proteomics has also rapidly expanded and includes technologies like protein microarrays. Protein microarrays described by Pawlak, M. et al 2002, are based on the innovative planar waveguide (PWR) technology for evanescent field detection. They are designed for robustness, sensitivity and high throughput in biological assays. They can be universally applied for the analysis of all prominent ligand - receptor pairs such as antigen - antibody, enzyme - substrate, and membrane receptor - ligand. Microarrays are available in the capture or the reverse array format.

Protein-protein interaction pathways and cell signalling networks, high-throughput protein structural studies using mass spectrometry, nuclear magnetic resonance, X-ray crystallography are also expanding fields.

Newly, biophotonics technologies combined with atomic force microscopy was reported by Kuznetzsov, V.Y., et al, 2002.

The field of proteomics covers also protein profiling. This is a very powerful tool in clinical medicine for identification of diagnostic markers. The proteinChip technology based on the surface enhanced laser desorption /ionization time- of- flight (SELDI-TOF) mass spectrometry was developed at Ciphergen Biosystems to determine protein expression patterns. Due to the wide dynamic range of protein expression in complex biological samples, initial sub-fractionation of the sample is carried out by retentate chromatography using Protein Chip arrays with differential surface chromatographic properties.

In large proteomics projects, the information on samples used is often recorded as part of a laboratory information management system (LIMS) and constitute the other components of proteomics. A more comprehensive list of technologies used in proteomics is given in chapter 9 by K.K. Jain.

5.   CONCLUSION

As new methodologies are being designed and implemented, proteomics will provide a large number of targets for drug discovery in ocular and other diseases. Within this chapter we have not attempted to provide an exhaustive overview of the entire proteomics landscape, but we have selected several high potential approaches ranging from identification of biomarkers and disease-causing proteins through in vivo protein pathways analysis to new technologies. Analytical developments have made a substantial contribution to the successful sequencing of the human genome. At this point, we expect that the most recent technical progresses will drive proteomics to the same level of achievement as genomics.

ACKNOWLEDGEMENTS

We should like to acknowledge Drs. Dieter Müller, Patrick Schindler, Hans Voshol, Sjouke Hoving, Angelique Augustin, William Phares, Urs Wirth and Harry Towbin for sharing their views on methods related to proteomics and many stimulating discussions on data interpretation and reliability issues.

REFERENCES

Aebersold, R., Hood, L.E., and Watts, J.D., 2000, Equipping scientists for the new biology. *Nat Biotech.* **18**: 359.
Battelle, B.A., 2002, Circadian efferent input to Limulus eyes: anatomy, circuitry, and impact. *Microsc. Res. Tech.* **58**: 345-55.
Battelle, B.A., Andrews, A.W., Calman, B.G., Sellers, J.R., Greenberg, R.M., and Smith, W.C., 1998, A myosin III from Limulus eyes is a clock-regulated phosphoprotein. *J. Neurosci.* **18**: 4548-59.
Creaser, C.S., Lill, J.R., Bonner, P.L., Hill, S.C., and Rees, R.C., 2000, Nano-electrospray and microbore liquid chromatography-ion trap mass spectrometry studies of copper complexation with MHC restricted peptides. *Analyst* **125**: 599-603.
Edwards, S.C., Wishart, A.C., Wiebe, E.M., and Battelle, B.A., 1989, Light-regulated proteins in Limulus ventral photoreceptor cells. *Vis. Neurosci.* **3**: 95-105.
Ehling, U.H., Favor, J., Kratochvilova, J., and Neuhauser-Klaus, A., 1982, Dominant cataract mutations and specific cross-locus mutations in mice induced by radiation or ethylnitrosurea. *Mutat. Res.* **92**: 181-192.
Faupel, M., Schindler, P., *Patent Pending* 2002 . GB 0010957.
Faupel, M., Righetti, P.G., U.S.*Patent* : 4,971,670, Nov.20, 1990.
Gygi, S.P., Rist, B., Griffin, T.J., Eng, J., and Aebersold, R., 2002, Proteome analysis of low-abundance proteins using multidimensional chromatography and isotope-coded affinity tags. *J. Proteome Res.* **1**: 47-54.
Hoving, S., Gerrits, B., Voshol, H., Muller, D., Roberts, R.C., and Van Oostrum, J., 2002, Preparative two-dimensional gel electrophoresis at alkaline pH using narrow range immobilized pH gradients. *Proteomics* **2**: 127-34.
Hoving, S., Voshol, H., Van Oostrum. J., 2000, Towards high performance two-dimentional gel electrophoresis using ultrazoom gels. *Electrophoresis*, **21**: 2617-2621.
Hyde, D.R., Mecklenburg, K.L., Pollock, J.A., Vihtelic, T.S., and Benzer, S., 1990, Twenty Drosophila visual system cDNA clones: one is a homolog of human arrestin. *Proc.Natl.Acad.Sci.USA* **87**: 1008-12.
Incerti, B., Cortese, K., Pizzigoni, A., Surace, E.M., Varani, S., Coppola, M., Jeffery, G., Seliger, M., Jaissle, G., Benett, D.C., Marigo, V., Schaiffino, M.V., Tachetti, C., and Ballabio, A., 2000, Oa1 knockout: new insights on the pathogenesis of ocular albinism type 1. *Hum. Mol. Genet.* **9**: 2781-2788.
Jimenez-Asensio, J., Colvis, C.M., Kowalak, J.A., Duglas-Tabor, Y., Datiles, M.B., Moroni, M., Mura, U., Rao, M., Balasubramanian, D., Janjani, A., and Garland, D., 1999, An atypical form of alphaB-crystallin is present in high concentration in some human

cataractous lenses. Identification and characterization of aberrant N- and C-terminal processing. *J. Biol. Chem.* **274:** 32827-94.

Jungblut, P.R., Otto, A., and Favor, J., 1998, Identification of mouse crystallins in 2D expression patterns by sequencing and mass spectrometry. Application to cataract mutants. *FEBS lett.* **435:** 131-137.

Komori, N., Usukura, J., Kurien, B., Shichi, H., and Matsumoto, H., 1994, Phosrestin I, an arrestin homolog that undergoes light-induced phosphorylation in dipteran photoreceptors. Insect Biochem. *Mol. Biol.* **24:** 607-17.

Koyama, T., Tago, K., Nakamura, T., Ohwada, S., Morishita, Y., Yokota, J., and Akiyama, T., 2002, Mutation and expression of the beta-catenin-interacting protein ICAT in human colorectal tumors. *Jpn. J. Clin. Oncol.* **32:** 358-62.

Kuznetsov, V.Y., Ivanov, Y.D., Bykov, V.A., Saunin, S.A., Fedorov, I.A., Lemeshko, S.V., Hoa, H.B., and Archakov, A.I., 2002, Atomic force microscopy detection of molecular complexes in multiprotein P450cam containing monooxygenase system. *Proteomics* **12:** 1699-705.

LeVine, H., 3rd, Smith, D.P., Whitney, M., Malicki, D.M., Dolph, P.J., Smith, G.F., Burkhart, W., Zuker, C.S., 1990, Isolation of a novel visual-system-specific arrestin: an in vivo substrate for light-dependent phosphorylation. *Mech. Dev.* **33:** 19-25.

Matsumoto, H., Kahn, E.S., and Komori, N., 1999, The emerging role of mass spectrometry in molecular biosciences: studies of protein phosphorylation in fly eyes as an example. *Novartis Found. Symp.* **224:** 225-44.

Matsumoto, H. and Komori N., 2000, Ocular proteomics: cataloging photoreceptor proteins by two-dimensional gel electrophoresis and mass spectrometry. *Methods Enzymol.* **316:** 492-511.

Matsumoto, H., Kurien, B.T., Takagi, Y., Kahn, E.S., Kinumi, T., Komori, N., Yamada, T., Hayashi, F., Isono, K., and Pak, W.L., 1994, Phosrestin I undergoes the earliest light-induced phosphorylation by a calcium/calmodulin-dependent protein kinase in Drosophila photoreceptors. *Neuron* **12:** 997-1010.

Matsumoto, H., O'Tousa, J.E., and Pak, W.L., 1982, Light-induced modification of Drosophila retinal polypeptides in vivo. *Science* **217:** 839-41.

Matsumoto, H., and Pak, W.L., 1984, Light-induced phosphorylation of retina-specific polypeptides of Drosophila in vivo. *Science* **223:** 184-6.

Nolan, P.M., 2000, Generation of mouse mutants as a tool for functional genomics. *Pharmacogenomics* **1:** 243-255.

Ottani V., Martini D., Franchi M., Ruggeri A. and Raspanti M.,2002, Hierarchical structures in fibrillar collagens. *Micron* **33:** 587-96.

Pawlak, M., Schick, E., Bopp, M.A., Schneider, M.J., Oroszlan, P., and Ehrat, M., 2002, Zeptosens' Protein Microarrays: A novel high performance microarray platform for low abundance protein analysis. *Proteomics* **2:** 383-393.

Pennesi, M.E., Cho, J.H., Yang, Z., Wu, S.M., and Tsai, M.J., 2003, BETA2\NeuroD1 null mice: a new model for transcription factor dependant photoreceptor degeneration. *J. Neurosci.* **23:** 453-61.

Perrault, I., Rozet, J..M., Calvas, P., Gerber, S., Camuzat, A., Dollfus, H., Chatelin, S., Souied, E., Ghazi, I., Leowski, C., Bonnemaison, M., Le Paslier, D., Frezal, J., Dufier, J.L., Pittler, S., Munnich, A., and Kaplan, J., 1996, Retinal-specific guanylate cyclase gene mutations in Leber's congenital amaurosis. *Nat Genet.* **14:** 461-4.

Ranish, J.A., Yi, E.C., Leslie, D.M., Purvine, S.O., Goodlett, D.R., Eng, J., and Aebersold, R., 2003, The study of macromolecular complexes by quantitative proteomics. *Nat. Genet.* **33:** 349-55.

Reifenberger, J., Knobbe, C.B., Wolter, M., Blaschke, B., Schulte, K.W., Pietsch, T., Ruzicka, T., and Reifenberger, G., 2002, Molecular genetic analysis of malignant melanomas for aberrations of the WNT signaling pathway genes CTNNB1, APC, ICAT and BTRC. *Int. J. Cancer.* **100:** 549-56.

Semple-Rowland, S.L., and Ulshafer, R.J., 1989, Analysis of proteins in developing rd, retinal degeneration, chick retina using two-dimensional gel electrophoresis. *Exp. Eye Res.* **49:** 665-75.

Semple-Rowland, S.L., Lee, N.R., Van Hooser, J.P., Palczewski, K., and Baehr, W., 1998, A null mutation in the photoreceptor guanylate cyclase gene causes the retinal degeneration chicken phenotype. *Proc.Natl.Acad.Sci.USA* **95:** 1271-6.

Sieving, P.A., Fowler, M.L., Bush, R.A., Machida, S., Calvert, P.D., Green, D.G., Makino, C.L., and McHenry, C.L., 2001, Constitutive "light" adaptation in rods from G90D rhodopsin: a mechanism for human congenital nightblindness without rod cell loss. *J. Neurosci.* **21:** 5449-5460.

Sivak, J.M. and Fini, M.E., 2002, MMP in the eye: emerging role for matrix metalloproteases in ocular physiology. *Prog. Ret. Eye Research* **21:** 1-14.

Schmitt, A. and Wolfrum, U., 2001, Identification of novel molecular components of the photoreceptor connecting cilium by immunoscreens. *Exp. Eye Res.* **73:** 837-49.

Smith, D.P., Shieh, B.H., and Zuker, C.S., 1990, Isolation and structure of an arrestin gene from Drosophila. *Proc.Natl.Acad.Sci.USA* **87:** 1003-7.

Steely, H.T., and Clark, A.F., 2000, The use of proteomics in ophthalmic research. *Pharmacogenomics* **1:** 267-280.

Thaung, C., West, K., Clark, B.J., McKie, L., Morgan, J.E., Arnold, K., Nolan, P.M., Peters, J., Hunter, A.J., Browm, S.D.M., Jackson, I.J., and Cross, S.H. , 2002, Novel ENU-induced eye mutations in the mouse: models for human eye disease. *Hum. Mol. Genet.* **11:** 755-767.

Wirth, U., Müller, D., Schindler, P., Lange, J., Van Oostrum, J., 2002, Post- translational modification detection using metastable ions in reflector matrix-assisted laser desorption / ionization-time of flight mass spectrometry. *Proteomics* **2:** 1445-1451.

Yamada, T., Takeuchi, Y., Komori, N., Kobayashi, H., Sakai, Y., Hotta, Y., and Matsumoto H., 1990, A 49-kilodalton phosphoprotein in the Drosophila photoreceptor is an arrestin homolog. *Science* **248:** 483-6.

Zhang, R., and Regnier, F.E., 2002, Minimizing resolution of isotopically coded peptides in comparative proteomics. *J. Proteome Res.* **1:** 139-47.

Zhang, R., Sioma, C.S., Thompson, R.A., Xiong, L., and Regnier, F.E., 2002, Controlling deuterium isotope effects in comparative proteomics. *Anal. Chem.* **74:** 3662-9.

Chapter 6

Proteomics in Oncology:
the Breast Cancer Experience

FRANCK VANDERMOERE*, IKRAM EL YAZIDI-BELKOURA*, ERIC
ADRIAENSSENS*, JEROME LEMOINE#, HUBERT HONDERMARCK*
*INSERM-ESPRI and #UMR-8576, IFR-118, University of Sciences and Technologies Lille,
Villeneuve d'Ascq, France

1.   INTRODUCTION

Until recently, it was difficult to sustain the idea that looking globally at proteins could be a productive way for understanding fundamental mechanisms of cancer cell growth or discovering molecules of clinical interest. Indeed, recent times have been dominated by the concepts and methods - the panacea - of molecular genetics. The near completion of human genome sequencing, and its limited number of direct outcomes for cancer, coupled with the technical progress made in protein identification, it gradually became reasonable to believe that looking directly at proteins could be fruitful. Today, it has become common to read that proteins are the main functional ouput of genes, and that proteomics will drive biology and medicine beyond genomics, particularly in the field of cancer (Hanash, 2003; Celis and Gromov, 2003). However, we are already at the end of the pioneer stage, and proteomics of breast cancer is now entering into a phase dominated by the ideas of large scale analyses and translation to the clinic. There are two main expected outcomes from proteomic analyses of breast cancer. The first is to discover new molecular markers for early diagnosis and profiling of breast tumors. The second is to decipher the intracellular signaling pathways leading to the initiation and progression of breast tumors. Such data should provide the knowledge base for the identification of new

H. Hondermarck (ed.), Proteomics: Biomedical and Pharmaceutical Applications, 139–161.

therapeutic targets and the development of innovative strategies against breast cancer.

2. CHALLENGES FACING RESEARCH ON BREAST CANCER

Current methods used to detect breast tumors, either benign or malignant, are based on mammography. However, there are intrinsic limitations to mammography. First, there are suggestions that X-rays can potentially induced carcinogenesis, and second, it is clear that to be detected in mammography, a breast tumor should be at least a few millimeters in size. However, a tumor of this size already contains several hundred million cells. From the cellular point of view, given the fact that a single cell can lead to the development of a whole tumor (clonal origin of cancer), it is already late when a breast tumor is detected by mammography. Thus, this constitutes an important limitation to mammography. In clinical practice, after the surgical removal of a tumor, its characterization as a malignant or benign tumor is made by histology. Such parameters as tumor size and inflammation, histoprognostic grading and node involvement are then used to decide treatment and prognosis. However, breast cancer is not an homogeneous disease and there are different types. Depending on the cellular and histological origin of the cancer cells and on the evolution of the disease, a broad range of breast tumor types have been described. Several studies have been performed in an attempt to identify factors associated with either the growth rate- or the metastatic-potential of breast tumors (i. e. prognostic factors) or factors related to sensitivity and/or resistance to therapeutic agents (i. e. predictive factors). The main biological markers recommended for routine use is the presence of estrogen (estradiol and progesterone) receptors, for the selection of patients potentially responding to treatment with anti-estrogen such as tamoxifen (Ciocca and Elledge, 2000).

On the therapeutic side, practical consequences for treatment derived from a better understanding of the molecular basis of cancer cell growth are now emerging. This is evidenced by the development of therapeutic strategies based, for example, on the inhibition of tyrosine kinase receptors. This approach is well illustrated with ErbB2, a tyrosine kinase receptor overexpressed in more than 20% of breast tumors. Specific inhibition of ErbB2 using Herceptin, a truncated blocking-antibody directed against it, has been successfully developed and has now entered into clinical practice (Colomer *et al.*, 2001). However, the growth of breast cancer cells can be regulated by other growth factors that either stimulate or inhibit their proliferation, migration and differentiation, thus acting in concert to promote tumor growth and metastasis (Ethier *et al.*, 1995, LeBourhis *et al.*, 2000). Probably related to this, the efficacy of Herceptin appears to be limited to a small proportion of breast cancers and the identification of other targets, and corresponding drugs, remains a major interest for the development of targeted therapeutic strategies.

Clearly, there is a critical need to find new molecular parameters not only for detection, but also for typing and treatment of breast cancer. As proteomics provides a global approach for the identification of protein regulation during pathological processes, the discovery of new markers and therapeutic targets as well as the corresponding drugs is a highly anticipated outcome (Figure 1). Such data would provide the knowledge base for the development of new strategies against breast cancer (Hondermarck *et al.*, 2001, Hondermarck *et al.*, 2003).

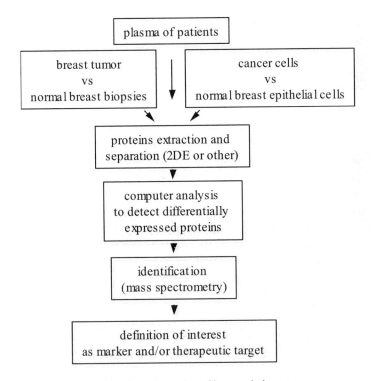

Figure 1. Strategy for identification of proteins of interest in breast cancer.

3.   ANALYSIS OF BREAST TUMORS BIOPSIES

Differential proteomic analysis, to detect and characterize polypeptides in disease versus normal tissue biospies is generally presented as an appropriate way to identify markers of pathologies. However, in case of breast cancer, but certainly not limited to it, the pertinence of this approach can seriously be questioned. As normal mammary gland tissues and breast tumors are dynamic and heterogenous cellular structures, a global differential analysis of their proteomes refers finally to a comparative study of cell mixtures with different proportions of epithelial and of other cell types including myoepithelial, fibroblast and endothelial cells, supplemented by adipocytes, nerve fibers, circulating cells and macrophages.

Several studies have reported differential proteomic analyses of breast tumor biopsies. As early as 1974, the use of 2DE to resolve serum proteins was reported, with differences being noted in the protein patterns of individuals suffering cancer (Wright, 1974) ; however this study was essentially descriptive, and no proteins identification were made. In 1984, it was found that most polypeptides were consistently present in both malignant and non-malignant breast tissues, as only 10 polypeptides differed out of 350 resolved (Stastny *et al.*, 1984). Subsequently, using better technology, it was shown that of approximately 1,000 silver-stained cytosolic polypeptides observed, the 2DE patterns from normal and malignant tissue differed in only 6 places qualitatively, and only 22 places quantitatively (Wirth *et al.*, 1987; Maloney *et al.*, 1989). A more precise characterization of such polypeptide differences was published in the early 90's with the demonstration of a defect in tropomyosin 1-2 and -3 expression in mammary carcinoma, suggesting that such abnormalities may play a role in breast neoplasia (Bhattacharya *et al.*, 1990; Franzen *et al.*, 1993). Differential distribution of heat shock protein (HSP) family members has also been described (Franzen *et al.*, 1996; Franzen *et al.*, 1997). Recently, a comparison of normal ductal/lobular units versus ductal carcinomas has been published with a total of fifty seven proteins found to be differentially expressed (Wulfkuhle *et al.*, 2002). From all of these data, it appears that the set of differentialy expressed proteins identified are different from one study to another, suggesting either a lack of experimental standardization or problems of heterogeneity between the biological material used in each study. Probably as a consequence of this, the relevance of these data for clinical practice has still to be established and indeed, none of the potential markers identified by proteomics of breast tumors so far are routinely used by clinicians for either diagnosis, treatment choice or prognosis. In our laboratory, we have also experienced the limit of analyzing breast tumor biopsies as a whole. Indeed, we described some of the modifications reported above for tropomyosin or HSP (Bhattacharya *et al.*, 1990; Franzen *et al.*, 1993; Franzen *et al.*, 1996; Franzen *et al.*, 1997), but these differences were only quantitative and no protein was found exclusively in cancer versus normal breast tissues (unpublished data). Considering the fact that such a global analysis is in fact describing the proteomes of about ten different cell types, the lack of specificity found in cancer samples might not be considered so unusual. A classical 2D gel of a breast tumor biopsy is shown in Figure 2 and reveals a further level of complication, namely the major protein that can be observed is serum albumin. Therefore, breast tumor biopsies not only contain a mixture of the different mammary cell types mentioned above, but they also contain blood compounds due to the vascular irrigation of breast tumors. As blood cannot be washed out of biospies, proteomic analyses of breast tumors also measure proteins from circulating cells and from the plasma, further confounding an already complex situation. On the other hand, blood should also be regarded as an important point to consider in any differential analysis of pathological versus normal situation. In terms of biomedical applications, the detection of proteomic markers in the biological fluids such as plasma and urine is of promising interest. To be able to detect breast cancer-or any other cancer - from a simple blood analysis would represent a major breakthrough and such an achievement provides impetus to the currently ongoing plasma project (Anderson and Anderson, 2002). However, even thought significant progress has already been made (Li *et al.*, 2002; Vejda *et al.*, 2002), searching for tumor -released proteins in plasma or their fragments in urine is like looking for a needle in a bale of hay, and it might turn out to be more productive to identify markers from tumor cells themselves before secondarily assaying them in the plasma.

Considering the cellular complexity of mammary tumors, microdissection is now regarded as a reasonable alternative for selectively isolating cell types to be analyzed. Techniques such as laser capture microdissection (Craven *et al.*, 2001) can be used for the isolation of malignant cells prior to sample preparation for proteomic analysis and might facilitate marker protein discovery. However, initial enthusiasm for microdissection has progressively been diminished by several limiting aspects. The first problem is quantitation : as no amplification technique can be applied to proteins (as PCR for DNA/RNA), it is necessary to start any proteomic analysis with a significant amount of material. In our hands, a minimum of 100, 000 cells are necessary to perform 2D gel analysis in a way that allows subsequent characterization of proteins by mass spectrometry; this has recently been illustrated (Wulfkuhle *et al.*, 2002). Alternative "off-gel" methods, such as Isotope-coded affinity tagging - ICAT (Aebersold and Mann, 2003) certainly allows one to start with a lower cell number, but tens of thousands cells are still required. Despite being at the upper-limit of the theoritical capabilities of current microdissection techniques, the preparation of such a number of cells represents a huge quantity of work with a constant problem in the choice of which cells to select. Indeed, cell morphology alone is not totally reliable to recognize cell types (specially cancer versus normal), this is the reason why we need to identify molecular markers and therefore subjectivity can seriously impair the quality of the cell preparation. The second problem of using microdissection for proteomic analysis is related to the biochemical quality of the dissected material. The use of fixatives must be avoided as the requisite compounds create artificial boundaries between amino acid residues. In addition, due to the time necessary to perform microdissection, protein modifications such as degradation or dephosphorylation are likely to take place. This last point is related to a more general concern about working with tissue biopies : the collecting procedure of the samples must be standardized to a maximum in order to avoid differential protein modification and/or degradation. Standardization of procedures and trackability of the samples, are crucial points to take under consideration. This issue is certainly more significant with proteomic analysis than it is with genomics because, in contrast to RNA which can easily be extracted from frozen tissues, protein resolubilization can be hampered after freezing. Ideally, protein extraction should be performed from fresh tissue to allow maximum solubilization and recovery of the proteins in SDS ; after that, samples can be frozen before performing proteomic analysis. When considering the setting up of large scale analyses of cancer proteomes, under way in several international programs, with tumor resection and protein extraction being performed in different clinical centers, the issue of standardization is obviously a sensitive point.

When comparing cancer tissues to normal tissues or the corresponding dissected cells, a fundamental physiological question also arises : what is a normal breast tissue ? Due to the changing hormone concentration (estrogens and progesterone) during the course of menstrual cycle, the status of normal breast tissue is clearly also changing and therefore it is difficult to define a reference state. Hormonal stimulation of breast epithelial cells will be different depending on hormonal status of the women at the time the sampling is made, and the same concern applies to resected breast tumor biopsies. The stimulation by hormones is known to modify gene expression in both normal and cancer cells and consequently the proteome of breast epithelial cells. Thus, the hormonal environment of biopsied tissue should ideally be evaluated, but this is practically very difficult if not impossible to implement. One way to avoid this intrinsic difficulty related to *in situ* endogenous hormonal stimulation of breast epithelial cells is to use cell cultures, in which cells can be kept in standard medium, can be regarded as a way to standardize the environment and the conditions applied to the cells.

Thus, despite its intrinsic promise, proteomic analysis of breast cancer is not so straight-forward. Clearly, "the devil is in the details" and differential proteomics of normal versus cancer breast biopsies is in fact associated with a series of difficult problems making the approach a potential pitfall. Unfortunately, this conclusion is probably not limited to breast cancer and the suitability of differential proteomics performed from biopsies should be questioned in any attempt to identify markers or therapeutic targets.

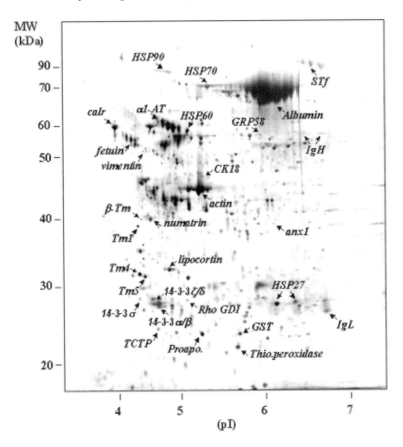

Figure 2. 2D gel of a human breast tumor. Proteins were identified using mass spectometry. Abbreviations : Anx, annexin; CK, cytokeratins; crtc, calreticulin; GRP, glucose-regulated protein; GST, glutathione S-transferase; HSP, heat-shock proteins; Ig, Immunoglobulins; PCNA, proliferating cell nuclear antigen; PGM, phosphoglycerate mutase; RhoGDI, Rho GTP-dissociation inhibitor 1; SODM, superoxide dismutase; STf, serotransferrine, TCTP, transcriptionaly controlled tumor protein; thio. peroxidase, thioredoxin peroxidase; TM, tropomyosin; vim,vimentin; 14-3-3: s, sigma; a, alpha; b, beta; d, delta; z, zeta.

4.   FROM CELLULAR MODELS TO TUMOR BIOPSIES

In contrast to the studies of breast tumor biospies, proteomic analysis starting from breast cancer cells in culture has already given significant results with the identification of proteins with clinical interest. In 1980, a secreted 46 kDa glycoprotein, induced by estrogens in human breast cancer cell lines, was identified with specific antibodies as being the protease cathepsin D (Westley and Rochefort, 1980). In 1989, a computer-based analysis of 2DE gels reported a total of 8 polypeptide differences between cancerous and normal breast epithelial cells in tissue culture (Worland *et al.,* 1989). More precise characterization of such polypeptide differences was published in the early 90's with the demonstration that normal breast epithelial cells produce keratins K5, K6, K7 and K17, whereas tumor cells produce mainly keratins K8, K18 and K19 (Trask *et al.,* 1990). This distribution was secondarily confirmed in tumor samples (Giometti *et al.,* 1995) and cytokeratin immunodetection is now eventually used to help discriminate benign from malignant cells on histopathological slides. Finally, the first characterization of normal breast epithelial and myoepithelial cell proteins and of breast cancer cells using mass spectrometry was published more recently and revealed a limited number of differences (Page *et al.,*1999). The study of cells, both normal and cancerous, in culture definitely offers the possibility to directly investigate proteins specifically produced, or not produced, by breast cancer cells. This has also been shown with proteinase inhibitors TIMP-1 and PAI-1, originally studied in breast cancer cells, which have recently been shown to be complementary in determining prognosis of breast tumors (Schrohl *et al.,* 2003). Cell fractionation prior to proteomic analysis has also been explored, as recently illustrated for plasma membranes of breast cancer cells (Adam *et al.*, 2003). Importantly, proteomic analysis from cell cultures can be realized in standardized medium conditions, providing a comparable and controlable status of hormone and growth factor stimulation. In addition, phenotype and behavior of the cells, i.e. proliferation/migration/differentiation/survival, can be experimentaly studied.

Interestingly, differences between the proteome of normal versus cancerous breast epithelial cells appear to be quite limited. In our laboratory, about 2,000 proteins were analysed in normal breast epithelial cells, compared to two cancer cell cultures placed in the same medium conditions. It can be noted that most detected proteins were present in similar quantities in the three samples. Importantly, only a limited number of proteins - about twenty - differ between normal and cancerous samples and therefore have the potential to be relevant markers. These values were obtained after computer analysis of 2D gel protein separations, but similar results have been obtained using off-gel approaches like ICAT (unpublished observation). The fact that only a limited number of individual protein modifications can be seen between normal and cancer cells is in fact not surprising as a similar situation has already been described at the genomic level. A limited number of molecular modifications, affecting oncogenes and suppressor genes, are required to transform normal cells into cancerous ones (Haber, 2000). This limited number of molecular modifications ultimately makes cancer cells quite similar to normal ones, and probably accounts for several important consequences encoutered with cancer cells. The first is the difficulty to detect and type cancer cells : so far, no one has been able to identify a universal cancer marker which would allow early detection of the pathology. It can also be postulated that for the same reason, the immune system encounters difficulties in efficiently recognizing cancer cells. The second consequence is that the definition of drugs specifically targeting breast cancer cells has so far been difficult and current treatments have therefore numbers of heavy side effects. In this context, the identification of proteins specifically expressed by cancer cells would be essential for drug targeting. Alternatively, proteins down-regulated in cancer cells would also be of interest as their level of expression could be used to discriminate and type breast cancer cells. In addition, if such a protein disappearance is germane to cancer cell growth and metastasis, its re-induction could potentialy lead to a reversion of the cancerous phenotype. To date this is only a distant possibility, but normalizing cancer cells by re-expressing specific proteins controling the normal state of growth would provide a theoritical basis for potential gene therapy of cancer.

A first step toward the identification of proteins down-regulated in cancer cells has recently been provided by the molecular chaperone 14-3-3 sigma, and interestingly, this is a good illustration of the usefullness of cellular models and of the complementarity of genomic- and proteomic-based approaches. 14-3-3 is a family of highly conserved protein forms (alpha, beta, delta, sigma, zeta) of 25- to 30-kDa, expressed in all eukaryotic cells, that plays a role in the regulation of signal transduction pathways implicated in the control of cell proliferation, differentiation and survival (Fu *et al.*, 2000). 14-3-3 proteins are known to associate directly or indirectly with signaling proteins such as the IGF-1 receptor, Raf, MEK kinases and PI3-kinase, but the precise molecular mechanism by which they activate or inhibit these elements remains unclear (Fu *et al.*, 2000). We have shown that the sigma form of 14-3-3 is easily detectable in 2DE gels of normal breast epithelial cells using low sensitivity Coomassie staining, whereas the spot was undetectable in breast cancer cell protein profiles (Vercoutter-Edouart *et al.*, 2001a). Nevertheless, 14-3-3 sigma was also present in breast cancer cells, but due to the very low levels, more sensitive silver staining was necessary for its detection (Figure 3). Therefore, 14-3-3 sigma down-regulation appears to be a major modification of the proteome associated with breast epithelial cell carcinogenesis and we subsequently have investigated its distribution in breast cancer biospies.

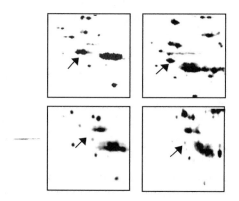

Figure 3. 2D-patterns showing the down-regulation of 14-3-3 sigma (indicated by an arrow). two normal breast samples (A, B) compared to 2 breast cancer samples (C, D).

Based on data gathered in cell culture, the analysis of breast tumors with 2DE was set-up with a narrow range pH gradient to provide an optimal view of the gel area containing 14-3-3 sigma. We showed that the level of 14-3-3 sigma is systematically down-regulated in tumor biopsies (Vercoutter-Edouart *et al.*, 2001a). This demonstrates that starting with a cellular proteome, which exhibits a lower level of complexity, facilitates the identification of relevant proteins that can secondarely be found and quantified in tumor biospies. At the mRNA level, it was shown that gene expression of 14-3-3 sigma is 7-10 time lower in breast cancer cells than in normal breast due to the high frequency of hypermethylation of the 14-3-3 sigma locus (Ferguson *et al.*, 2000). Interestingly, the mRNA for 14-3-3 sigma was undetectable by Northern blot analyses in 45/48 primary breast carcinomas studied; in contrast, we have detected the 14-3-3 sigma protein in 30/35 primary tumor samples, indicating the high sensitivity provided by proteomic analysis as well as the complementarity of both strategies for identifying cancer markers. From the clinical point of view, it has been reported that down-regulation of 14-3-3 sigma gene expression is an early event in breast carcinogenesis (Umbricht *et al.*, 2001) and we have recently observed that overexpression of 14-3-3 sigma in cancer cells reverses the proliferative phenotype (Adriaenssens E. and Hondermarck H., unpublished observation). This emphasizes the invaluable use of cell cultures, not only for the detection of markers directly from cancer cells, but also because *in vitro* models allow experimental work necessary for the understanding of protein function.

5. EXPLORING THE MECHANISM OF CARCINOGENESIS : FROM ONCOGENES TO GROWTH FACTOR SIGNALING

Molecular events leading to breast epithelial cell carcinogenesis involve modification in the structure and expression of both oncogenic and tumor suppressive genes (such as *myc, ErbB2* and *p53*), leading to an unbalanced growth characterized by high rates of cell proliferation and eventually cell migration (Haber, 2000). These genetic modifications also confer an ability of cancer cells to survive environmental stresses, which would otherwise lead to apoptosis. Moreover, mammary tumors develop slowly, with an estimated time of 6 to 8 years for the attainment of a tumor/0.5 cm diameter, from one originating cell. This fact emphasizes the crucial importance played by endocrine and paracrine regulation of breast cancer cell growth in tumor development. Therefore, breast carcinogenesis is clearly based on both genetic and epigenetic parameters; and interestingly proteomics can provide useful information for the comprehension of these two aspects.

5.1 Mechanism of oncogene activity

Despite the fact that modifications of oncogenes and tumor suppressor gene expression are recognized to be central to cancer cell development and metastasis, the mechanism of several of these remains to be determined. This is the case for *H19*, an oncofetal gene, which encodes an untranslated mRNA (Brannan *et al.*, 1990) that has been shown to stimulate cancer cell growth (Lottin et al., 2002a). *H19*-transfected breast epithelial cells appear to grow faster (Lottin *et al.*, 2002a), but no molecular target has been described. Recently, it was shown that growth factor stimulation of breast cancer cells, for example by FGF-2, results in a strong and sustained stimulation of *H19* gene transcription (Adriaenssens *et al.*, 2002). In order to identify molecular events involved in H19 oncogenic activity, we have developed a proteomic-based strategy (Lottin *et al.*, 2002b). Breast mammary cells were transfected with *H19* and the resulting proteomic profile established using 2DE. Changes in protein synthesis were determined by computerized analysis and revealed an essential modification in the intensity of one spot induced by *H19* overexpression. Mass spectrometric analysis of this spot, performed by MALDI-TOF and MS-MS, allowed the identification of the protein up-regulated by *H19* gene as thioredoxin (Lottin *et al.*, 2002b), one of the major proteins regulating intracellular redox metabolism. The thioredoxin system is a general protein disulfide-reducing complex and includes the NADPH-dependent flavoprotein thioredoxin reductase (Nakamura *et al.*, 1997). These data do not fully elucidate the mechanism of action of H19, but provide a first identified target, demonstrating the value of proteomic analysis for the understanding of the molecular mechanisms involved in oncogene activity. More generally, this shows that the transfer of information is not only from genomics to proteomics, but also from proteomics to genomics. It is usually said that proteomics can be performed because the human genome was previously sequenced, allowing the possibility to perform protein identification through database searching. On the other hand, as shown with the oncogene *H19*, proteomics can provide useful informations for the understanding of the genome and its functioning in a physiopathological context.

5.2   Growth factors signaling

The growth of breast cancer cells is regulated not only by estrogenic hormones (estradiol and progesterone), but also by different growth factors (Ethier *et al.*, 1995, LeBourhis *et al.*, 2000). For example, insulin-like growth factor I (IGF-I) or epidermal growth factor (EGF) stimulate the proliferation of breast cancer cells, whereas other factors like mammary-derived growth factor inhibitor (MDGI) inhibit their growth. Hepatocyte growth factor/scatter factor (HGF/SF) has been shown to stimulate the migration of breast cancer cells, and thus metastasis. We and others have shown that fibroblast growth factors (FGFs), which are pleiotropic polypeptides involved in the control of cell growth, are stimulators of both breast tumor growth and metastasis (Rahmoune *et al.*, 1998; Nurcombe *et al.*, 2000). More recently, it has also been demonstrated that nerve growth factor (NGF), well known as the archetypal neurotrophin, is able to stimulate both proliferation and survival of breast cancer cells through distinct signaling pathways initiated by different NGF receptors (Descamps *et al.*, 1998; Descamps *et al.*, 2001; El Yazidi-Belkoura *et al.*, 2003). NGF stimulation of the tyrosine kinase receptor trkA leads to the activation of the MAP-kinases pathway resulting in cell proliferation (Descamps *et al.*, 2001). On the other hand the activation of the common neurotrophin receptor p75[NTR] (a Tumor Necrosis Factor-receptor family member) induces a nuclear translocation of the transcription factor NF-κB, resulting in increased survival of breast cancer cells (Descamps *et al.*, 2001; El Yazidi-Belkoura *et al.*, 2003). The hypothesis of a recruitment and cooperation between trkA and Her-2 (also known as c-ErbB2 or neu) has been proposed for the induction of mitogenesis by NGF (Tagliabue *et al.*, 2000) and it has also been shown that the drug tamoxifen induces an inhibition of NGF signaling in breast cancer cells (Chiarenza *et al.*, 2001). Interestingly, it has also been shown that NGF is overexpressed by breast cancer cells, resulting in an autocrine stimulation (Dollé *et al.*, 2003), and NGF increasingly appears as an essential regulator of breast tumor growth. Figure 4 summarizes the signaling pathway of NGF in breast cancer cells.

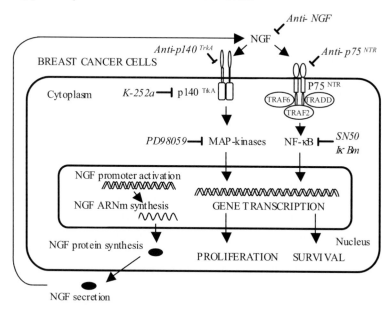

Figure 4. NGF signaling pathways in breast cancer cells. Two signaling pathways are mediated via the two distinct receptors p140ᴛᴿᴬ and p75ɴᴛᴿ. The effects of NGF can be modulated by pharmacological agents in several ways, as indicated (anti-NGF, anti NGF-receptors, tyrosine kinase inhibitor K-252a, MEK inhibitor PD98059, NF-kB inhibitors IkBm and SN50).

Many growth factors initiate intracellular signaling through tyrosine kinase-membrane receptors, which in turn induce intracellular protein-protein interaction and phosphorylation cascades involving a variety of signaling proteins such as the mitogen activated protein-kinases (MAP-kinases). These cascades of protein phosphorylation ultimately induce changes in gene expression, with consequent modifications in protein synthesis, leading to either cell survival, proliferation, differentiation or migration. The development of Herceptin, a truncated antibody directed against the tyrosine kinase recptor Erb-B2, and its succesfull use in clinical practice shows that deciphering growth signaling in breast cancer cells can potentially lead to the development of realistic therapeutic strategies. In addition, modifications of the proteome induced by expression of the oncogene Erb-B2 have also been described (Gharbi *et al.*, 2002). Nevertheless, signaling proteins - such as kinases and phosphatases - are regulatory proteins present in low quantity in the cell and their activation-deactivation, by phosphorylation-dephosphorylation, renders these difficult to study. Classical methods to investigate these processes are based on the use of specific antibodies to purify known signaling proteins, followed by SDS-PAGE separation and Western blot analysis of tyrosine phosphorylation. Alternatively, phosphotyrosine-containing proteins can be purified and signaling proteins immunodetected. However, the main drawback of these traditional protocols is their difficulty in identifying proteins with no previously described function in signal transduction. In contrast, proteomics is providing a way to *denovo* identify signaling proteins and it is certainly one of the most exciting challenge of functional proteomics to shed new light on the signaling pathways leading to cancer cell development.

Before the word proteomics appeared, the use of 2DE for studying growth factors mechanism of action was initiated with pheochromocytoma PC12 cells (Hondermarck *et al.*, 1994). In these cells, early protein synthesis induced by the differentiating activity of NGF and FGF-2 has been compared to the mitogenic activity of epidermal growth factor (EGF) using 2-DE separation of ^{35}S metabolically-labeled amino acids. The results revealed an initial modulation of protein synthesis induced by the three growth factors, and provided some of the early evidence for a specificity of the differentiative versus proliferative pathways induced by tyrosine kinase growth factor receptors. Several years after, proteomics has been used succesfully for the study of protein phosphorylation cascades (Soskic *et al.*, 1999; Pandey *et al.*, 2000). Recently, a major breakthough in the use of proteomics for studying signal transduction has been achieved with Phosphoaminoacid Ion Scanning, allowing the localization of phosphorylated aminoacid residues in a protein, and its first application for EGF and FGF-2 signaling (Blagoev *et al.*, 2003; Hinsby *et al.*, 2003).

In breast cancer cells, the increase in tyrosine phosphorylation of several proteins induced by FGF-2 have been characterized and include the FGF receptor, the FGF receptor substrate (FRS2), the oncogenic protein Src and the MAP kinases p42/p44. Interestingly, it was also shown that such stimulation induces the tyrosine phosphorylation of cyclin D2 (Vercoutter-Edouart *et al.*, 2000). Phosphorylation of cyclin D2 had not been previously described in growth factor signaling, although it was known that progression through the cell cycle is under the strict control of cyclins and their catalytic subunits, the Cdks (cyclin-dependent kinases). Modifications in protein synthesis induced by FGF-2 have also been studied (Vercoutter-Edouart *et al.*, 2001b) and revealed that this growth factor induced an increase in the level of stress proteins heat-shock HSP90 and HSP70 as well as of the proliferating cell nuclear antigen (PCNA, a regulator of the polymerase delta) and the transcriptionally controled tumor protein (TCTP), a protein of unknown function which is found associated with cell proliferation.

5.3　Toward single cell proteomics ?

As described above, proteomics is now an efficient way to identify growth signaling proteins. Sensitivity of the methods of detection have been improved and functional aspects as protein-protein interaction as well as post-translational modifications can reliably be studied. However, the most challenging aspect of studying intracellular signaling has now shifted to the understanding of global signaling networks. A living cell, in a physiological or pathological situation, is under stimulation of different factors at the same time and the activated signaling pathways are succeptible to interact one with another. For example it is known that the activation of TGFb receptors leads to stimulation of Smad signaling proteins, which are translocated to the nucleus to modify gene expression; interestingly, Smad proteins can also lead to the phosphorylation of the MAP-kinases leading to the inactivation of their translocation to the nucleus (Piek and Roberts, 2001). Therefore the signaling pathways of growth factors and hormones are definitely not independent ways that regulate cell behavior and their functioning increasingly appears to depend on inter-relations and cross-talks. A summary of known cellular interaction of breast cancer cells is presented in Figure 5. In this regard, kinetics of signal activation is also subject of increasing interest, and to tackle these new challenges in the field of signal transduction, further improvement of the proteomic technologies are necessary. Indeed, current analysis of protein-protein modifications and post-translational modifications requires a certain amount of proteins which have therefore to be prepared from whole cell cultures. As mentioned before, tens of thousands cells are required for any proteomic analysis and it is now well established that all individual cells in culture will have a different response profile to growth stimulators. For example, cancer cells are hardly synchronized in culture and stimulation will be obtained only on cells in G0/G1 phase. Therefore not all cells will respond with the same time course to growth factor stimulation and the kinetic of signaling pathway activation will consequently vary from a cell to another. Ideally, a precise investigation of intracellular signaling would require to be able to work on a single cell, allowing the spatio-temporal integration of multiple pathways. The concept of single cell proteomics and its implementation for the study of intracellular signaling would definitely be a major advancement to investigate the complexity of intracellular signaling in cancer cells - and consequently to define future therapeutic strategies.

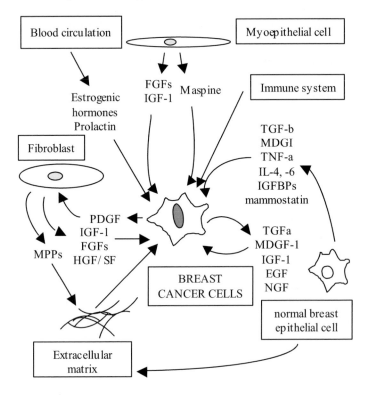

Figure 5. Breast cancer cells are under a network of cellular interactions mediated by hormones and growth factors. (+) indicates a stimulation of cellular growth and (-) an inhibition. EGF, epidermal growth factor; FGFs, fibroblast growth factors; IGF, insulin-like growth factor; IL, interleukin; MPP, metalloprotease; NGF, nerve growth factor; MDGI, mammary derived growth inhibitor; PDGF, platelet derived growth factor; TGF, transforming growth factor; HGF/SF: Hepathocyte growth factor/ Scatter Factor; TNF: Tumor Necrosis Factor alpha; IGFBPs: Insulin-like growth factor binding proteins; MDGF-I: mammary derived growth factor I.

6.   STRATEGY FOR DRUG TARGET IDENTIFICATION USING PROTEOMICS

Treatment derived from a better understanding of the molecular basis of breast cancer cell growth are now emerging, as evidenced by the development of therapeutic strategies basedon the inhibition of tyrosine kinase receptors. This approach is well illustrated with ErbB2. ErbB2 overexpression in breast cancer cells has been shown at both the mRNA and protein levels. Specific inhibition of ErbB2 using monoclonal antibodies leads to a decreased level of cellular proliferation. Based on these *in vitro* experiments, Herceptin, a truncated blocking-antibody directed against ErbB2 has been successfully developed and is now used in clinical practice (Colomer *et al.*, 2001). Since proteomics is providing a global way for the identification of protein regulation during pathological processes, the discovery of new therapeutic targets and corresponding drugs is a highly awaited outcome.

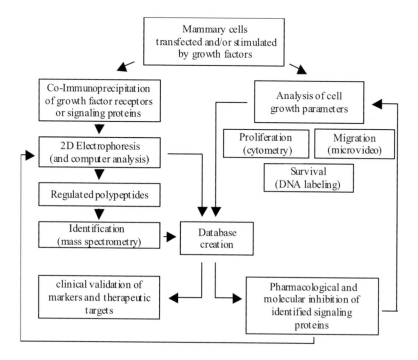

Figure 6. Protocol of proteomic analysis for the definition of drug targets in breast cancer cells.

Alternatively to the systematic analysis of genome and proteome, more focused approaches are proposed to define therapeutic targets in cancer cells. Figure 6 reports such an approach combining the use of proteomics with the analysis of cell growth parameters. Differential analysis of proteomes of cells stimulated or not stimulated by growth factors or the analysis of cells transfected or not transfected by oncogenes can provide the experimental basis for the identification of crucial signaling proteins. Additionally, the testing of existing pharmacological inhibitors or the development of new drugs can also be monitered by using proteomic analysis. The application of such a protocol has, for example, led to the identification of HSP90 as a target in breast cancer. Previous studies using mammalian cells have shown that mitogens such as insulin-like growth factor and epidermal growth factor increase cellular synthesis and accumulation of HSP90 and HSP70 (Jolly and Morimoto, 2000). Using proteomic analysis, it has been shown that synthesis of HSP90 is up-regulated after FGF-2 stimulation in breast cancer cells (Vercoutter-Edouart *et al.*, 2001b). HSP90 is up-regulated in cancer cells and after stimulation by growth factors. HSP90 can form complexes with many components of growth factor signaling pathways (Pearl and Prodromou, 2000), and these interactions permit their protection and correct conformational folding. Interestingly, geldanamycin, an inhibitor of HSP90 activity, totally blocks FGF-2-induced growth of breast cancer cells, suggesting that appropriate protein folding and trafficking are essential for the stimulation of breast cancer cell growth. Therefore, the overexpression of HSP90 appears to be essential to breast epithelial cell tumorigenesis, and this protein is a potential therapeutic target in breast cancer.

7.   PERSPECTIVES: TRANSITION FROM BENCHSIDE TO BEDSIDE

   Future developments in the field of breast cancer proteomics are likely to depend on technological innovations. Currently, a typical 2DE gel can reliably separate and detect about 2,000 proteins. The human genome may contain as many as 30,000 genes; as a result of mRNA splicing, post-translational modifications and proteolysis, the number of proteins is clearly higher than the number of spots detectable on a 2-DE gel. Improvement in resolution and capacity of gels may be achieved, for instance by the use of a series of immobilized pH gradient gels. Improving protein detection can also increase the capacity of 2DE. Although proteins separated on a gel can be stained using Coomassie blue dye, silver staining is more routinely used because it offers detection limits down to about 1 ng. However, silver is hampered by its non-stoichiometric reactivity, and other staining techniques that employ fluorescent dyes have recently been developed. In addition, although 2DE provides the highest resolution for protein separation, some proteins, such as hydrophobic membrane proteins, cannot be studied. Clearly, 2DE has serious limits that are now hampering its use for the sort of large scale/high throughput analysis and for the generation of future, clinically relevant data. Alternative approaches for proteomic analysis are currently being developped and used. The Isotope affinity tagging of cysteine residues (ICAT) method in conjunction with liquid chromatography and mass spectrometry (Gygi *et al.*, 1999) is a promising way to access less soluble proteins. Multi-dimensional chromatography is also being developped (Issaq, 2001), combining, for example, cation exchange high performance liquid chromatography as a first dimension and reverse phase HPLC for a second separation.

The use of protein chip technologies is certainly the most promising concept now avalaible for large scale/high throughput proteomic analysis, and its practical implementation is now emerging (Weinberg *et al.*, 2000; Jain, 2002). The application of protein chips depends on being able to flush potential ligands over arrays impregnated with different types of affinity probes (such as antibodies), and the subsequent analysis of differential protein binding (Nelson *et al.*, 2000; Fung *et al.*, 2001). The use of protein arrays has great potential for diagnostics. Once a specific set of biomarkers or a drug target is determined for a particular tumor type, its use for diagnosis requires a large scale routine analysis of the protein profile of tumors that could not be achieved by 2DE. Then, automated chip-based technologies for analyzing hundreds or thousands of proteins simultaneously, similarly to the cDNA based technologies, could provide a powerful tool for cancer diagnosis, as well as for treatment choice and subsequent monitoring. Protein chips and microarrays probably constitute the most promising technologies for the future development of breast cancer proteomics, especially for the routine assays that will have to be carried out at the clinical level. In addition, given that the expression profiles obtained by genomics and proteomics are complementary, a combined approach is expected to become the basis for the development of new research and diagnostic tools (Hanash, 2003; Celis, 2003).

8. CONCLUDING REMARKS

Proteomics of breast cancer is still in a pioneer phase of exploration. Using different technological approaches, the quest for specific targets that would provide the basis for future treatments is under way. However, the application of proteomic technologies for clinical purposes, for prognosis, diagnosis and treatment monitoring is still to be developed. A major requirement for the clinical application of proteomics will certainly be to integrate the mass of data issuing forth from genomics, proteomics and from the clinic. This integration will require the development of bioinformatics in a way that can make proteomics clinically valuable. The most exciting challenge of the coming years is to move breast cancer proteomics from the bench to the "bedside" and to bridge the existing gap between our advanced molecular understanding of oncogenesis and practical outcomes.

ACKNOWLEDGEMENTS

This work is supported by the "Ligue Nationale Contre le Cancer" (Comité Departemental), the "Fondation pour la Recherche Médicale" (FRM, Comité du Nord), the "Association pour la Recherche contre le Cancer" (ARC, contract 3305), the "Ministère de la Recherche et de l'Education Nationale", the "Genopole de Lille" and the "Région Nord-pas-de-Calais".

REFERENCES

Adam P.J., Boyd R., Tyson K.L., Fletcher G.C., Stamps A., Hudson L., Poyser H.R., Redpath N., Griffiths M., Steers G., Harris A.L., Patel S., Berry J., Loader J.A., Townsend R.R., Daviet L., Legrain P., Parekh R., Terrett J.A., 2003, Comprehensive proteomic analysis of breast cancer cell membranes reveals unique proteins with potential roles in clinical cancer. *J. Biol. Chem.* **278**: 6482-6489.

Adriaenssens E., Lottin S., Berteaux N., Hornez L., Fauquette W., Fafeur V., Peyrat J.P., Le Bourhis X., Hondermarck H., Coll J., Dugimont T., Curgy J.J., 2002, Cross-talk between mesenchyme and epithelium increases H19 gene expression during scattering and morphogenesis of epithelial cells. *Exp. Cell. Res.* **275**: 215-29.

Anderson, N.L., Anderson, N.G., 2002, The human plasma proteome: history, character, and diagnostic prospects. *Mol. Cell. Proteomics* **1**: 845-67.

Bhattacharya, B., Prasad, G.L., Valverius, E M., Salomon, D.S., and Cooper, H.L., 1990, Tropomyosins of human mammary epithelial cells: consistent defects of expression in mammary carcinoma cell lines. *Cancer Res.* **50**: 2105-2112.

Blagoev B., Kratchmarova I., Ong S.E., Nielsen M., Foster L.J., Mann M., 2003, A proteomics strategy to elucidate functional protein-protein interactions applied to EGF signaling.*Nat Biotechnol.* **21**: 315-318.

Brannan C.I., Dees E.C., Ingram R.S., Tilghman S.M., 1990, The product of the H19 gene may function as an RNA. *Mol. Cell. Biol.* **10**: 28-36.

Celis J.E., Gromov P., 2003, Proteomics in translational cancer research: Toward an integrated approach. *Cancer Cell* **3**: 9-15.

Chiarenza A., Lazarovici P., Lempereur L., Cantarelle G., Bianchi A., Bernardini R., 2001, Tamoxifen inhibits nerve growth factor-induced proliferation of the human breast cancerous cell line. *Cancer Res.* **61**: 3002-3008.

Ciocca D.R., Elledge R., 2000, Molecular markers for predicting response to tamoxifen in breast cancer patients. *Endocrine* **13**: 1-10.

Colomer R., Shamon L.A., Tsai M.S., Lupu R., 2001, Herceptin : from the bench to the clinic. *Cancer Invest.* **19**: 49-56.

Craven R.A., Banks R.E., 2001, Laser capture microdissection and proteomics: possibilities and limitation. *Proteomics* **1**: 1200-4.

Descamps S., Lebourhis X., Delehedde M., Boilly B., Hondermarck H., 1998, Nerve growth factor is mitogenic for cancerous but not normal human breast epithelial cells. *J. Biol. Chem.* **273**: 16659-16662.

Descamps S., Toillon R.A., Adriaenssens E., Pawlowski V., Cool S.M., Nurcombe V., Le Bourhis X., Boilly B., Peyrat J.P., Hondermarck H., 2001, Nerve growth factor stimulates proliferation and survival of human breast cancer cells through two distinct signaling pathways. *J. Biol. Chem.* **276:** 17864-17870.

Dollé L., El Yazidi-Belkoura I., Adriaenssens E., Nurcombe V., Hondermarck H., (2003), Nerve growth factor overexpression and autocrine loop in breast cancer cells. *Oncogene* **22:** 5592-601.

El Yazidi-Belkoura I., Adriaenssens E., Dollé L., Descamps S., Hondermarck H., 2003, Tumor Necrosis Factor Receptor-Associated Death Domain Protein (TRADD) is Involved in the Neurotrophin Receptor (p75NTR)-Mediated Anti-Apoptotic Activity of Nerve Growth Factor in Breast Cancer Cells. *J.Biol.Chem.* **278:** 16952-16956.

Ethier S.P., 1995, Growth factor synthesis and human breast cancer progression. *J. Natl. Cancer Inst.* **87:** 964-973.

Ferguson A.T., Evron E., Umbricht C.B., Pandita T.K., Chan T.A., Hermeking H., Marks J.R., Lambers A.R., Futreal P.A., Stampfer M.R. and Sukumar, S., 2000, High frequency of hypermethylation at the 14-3-3 sigma locus leads to gene silencing in breast cancer. *Proc. Natl. Acad. Sci. USA* **97:** 6049-6054.

Franzen, B., Linder, S., Alaiya, A.A., Eriksson, E., Fujioka, K., Bergman, A.C., Jornvall, H., and Auer, G., 1997, Analysis of polypeptide expression in benign and malignant human breast lesions. *Electrophoresis* **18:** 582-587.

Franzen, B., Linder, S., Alaiya, A.A., Eriksson, E., Uruy, K., Hirano, T., Okuzawa, K., and Auer, G., 1996, Analysis of polypeptide expression in benign and malignant human breast lesions: down-regulation of cytokeratins. *Br. J. Cancer* **74:** 1632-1638.

Franzen, B., Linder,S., Okuzawa, K., Kato, H., and Auer, G., 1993, Non enzymatic extraction of cells from clinical tumor material for analysis of gene expression by two-dimensional polyacrylamide gel electrophoresis. *Electrophoresis* **14:** 1045-1053.

Fu H., Subramanian R.R., and Masters S.C., 2000, 14-3-3 proteins: structure, function, and regulation. *Annu. Rev. Pharmacol. Toxicol.* **40:** 617-647.

Fung E.T., Thulasiraman V., Weinberger S.R., Dalmasso E.A., 2001, Protein biochips for differential profiling. *Curr. Opin. Biotechnol.***12:** 65-69.

Gharbi S., Gaffney P., Yang A, Zvelebil M.J., Cramer R., Waterfield M.D., Timms J.F. , 2002, Evaluation of two-dimensional differential gel electrophoresis for proteomic expression analysis of a model breast cancer cell system. *Mol. Cell. Proteomics* **1:** 91-8.

Giometti C.S., Tollaksen S.L., Chubb C., Williams C., Huberman E., 1995, Analysis of proteins from human breast epithelial cells using two-dimensional gel electrophoresis. *Electrophoresis* **16:** 1215-1224.

Gygi S.P., Rist B., Gerber S.A., Turecek F., Gelb M.H., Aebersold R., 1999, Quantitative analysis of complex protein mixtures using isotope-coded affinity tags. *Nat. Biotechnol.* **17:** 994-999.

Haber D., 2000, Roads leading to breast cancer. *N. Engl. J. Med.* **343:** 1566-1568.

Hanash S., 2003, Disease proteomics. *Nature* **422:** 226-232.

Hinsby A..M., Olsen J.V., Bennett K.L., Mann M., 2003, Signaling initiated by overexpression of the fibroblast growth factor receptor-1 investigated by mass spectrometry. *Mol. Cell. Proteomics* **2:** 29-36.

Hondermarck H., 2003, Breast cancer: when proteomics challenges biological complexity. *Mol. Cell. Proteomics* **2:** 281-91.

Hondermarck H., Vercoutter-Edouart A.S., Revillion F., Lemoine J., El-Yazidi-Belkoura I., Nurcombe V., Peyrat J.P., 2001, Proteomics of breast cancer for marker discovery and signal pathway profiling. *Proteomics* **1:** 1216-32.

Hondermarck H., McLaughlin C.S., Patterson S.D., Bradshaw R.A., 1994, Early changes in protein synthesis induced by basic fibroblast growth factor, nerve growth factor, and epidermal growth factor in PC12 pheochromocytoma cells. *Proc. Natl. Acad. Sci. U S A.* **91:** 9377- 9381.

Issaq H.J., 2001, The role of separation science in proteomics research. *Electrophoresis* **22:** 3629-3638.

Jain K.K., 2002, Post-genomic applications of lab-on-a-chip and microarrays. *Trends Biotechnol.* **20:** 184-5.

Jolly C., Morimoto R.I., 2000, Role of the heat shock response and molecular chaperones in oncogenesis and cell death. *J. Natl.CancerInst.* **92:** 1564-1572.

LeBourhis X., Toillon R.A., Boilly B., Hondermarck H., 2000, Autocrine and paracrine growth inhibitors of breast cancer cells. *Breast Cancer Res. Treat.* **60:** 251-258.

Li J, Zhang Z, Rosenzweig J, Wang YY, Chan DW., 2002, Proteomics and bioinformatics approaches for identification of serum biomarkers to detect breast cancer. *Clin. Chem.* **48:** 1296-304.

Lottin S., Adriaenssens E., Dupressoir T., Berteaux N., Montpellier C., Coll J., Dugimont T., Curgy J.J., 2002, Overexpression of an ectopic *H19* gene enhances the tumorigenic properties of breast cancer cells. *Carcinogenesis* **23:** 1885-1895.

Lottin S., Vercoutter-Edouart A.S., Adriaenssens E., Czeszak X., Lemoine J., Roudbaraki M., Coll J., Hondermarck H., Dugimont T., Curgy J.J., 2002, Thioredoxin post-transcriptional regulation by *H19* provides a new function to mRNA-like non-coding RNA. *Oncogene* **21:** 1625-31.

Maloney T.M., Paine P.L., Russo J., 1989, Polypeptide composition of normal and neoplastic human breast tissues and cells analyzed by two-dimensional gel electrophoresis. *Breast Cancer Res. Treat.* **14:** 337-348.

Nakamura H., Nakamura K., Yodoi J., 1997, Redox regulation of cellular activation. *Annu. Rev. Immunol.* **15:** 351-69.

Nelson R.W., Nedelkov D., Tubbs K.A., 2000, Biosensor chip mass spectrometry: a chip-based proteomics approach. *Electrophoresis* **21:** 1155-63.

Nurcombe V., Smart C.E., Chipperfield H., Cool S.M., Boilly B., Hondermarck, H., 2000, The proliferative and migratory activities of breast cancer cells can be differentially regulated by heparan sulfates. *J. Biol. Chem.* **275:** 30009-30018.

Page M.J., Amess B., Townsend R.R., Parekh R., Herath A., Brusten L., Zvelebil M.J., Stein R.C., Waterfield M.D., Davies S.C., and O'Hare M.J., 1999, Proteomic definition of normal human luminal and myoepithelial breast cells purified from reduction mammoplasties.*Proc. Natl. Acad. Sci. U S A* **96:** 12589-12594.

Pandey A., Podtelejnikov A.V., Blagoev B., Bustelo X.R., Mann M., Lodish H.F., 2000, Analysis of receptor signaling pathways by mass spectrometry: identification of vav-2 as a substrate of the epidermal and platelet-derived growth factor receptors. *Proc. Natl. Acad. Sci. U S A* **97:** 179-184.

Pearl L.H., Prodromou C., 2000, Structure and in vivo function of HSP90. *Curr. Opin. Struct. Biol.* **10:** 46-51.

Piek E., Roberts A.B., 2001, Suppressor and oncogenic roles of transforming growth factor-beta and its signaling pathways in tumorigenesis. *Adv. Cancer Res.* **83:** 1-54.

Rahmoune H., Chen H.L, Gallagher J.T., Rudland P.S., Fernig D.G., 1998, Interaction of heparan sulfate from mammary gland cells with acid fibroblast growth factor and basic FGF. Regulation of the activity of bFGF by high and low affinity binding sites in heparan sulfate. *J. Biol. Chem.* **273:** 7303-7310.

Schrohl A.S., Pedersen A.N., Jensen V., Mouridsen H., Murphy G., Foekens J.A., Brunner N., Holten-Andersen M.N., 2003, Tumour tissue concentrations of the proteinase inhibitors TIMP-1 and PAI-1 are complementary in determining prognosis in primary breast cancer. *Mol. Cell. Proteomics* **3:** 164-72.

Soskic V., Gorlach M., Poznanovic S., Boehmer F.D., Godovac-Zimmermann J., 1999, Functional proteomics analysis of signal transduction pathways of the platelet-derived growth factor beta receptor. *Biochemistry* **38:** 1757-1764.

Stastny J., Prasad R., Fosslien E., 1984, Tissue proteins in breast cancer, as studied by use of two-dimensional electrophoresis. *Breast Cancer Res. Treat.* **30:** 1914-1918.

Tagliabue E., Castiglioni F., Ghirelli C., Modugno M., Asnaghi L., Melani C., Menard S., 2000, Nerve growth factor cooperates with p185HER-2 in activating growth of human breast carcinoma cells. *J. Biol. Chem.* **275:** 5388-5394.

Trask D.K., Band V., Zajchowski D.A., Yaswen P., Suh T., Sager. R., 1990, Keratins as markers that distinguish normal and tumor-derived mammary epithelial cells. *Proc. Natl. Acad. Sci. U S A.* **87:** 2319-2323.

Umbricht C.B., Evron E., Gabrielson E., Ferguson A., Marks J., Sukumar S., 2001, Hypermethylation of 14-3-3 sigma (stratifin) is an early event in breast cancer. *Oncogene* **20:** 3348-53.

Vejda S, Posovszky C, Zelzer S, Peter B, Bayer E, Gelbmann D, Schulte-Hermann R, Gerner C., 2002, Plasma from cancer patients featuring a characteristic protein composition mediates protection against apoptosis. *Mol.Cell.Proteomics* **1:** 387-93.

Vercoutter-Edouart A.S., Czeszak X., Crepin M., Lemoine J., Boilly B., Le Bourhis X., Peyrat J.P., Hondermarck H., 2001, Proteomic detection of changes in protein synthesis induced by fibroblast growth factor-2 in MCF-7 human breast cancer cells. *Exp. Cell Res.* **262:** 59-68.

Vercoutter-Edouart A.S., Lemoine J., Le Bourhis X., Hornez L., Boilly B., Nurcombe V., Revillion F., Peyrat J.P. and Hondermarck H., 2001, Proteomic analysis reveals that 14-3-3 sigma is down-regulated in human breast cancer cells. *Cancer Res.* **61:** 76-80.

Vercoutter-Edouart A.S., Lemoine J., Smart C.E., Nurcombe V., Boilly B., Peyrat J.P., Hondermarck H., 2000, The mitogenic signaling pathway for fibroblast growth factor-2 involves the tyrosine phosphorylation of cyclin D2 in MCF-7 human breast cancer cells. *FEBS Lett.* **478:** 209-215.

Weinberger S.R., Morris T.S., Pawlak M., 2000, Recent trends in protein biochip technology. *Pharmacogenomics* **1:** 395-416.

Westley B. and Rochefort H., 1980, A secreted glycoprotein induced by estrogen in human breast cancer cell lines. *Cell* **20:** 353-362.

Westley B. and Rochefort H., 1980, A secreted glycoprotein induced by estrogen in human breast cancer cell lines. *Cell* **20:** 353-362.

Wirth P.J., Egilsson V., Gudnason V., Ingvarsson S., Thorgeirsson S.S., 1987, Specific polypeptide differences in normal versus malignant human breast tissues by two-dimensional electrophoresis. *Breast Cancer Res. Treat.* **10:** 177-189.

Wirth P.J., 1989, Specific polypeptide differences in normal versus malignant breast tissue by two-dimensional electrophoresis. *Electrophoresis* **10:** 543-54.

Worland P.J., Bronzert D., Dickson R.B., Lippman M.E., Hampton L., Thorgeirsson S.S., Wirth. P.J., 1989, Secreted and cellular polypeptide patterns of MCF-7 human breast cancer cells following either estrogen stimulation or v-H-ras transfection. *Cancer Res.* **49:** 51-57.

Wright G.L., 1974, Two-dimensional acrylamide gel electrophoresis of cancer-patient serum proteins. *Annu. Clin. Lab. Sci.* **4:** 281-293.

Wulfkuhle JD, Sgroi DC, Krutzsch H, McLean K, McGarvey K, Knowlton M, Chen S, Shu H, Sahin A, Kurek R, Wallwiener D, Merino MJ, Petricoin EF 3rd, Zhao Y, Steeg PS., 2002, Proteomics of human breast ductal carcinoma in situ. *Cancer Res.* **62:** 6740-9.

Chapter 7

Proteomics of Hepatocellular Carcinoma : Present Status and Future Prospects

MAXEY C.M. CHUNG [*], ROSA C.M.Y. LIANG[#], TECK KEONG SEOW[#], JASON C.H. NEO[#], SIAW LING LO[#], GEK SAN TAN[#]

[*]Department of Biochemistry, Faculty of Medicine, [#]Department of Biological Science, Faculty of Science, National University of Singapore, Singapore 119260, Republic of Singapore

1. INTRODUCTION

Hepatocellular carcinoma (HCC or hepatoma) is the most common primary cancer of the liver. It represents more than 5% of all cancers in the world, and affects approximately 550,000 people in 2000 (Parkin *et al.*, 2001). Persistent viral infection by the hepatitis B or C virus (HBV or HBV) is probably the most important cause of HCC worldwide. Together with dietary exposure to aflatoxin B1, they are responsible for about 80% of all HCCs in humans (Thorgeirsson and Grisham, 2002). In addition, males are afflicted at least twice as often as females, and its occurrence also increases progressively with age (Schafer and Sorrell, 1999; McGlynn *et al.*, 2001).

HCC has a high incidence rate in the underdeveloped and developing countries of sub-Saharan Africa, China and the Far East where the patients are often diagnosed with infiltrative or massive tumours (Llovet and Beaugrand, 2003). The incidence of HCC in the developed countries is also increasing (McGlynn *et al.*, 2001). For example, approximately 15,000 new cases are reported each year in the United States (El-Serag *et al.*, 1999), while in France, it has increased markedly to about 4000 cases per year (Deuffic *et al.*, 1999). In these countries, however, HCC is mostly diagnosed at an asymptomatic stage by routine ultrasonography since the patient's

163

H. Hondermarck (ed.), Proteomics: Biomedical and Pharmaceutical Applications, 163–181.
© 2004 *Kluwer Academic Publishers. Printed in the Netherlands.*

underlying liver disease has been detected and regularly followed (Llovet and Beaugrand, 2003)

For most patients who suffer from HCC, long-term survival is rare. They usually die within a year after diagnosis. The options for treatment are surgery, systematic chemotherapy, loco-regional treatment, and symptomatic relief, and of these, only surgery has the potential to cure. For example, a 5-year survival rate in surgical patients may exceed 70% in Western countries (Llovet and Beaugrand, 2003). Unfortunately, liver resection is only feasible for 10 – 15% of the patients as they are often presented late. While primary preventive measures such as vaccination against the hepatitis B virus can prevent the onset of HCC (Kao and Chen, 2002), we are still faced with the problem as how best to detect this malignant disease *early* for those who have been infected with the virus or are at risk of developing the disease. While tumours of 1 cm in size may be detected by improved image technology, it is also essential to develop a molecular diagnosis of HCC by establishing a molecular profile of the preneoplastic and *carcinoma-in-situ* lesions (Llovet and Beaugrand, 2003). Proteomics is a promising approach in this respect as by analyzing the proteome of HCC, it is possible to identify novel diagnostic biomarkers for the early detection of HCC, and specific disease-associated proteins as potential therapeutic targets in the treatment of HCC.

2.   AETIOLOGY

The aetiological and other clinical aspects of HCC has been reviewed recently by several authors (Schafer and Sorrell, 1999; Seow *et al.*, 2001; Llovet and Beaugrand, 2003), so we will highlight only some of the salient points here. As mentioned above, persistent viral infection by HBV and HCV is the main causative agent for HCC in the world. This viral infection results in liver cirrhosis, which predisposes to development of HCC. It is believed that the risk of developing HCC in a chronic HBV carrier is increased by a factor of 100 as compared to a non-infected individual (Beasley *et al.*, 1981). Since the development of HCC in humans may take up to 30 years after it was first diagnosed with chronic HBV or HCV infection (see Fig. 1) (Thorgeirsson and Grisham, 2002), the ability to prevent liver diseases to progress to cirrhosis is the most effective mode of preventing the cancer. For the prevention of HBV infection, vaccinating children in Taiwan against the virus (Kao and Chen, 2002) showed a drastic decrease in the incidence of HCC. On the other hand, the primary prevention in the Western countries hinges on the screening of the general population for liver diseases to detect advanced fibrosis or cirrhosis. Unfortunately, there is as yet a reliable non-invasive method for the detection of cirrhosis (Llovet and Beaugrand, 2003).

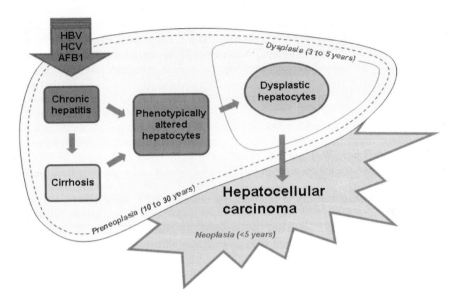

Figure 1. Chronological events leading to the development of hepatocellular carcinoma in human.

3.   SCREENING AND DIAGNOSIS

Since HCC is prevalent in a certain population (such as those with chronic hepatitis and cirrhosis), and geographical region of the world (Fig. 2), it is reasonable to target the screening for cirrhotic patients so that they can receive early and successful therapy.

Regular screening by ultrasonography and measurement of serum α-fetoprotein (AFP) levels is recommended for high risk individuals. This will facilitate the early detection of HCC and thus increase the tumour resection rate. However, the overall survival rate of HCC patients is still low even after resection because of frequent recurrence. It is therefore important to search for reliable and specific biomarkers that can identify patients at high risk for early recurrence so that proper treatment regime, including re-resection, can be implemented.

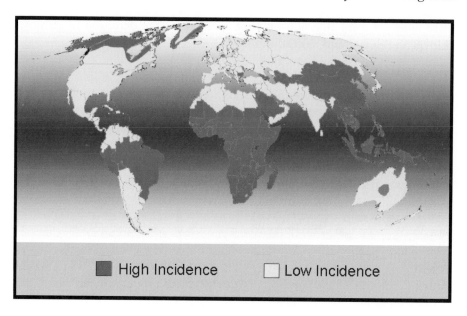

Figure 2. Geographical correlation between the incidence of HCC and prevalence of HBV infection.

Although AFP is the best available marker for the diagnosis and monitoring of HCC, it suffers from a number of disadvantages. For example, a serum AFP concentration of >500 ng/ml in a high risk individual is usually diagnostic of HCC. However, about 70% of HCC patients have levels above the reference range of >10 ng/ml but below the diagnostic level of >500 ng/ml. Furthermore, individuals without HCC but with benign chronic liver disease can have elevated AFP in this 'grey zone' of elevated AFP values (Seow *et al.*, 2001). The use of AFP glycosylated variants for diagnosis of HCC has been proposed to overcome this inadequacy, but the results are not very reproducible (Du *et al.*, 1991). However, a recent report suggested that a *Lens culinaris* agglutinin-reactive AFP, named AFP-L3, has the potential to detect small HCC tumours (<2cm) (Li *et al.*, 2001). As an alternative to testing for lectin-affinities, isoelectric focusing (IEF) was used to separate and detect HCC-specific AFP isoforms (Ho *et al.*, 1996), of which Band +II was relatively specific for HCC (Johnson *et al.*, 1999; Poon *et al.*, 2002).

The other useful biomarker for the diagnosis of HCC is serum des-γ-carboxy prothrombin (DCP), where a high positive rate is reported for patients with HCV infection (Saitoh *et al.*, 1994). However, DCP is also elevated in benign liver diseases such as cirrhosis, although the concentration is significantly lower than in HCC. Recently, Naraki *et al.* (2002) examined the difference in the γ-carboxyglutamic acid (Gla) content of HCC-associated DCP and benign liver disease DCP using a monoclonal antibody approach. It was concluded that HCC-associated DCP has less than four Gla residues while the DCP variants in benign liver diseases have more than five Gla residues (Naraki *et al.*, 2002).

4. PROTEOMICS OF HEPATOCELLULAR CARCINOMA : PRESENT STATUS

A review of the recent literature revealed that there were only a few publications dedicated to the proteome analysis of HCC. Two groups reported the proteomics of hepatoma cell lines (Seow *et al.*, 2000; Ou *et al.*, 2001; Yu *et al.*, 2000, 2001) while the other two on the proteome analysis of HCC from human liver tissues (Lim *et al.*, 2002, Kim *et al.*, 2002). By comparison, there were nine publications in the last two years on the gene expression analysis of HCC from human cancer tissues alone using cDNA arrays (Lau *et al.*, 2000; Shirota *et al.*, 2001; Okabe *et al.*, 2001; Tackels-Horne *et al.*, 2001; Kim *et al.*, 2001; Xu *et al.*, 2001; Goldenberg *et al.*, 2002; Delpuech *et al.*, 2002; Chen *et al.*, 2002). This showed that proteome analysis of HCC is still in its infancy, and thus more effort must be expended to speed up the identification of the disease associated proteins in HCC to compare, confirm and complement the published gene expression results. This is essential since the level of mRNA expression does not always correlate with the amount of active protein in the cell (Gygi *et al.*, 1999a).

4.1 Proteome analysis of hepatoma cell lines

4.1.1 HCC-M : A HBV surface antigen-positive derived cell line

Although there was an earlier report on the expression profiling of several hepatoma cell lines by 2-DE by Wirth *et al.* (1995), the most comprehensive proteomic analysis of a hepatoma cell line, HCC-M, was carried out by our group (Seow *et al.*, 2000). HCC-M was chosen as the model cell line for our work because it was a HBV surface antigen-positive derived cell line (Watanabe *et al.*, 1983). In this study, 192 proteins separated by 2-DE were identified by matrix-assisted laser desorption/ionization-time of flight mass spectrometry (MALDI-TOF MS), and another 29 identified by nanoelectrospray ionization-tandem mass spectrometry (nESI-MS/MS) (Ou *et al.*, 2001). It is noteworthy that most of the proteins identified by tandem MS were located in the lower molecular mass region of the 2-DE gel (Fig. 3). It was suggested that these lower molecular mass proteins may not have generated sufficient tryptic peptides for unambiguous identification by MALDI-TOF MS analysis.

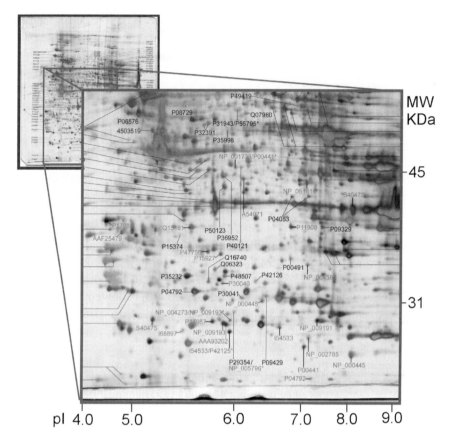

Figure 3. HCC-M proteins identified by nanoelectrospray ionization tandem MS/MS
(nESI MS/MS).

In the original publication, the proteins were listed in the 2-DE
database based on the number assigned to a given spot followed by the
NCBI/SWISS-PROT accession number (Seow *et al*, 2000). Subsequently, it
was decided to reorganize the database by grouping the proteins into fifteen
different functional categories. In addition, a separate list of proteins that
have been implicated in HCC and other types of cancer was also compiled.
We believe that such a categorization and grouping of HCC-M proteins will
simplify the database of HCC proteins and facilitate the rapid identification
of novel proteins involved in hepatocarcinogenesis (Liang *et al.*, 2002).

4.1.1.1   Integrated proteomics approach for the discovery of a novel protein, Hcc-1

An integrated approach consisting of 2-DE, MS (MALDI-TOF and tandem MS), bioinformatics, and molecular biology was used to characterize one of the expressed proteins, Hcc-1, of this cell line (Choong *et al.*, 2001). Briefly, Hcc-1 is a protein spot with a pI of 6.2 and a relative molecular mass of 35 kDa, and did not have any match with proteins from the database search. Three tryptic peptides were then sequenced, and these sequences were used to search the EST database. The CAP3 (contig assembly programme) software (TigemNET) was used to assemble the putative full sequence of the protein, which was confirmed subsequently by 3 additional peptides from the MALDI-TOF spectrum as well as by rapid amplification of cDNA ends (RACE). Subsequently standard molecular techniques were used to characterize the protein further (Choong *et al.*, 2001). It demonstrates clearly that an integrated proteomics approach is a very powerful means of discovering novel proteins, the complete sequence of which is not yet available from the database (Gray, 2001). This procedure is summarized in Fig. 4.

Interestingly, Fukuda *et al.* (2002) reported recently the cloning and characterization of a proliferation-associated cytokine-inducible protein, named CIP29, that has the same amino acid sequence as Hcc-1. This protein was present in the 2-DE gels of the lysate of erythropoietin (Epo)-stimulated UT-7/Epo cells, and its expression was also found to be higher in cancer and foetal tissues than in normal tissues. It was also concluded that up-regulation of CIP29 in these cells was associated with cell cycle progression but not antiapoptosis (Fukuda *et al.*, 2002).

4.1.1.2   HCC-M database

Recently, an interactive protein database that integrates the spots with the 2-DE map for HCC-M has been created. It is now freely accessible on the World Wide Web at http://proteome.btc.nus.edu.sg/hccm/ (Liang *et al.*, 2002). There are two options to query the HCC-M database: either (1) protein search by NCBI/SWISS-PROT Accession number, Protein name and Protein ID (ID as published in Seow *et al.*, 2000) ; or (2) interactive protein spots query on the original 2-DE image maps (Fig. 5). The availability of this protein database and the integrated proteome database of Cho *et al.* (2002) should serve as a useful resource for other groups involved in HCC research. They will now join a list of databases dedicated to cancer proteomics (Simpson and Dorow, 2001).

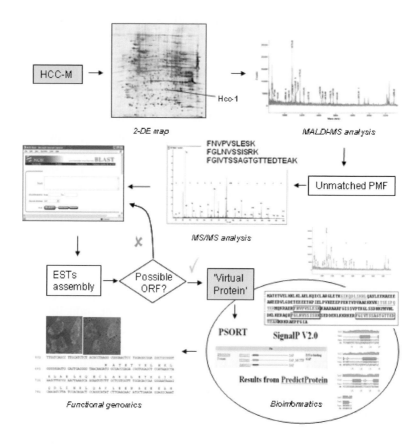

Figure 4. Integrated proteomics approach for the discovery of Hcc-1.

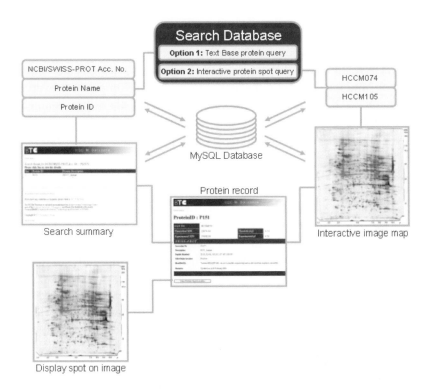

Figure 5. Summary of the procedure used to query the HCC-M proteome database.

4.1.2 Hepatoma cell line BEL-7404 and normal liver cell line L-02

The identification of differentially expressed proteins in a human hepatoma cell line, BEL-7404 and the normal human liver cell line, L-2, was reported by Yu *et al.*(2001). Image analysis showed that 99 protein spots showed quantitative and qualitative differences, but only 12 of these proteins were identified. Out of these, 7 protein spots were up-regulated in BEL-7404 while the remaining 5 showed higher levels of expression in the normal cell line, L-02. These proteins had been identified by liquid chromatography-ion trap-mass spectrometry (LC-IT-MS), and on the basis of their identities, their functional implication in hepatocarcinogenesis was discussed. For example, IMPDH type II, which is up-regulated in BEL-7404 cells, is commonly over-expressed in malignant cells. The over-expression (18-fold) of glutathione-S-transferase P (GSTP1-1) (a member of GST pi family) has been associated with carcinogenesis and the development of many cancers, including testis, ovarian, and colorectal (Yu *et al.*, 2000).

The same group of workers (Yu *et al.*, 2001) also investigated the alterations in the proteome of human hepatoma cells transfected with antisense epidermal growth factor receptor sequence. For these experiments, the cell strain JX-1 was established by transfecting the complete human EGFR cDNA into the hepatoma cell line BEL-7404 in the antisense orientation, and the cell strain, JX-0 was the control. Comparative expression profiling by 2-DE followed by image analysis of the silver stained gel showed 40 proteins were differentially expressed. Only 5 proteins (maspin, two isoforms of Hsp27, glutathione peroxidase and 14-3-3-sigma) have been identified by LC-IT-MS (Yu *et al.*, 2001).

4.2   Proteome analysis of human liver tumour tissues

Two recent papers on the proteome analysis of HCC from human liver tissues were published by Korean researchers (Lim *et al.*, 2002 ; Kim *et al.*, 2002). Using normal, cirrhotic and tumorous liver tissues, Lim *et al.* (2002) found that 21 proteins had a significant change in expression. Of these proteins, lamin B1 was proposed as a marker for cirrhosis as this protein was highly expressed in the cirrhotic tissue when compared to the normal liver tissue. However, a non-invasive method for its detection in cirrhosis, if any, was not proposed.

Kim *et al.* (2002), on the other hand, created an expanded reference map for human liver and also carried out differential profiling of proteins for normal and HCC tissues. For the expression profiling experiments, 127 different proteins were identified by peptide mass fingerprinting. On the basis of a 3-fold difference in expression level, 37 protein spots (9 over- and 28 under-expressed) were found on the 2-DE gel, but only 16 of these spots (1 over- and 15 under-expressed) corresponding to 11 proteins were identified by MALDI-TOF MS. Among the proteins identified were acyl CoA dehydrogenase and glycerol-3-phosphate dehydrogenase which are enzymes involved in lipid catabolism, aldehyde dehydrogenase, cathepsin D and growth factor receptor-bound protein-2 (Grb2). The implication of the differential expression of these proteins were discussed in the context of their possible role in tumorigenesis (Kim *et al.*, 2002). It was of interest to note that comparison of the list of differentially expressed proteins provided by the two groups showed that there were only two proteins, aldehyde dehydrogenase and protein disulphide isomerase, that were common to each list, even though the samples were from the same country.

In a separate paper, Park *et al.* (2002) reported that they had identified more than 250 proteins in the 2-DE gels of normal human liver tissues using

MALDI-TOF MS, but only a few representative proteins were listed in the figure caption of their 2-DE gel. However, their focus was to examine the differences in the electrophoretic pattern of aldehyde dehydrogenase (ALDH) between normal and tumorous tissues on 2-DE gels. They showed that alterations of the aldehyde dehydrogenase isozyme variants, ALDH-3 and ALDH-2, are closely correlated to HCC. This suggests that a proteomic approach could be used to identify potential biomarkers of HCC, and to elucidate molecular changes in hepatocarcinogenesis (Park *et al.*, 2002).

4.3 Proteomic approach for the discovery of serological biomarkers for hepatocellular carcinoma

Proteomics is well suited for the discovery of new HCC serological biomarkers by profiling the proteins found in the sera of control subjects versus HCC patients by 2-DE, and to screen for any protein spots that are either absent or present, or altered in abundance or location prior to identification by mass spectrometry. If other serum proteins, other than α-fetoprotein (AFP) can be added to a panel of biomakers, the confidence of predicting the onset of HCC will be increased considerably.

This approach was suggested and adopted by two groups of researchers (Steel *et al.*, 2001; Poon and Johnson, 2001). However, one of the problems of this method is due to the presence of a few highly abundant proteins such as albumin, transferrin, and immunoglobulin G (IgG) in the serum proteome profile, which may distort the separation and/or mask the minor proteins present. Thus selective removal of these proteins, especially albumin (representing 30-50% of the total serum protein) is recommended prior to fractionation by 2-DE. In many cases, the differences in the protein of interest in the serum protein profile may be due to post-translational modifications, especially glycosylation. In fact, different glycosylated forms of AFP were detected in the sera of HCC patients (Johnson *et al.*, 1999).

4.4 Identification of tumour-associated antigens in hepatocellular carcinoma by proteomics

Besides using proteome analysis to identify disease-associated proteins directly by differential display, one can also use the proteomics approach in combination with serology to screen for autoantibodies against tumour associated antigens (Selinger and Kellner, 2002). This is because autoantibodies against intracellular antigens have been found in the sera of a number of malignancies, including cancers and autoimmune diseases (Old and Chen, 1998). The procedure entailed the detection of immunoreactive proteins on the electroblots of 2-DE gels by incubating the membranes with sera from cancer or normal healthy volunteers. It is termed either SERPA or PROTEOMEX (Selinger and Kellner, 2002). In contrast, the genetic method is termed SEREX (<u>ser</u>ological analysis of recombinant cDNA <u>ex</u>pression libraries). This method involves monitoring the autoantibodies in serum from cancer patients by screening against an expression library prepared from the tumour tissues (Sahin *et al.*, 1995).   The identified antigens obtained either by PROTEOMEX or SEREX can be developed for cancer screening, diagnosis, prognosis as well as for immunotherapy.

Recently Le Naour *et al.* (2002) used the PROTEOMEX method to identify autoantibodies found in the sera of HCC patients. They have identified eight proteins that elicited a humoral response. However, only one of these autoantibodies, which is reactive against a truncated form of calreticulin, Crt32, is largely restricted to HCC. On the other hand, two earlier studies using the SEREX approach have reported the identification of numerous tumour-associated antigens of HCC (Steiner-Liewen *et al.*, 2000; Wang *et al.*, 2002). It is worth noting that none of the antigens identified in the two SEREX analyses, and the PROTEOMEX method is identical. This suggests that much more work needs to be done to resolve this apparent anomaly.

5.   GENE EXPRESSION ANALYSIS OF HEPATOCELLULAR CARCINOMA

Recently there were two excellent published reviews on the molecular pathogenesis of HCCs (Feitelson *et al.*, 2002; Thorgeirson and Grisham, 2002).   Coupled with this are nine publications on the gene expression analysis of HCC from human liver cancer tissues using cDNA arrays in the last two years (Lau *et al.*, 2000; Shirota *et al.*, 2001; Okabe *et al.*, 2001; Tackels-Horne *et al.*, 2001; Kim *et al.*, 2001; Xu *et al.*, 2001; Goldenberg *et al.*, 2002; Delpuech *et al.*, 2002; Chen *et al.*, 2002), and one on the profiling of the adult human liver transcriptome (Yano *et al.*, 2001). This clearly indicates that there is a great deal of efforts being directed at the elucidation of the molecular mechanisms of hepatocarcinogenesis, and the unravelling of the genetic factors that participate in the development of HCC.

Various types of HCC samples were used for these gene expression experiments. They included (a) HBV-associated HCC (Kim *et al.*, 2001; Xu *et al.*, 2001), (b) HBV- and HCV-associated HCC (Okabe *et al.*, 2001; Delpuech *et al.*, 2002), (c) cirrhotic and non-cirrhotic liver tissues, and HCC tumours of different degree of differentiation (Delpuech *et al.*, 2002), and (d) primary liver and metastatic carcinoma without associated viral infection (Shirato *et al.*, 2001; Tackels-Horne *et al.*, 2001; Chen *et al.*, 2002). The number of genes used in the various types of arrays range from 600 – 41,000. Each of this group produced a list of genes and expressed sequence tags in their gene expression profiles, but upon comparison of their results, the number of common genes is very limited (Goldenberg *et al.*, 2002). This is in spite of the fact that four of these samples were from Asian patients. This anomaly is a challenge for scientists working in this area, and efforts must be made to address this problem. However, one important conclusion arising from these studies is that despite the heterogeneity in gene expression profiles in HCC, there seems to be a link between gene expression patterns and pathological or virological features (Delpuech *et a*l., 2002).

6. PROTEOMICS OF HEPATOCELLULAR CARCINOMA : FUTURE PROSPECTS

It is apparent from the earlier discussion that there is much to be done in the proteome analysis of HCC if we were to compare with the detailed and extensive results of the gene expression analysis of HCC using cDNA microarrays. As many changes may occur in proteins that are not reflected at the RNA level, it is important that mRNA expression studies must be validated by proteomic information. However, proteome analysis still faces many technical challenges especially in terms of speed/throughput and sensitivity, so new proteome profiling technologies must be evaluated for implementation to avoid this bottleneck. The future prospects on the proteome analysis of HCC will depend on the adoption of some of these new technologies to complement the available cDNA microarray results, with the aim to identify novel disease-associated proteins for future drug development.

6.1 Differential gel electrophoresis (DIGE)

All the proteome profiling of HCC cell line and tumour tissue extracts have so far been carried out using 2-DE (Seow *et al.*, 2000; Yu *et al.*, 2000; Ou *et al.*, 2001; Lim *et al.*, 2002; Kim *et al.*, 2002). A recent innovation in 2-DE gels is termed differential gel electrophoresis (DIGE) where the tissue samples are prelabelled with fluorescent dyes (Cy3, Cy5) first, mixed, and then ran on the same 2-D gel (Unlu *et al.*, 1997; Tonge *et al.*, 2001). DIGE overcomes many of the current problems of 2-DE, such as minimizing between run-variability and cutting down the number of 2-D gels at least by half. In this respect, adopting DIGE will be an advantage in the proteome analysis of HCC and other cancer tissues. We have preformed a preliminary DIGE experiment on a pair of normal and HCC tumour tissues, and the result is shown in Fig.6. The result looks very promising, and we are now in the process of extending our studies using DIGE for differential expression analysis of tumorous samples of HCC.

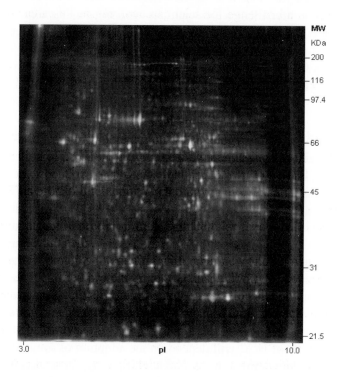

Figure 6. Differential gel electrophoresis of hepatocellular carcinoma. Cy3 (green), normal; Cy5 (red), tumorous tissue.

(See also Colour Plate Section, page 388)

6.2   Isotope-Coded Affinity Tag (ICAT) Labelling

The other approach to quantitative protein profiling is based on stable isotope labelling of proteins and peptides and automated tandem mass spectrometry (MS/MS) (Gygi *et al.*, 1999b). This method, termed isotope-coded affinity tag labelling or ICAT, has been shown to be robust, reproducible and amenable to high throughput automation (Ideker *et al.*, 2001). Compared to 2-DE gels, it has several advantages: it is (a) automated, (b) scalable and (c) quantitative. Application of this technology should facilitate the rapid profiling of more HCC samples, but to-date, there has been no published reports on the application of ICAT labelling for the proteome analysis of HCC yet.

6.3   Subproteomics

The other important area of proteome analysis is sample preparation before analysis, and this is true for either 2-DE or non-gel based approaches. In view of the great dynamic range of proteins in a cell (more than 10^6-fold difference), it is advantageous to analyze proteins from the subcellular organelles separately (subproteomics) so that lower abundance proteins may be detected by either 2-DE gels or other separation techniques. In addition the identification of differentially expressed proteins in subcellular organelles often shed light on the functional roles of these proteins (Dreger, 2003). This approach should be applicable to all expression proteomics projects, including HCC. However, a subproteomics strategy involves a lot of extra experimental work as well as requiring higher sample amounts for the prefractionation steps. This could be limiting for clinical samples, but should be applicable to the various hepatoma cell lines that are being used to study hepatocarcinogenesis.

6.4   HUPO human liver proteome project

The recent initiative by the Human Proteome Organization (HUPO) to establish a human liver proteome project (HLPP) with the objective to identify proteins expressed in the liver will undoubtedly accelerate the progress towards the understanding of liver diseases, including HCC, at the genomics and proteomics level.   With the anticipated participation of international scientists having a wide range of technical expertise and knowedge in the HLPP, this ambitious initiative may serve as a model as to how future international proteome projects could be structured.

7.   CONCLUSIONS

Proteome analysis of HCC is an emerging area of research in the discovery of novel proteins as early biomarkers for diagnosis and potential therapeutic targets for the treatment of HCC. Although numerous publications have appeared on the gene expression analysis of HCC by cDNA microarrays, there is still a need to analyze the differentially expressed proteins in the cellular proteome as many protein modifications are not reflected by changes at the RNA level. New and improved proteomic technologies such as DIGE, and ICAT labelling in combination with mass spectrometric analysis will drive and accelerate the next phase of expression profiling of HCC. There will also be a concerted effort in performing proteome analysis at the subcellular level (subproteomics) in order to detect and identify low abundance proteins which may be missed at the global expression profile of the whole cell. In addition the localization sites of these identified proteins will provide clues to their functions. Finally, it is hoped that integration of proteome-based approaches with data from genomic profiling will lead to a better understanding of hepatocarcinogenesis which will then contribute to the direct translation of the research findings into clinical practice (Selinger and Kellner, 2002).

ACKNOWLEDGEMENTS

We wish to thank NMRC (grant # 0594/2001 to MCMC) and the Lee Hiok Kwee Fund (MCMC, RCMYL, JCHN, SLL and GST) for financial assistance in this project. Earlier work was supported by and performed in the Proteomics Laboratory of the Bioprocessing Technology Centre (BTC).

REFERENCES

Beasley, R.P., Hwang, L.Y., Lin, C.C., and Chien, C.S., 1981, Hepatocellular carcinoma and hepatitis B virus: a prospective study of 22707 men in Taiwan. *Lancet*, **2:** 1129 – 1133.
Chen, X., Cheung, S.T., So, S., Fan, S.T., Barry, C., Higgins, J., Lai, K.-M., Ji, J., Dudoit, S., Ng, I.O.L., van de Rijn, M., Botstein, D., and Brown , P.O., 2002, Gene expression patterns in human liver cancers. *Mol. Biol. Cell*, **13:** 1929 – 1939.
Cho, S.Y., Park, K.-S., Shim, J.E., Kwon, M.-S., Joo, K.H., Lee, W.S., Chang, J., Kim, H., Chung, H.C., Kim, H.O., Paik, Y.-K., 2002, An integrated proteome database for two-dimensional electrophoresis data analysis and laboratory information management system. *Proteomics*, **2:** 1104 – 1113.
Choong, M.L., Tan, L.K., Lo, S.L., Ren, E.C., Ou, K., Ong, S.-E., Liang, R.C.M.Y., Seow, T.K., and Chung, M.C.M., 2001, An integrated approach in the discovery and characterization of a novel nuclear protein over-expressed in liver and pancreatic tumours. *FEBS Letts.*, **496:** 109 – 116.
Delpuech, O., Trabut, J.-P., Carnot, F., Feuillard, J., Brechot, C., and Kremsdorf, D., 2002, Identification, using cDNA macroarray analysis, of distinct gene expression profiles associated with pathological and virological features of hepatocellular carcinoma. *Oncogene*, **21:** 2926 – 2937.
Deuffic, S., Buffat, L., Poynard, T., and Valleron, A.J., 1999, Modeling the hepatitis C virus epidemic in France. *Hepatology,* **29:** 1596 – 1601.

Dreger, M., 2003, Proteome analysis at the level of subcellular structures. *Eur. J. Biochem.*, **270:** 589 – 599.

Du, M.Q., Hutchinson, W.L., Johnson, P.J., and Williams, R., 1991, Differential α-fetoprotein lectin binding in hepatocellular carcinoma. *Cancer*, **61:** 476 – 480.

El-Serag, H.B. and Mason, A.C., 1999, Rising incidence of hepatocellular carcinoma in the United States. *N. England J. Med.*, **340:** 745 – 750.

Feitelson, M.A., Sun, B., Tufan, N.L.G., Liu, J., and Lian, Z., 2002, Genetic mechanisms of hepatocarcinogenesis. *Oncogene*, **21:** 2593 – 2604.

Fukuda, S., Wu, D.W., Stark, K., and Pelus, L.M., 2002, Cloning and characterization of a proliferation-associated cytokine-inducible protein, CIP29. *Biochem. Biophys. Res. Commun.*, **292:** 593 – 600.

Goldenberg, D., Ayesh, S., Schneider, T., Pappo, O., Jurim, O., Eid, A., Fellig, Y., Dadon, T., Ariel, I., de Groot, N., Hochberg, A., and Galun, E., 2002, Analysis of differentially expressed genes in hepatocellular carcinoma using cDNA arrays. *Molec. Carcinog.*, **33:** 113 – 124.

Gray, S.G., 2001, Proteomic payoff. *Trends in Genetics*, **17:** 382 – 382.

Gygi, S.P., Rochon, Y., Franza, B.R., and Aebersold, R., 1999a, Correlation between protein and mRNA abundance in yeast. *Mol. Cell. Biol.*, **19:** 1720 – 1730.

Gygi, S.P., Rist, B., Gerber, S.A., Turecek, F., Gelb, M.H., and Aebersold, R., 1999b, Quantitative analysis of complex protein mixtures using isotope-coded affinity tags. *Nature Biotechnol.*, **17:** 994 – 999.

Ho, S., Cheng, P., Chan, A., Leung, N., Yeo, W., Leung, T., Lau, W.Y., Li, A.K.C. and Johnson, P.J., 1996, Isoelectric focusing of alpha-fetoprotein in patients with hepatocellular carcinoma – frequency of specific banding patterns at non-diagnostic serum levels. *Brit. J. Cancer*, **73:** 985 – 988.

Ideker, T., Thorsson, V., Ranish, J.A., Christmas, R., Buhler, J., Eng, J.K., Bumgarner, R., Goodlett, D.R., Aebersold, R., and Hood, L., 2001, Integrated genomic and proteomic analyses of a systematically perturbed metabolic network. *Science*, **292:** 929 – 934.

Johnson, P.J., Leung, N., Cheng, P., Welby, C., Leung, T., Lau, W.Y., Yu, S., and Ho, S., 1997, 'Hepatoma-specific' alpha-fetoprotein may perrmit preclinical diagnosis of malignant change in patients with chronic liver disease. *Brit. J. Cancer*, **75:** 236 – 242.

Kao, J.-H. and Chen, D.-S., 2002, Recent updates in hepatitis vaccination and the prevention of hepatocellular carcinoma. *Int. J. Cancer*, **97:** 269 – 271.

Kim, J., Kim, SA.H., Lee, S.U., Ha, G.H., Kang, D.G., Ha, N.-H., Ahn, J.S., Cho, H.Y., Kang, S.-J., Lee, W.-J., Hong, H.C., Ha, W.S., Bae, J.M., Lee, C.-W., and Kim, J.W., 2002, Proteome analysis of human liver tissue by two-dimensional gel electrophoresis and matrix assisted laser desorption/ionization mass spectrometry for identification of disease-related proteins. *Electrophoresis*, **23:** 4142 – 4156.

Kim, M.-Y., Park, E., Park, J.-H., Park, D.-H., Moon, W.-S., Cho, B.-H., Shin, H.-S., and Kim, D.-G., 2001, Expression profile of nine novel genes differentially expressed in hepattis B virus-associated hepatocellular carcinomas. *Oncogene*, **20:** 4568 – 4575.

Lau, W.-Y., Lai, P.B.S., Leung, M.-F., Leung, B.C.S., Wong, N., Chen, G., Leung, T.W.T., and Liew, C.-T., 2000, Differential gene expression of hepatocellular carcinoma using cDNA microarray analysis. *Oncology Res.*, **12:** 59 – 69.

Le Naour, F., Brichory, F.B., Misek, D.E., Brechoy, C., Hanash, S.M., and Beretta, L., 2002, A distinct repertoire of autoantibodies in hepatocellular carcinoma identified by proteomic analysis. *Mol.Cell.Proteomics*, **1.3:** 197 – 203.

Li, D., Mallory, T., and Satomura, S., 2001, ALF-L3: a new generation of tumour marker for hepatocellular carcinoma. *Clin. Chim. Acta*, **313:** 15 – 19.

Liang, R.C.M.Y., Neo, J.C.H., Lo, S.L., Tan, G.S., Seow, T.K., and Chung, M.C.M., 2002, Proteome database of hepatocellular carcinoma. *J. Chrom. B*, **771**: 303 – 328.

Lim, S,O., Park, S.-J., Kim, W., Park, S.G., Kim, H.-J., Kim, Y.-I., Sohn, T.-S., Noh, J.-H., and Jung, G., 2002, Proteome analysis of hepatocellular carcinoma. *Biochem. Biophys. Res. Comm.*, **291**: 1031 – 1037.

Llovet, J.M. and Beaugrand, M., 2003, Hepatocellular carcinoma : present status and future prospects. *J. Hepatology*, **38**: S136 – S149.

McGlynn, K.A., Tsao, L., Hsing, A.W., Devesa, S.S., and Fraumeni, Jr., J.F., 2001, International trends and patterns of primary liver cancer. *Int. J. Cancer*, **94**: 290 – 296.

Naraki, T., Kohno, N., Saito, H., Fujimoto, Y., Ohhira, M., Morita, T. and Kohgo, Y., 2002, γ-Carboxyglutamic acid content of hepatocellular carcinoma-associated des-γ-carboxy prothrombin. *Biochim. Biophys. Acta*, **1586**: 287 – 298.

Okabe, H., Satoh, S., Kato, T., Kitahara, O., Yanagawa, R., Yamaoka, Y., Tsunoda, T., Furukawa, Y., and Nakamura, Y., 2001, Genome-wide analysis of gene expression in human hepatocellular carcinomas using cDNA microarray: identification of genes involved in viral carcinogenesis and tumour progression. *Cancer Res.*, **61**: 2129 – 2137.

Old, L.J., and Chen, Y.T., 1998, New paths in human cancer serology. *J. Exp. Med.*, **187**: 1163 – 1167.

Ou, K., Seow, T.K., Liang, R.C.M.Y., Ong, S.-E., and Chung, M.C.M., 2001, Proteome analysis of a human hepatocellular carcinoma cell line, HCC-M: An update. *Electrophoresis*, **22**: 2804 – 2811.

Park, K.-S., Cho, S.-Y., Kim, H., and Paik, Y.-K., 2002, Proteomic alterations of the variants of human aldehyde dehydrogenase isozymes correlate with hepatocellular carcinoma. *Int. J. Cancer*, **97**: 261 – 265.

Parkin, D.M., Bray, F., Ferlay, J., and Pisani, P., 2001, Estimating the world cancer burden: Globocan 2000. *Int J. Cancer*, **94**: 153 – 156.

Poon, T.C.W., and Johnson, P.J., 2001, Proteome analysis and its impact on the discovery of serological tumour markers. *Clin. Chim. Acta*, **313**: 231 – 239.

Poon, T.C.W., Mok, T.S.K., Chan, A.T.C., Chan, C.M.L., Leong, V., Tsui, S.H.T., Leung, T.W.T., Wong, H.T.M., Ho, S.K.W. and Johnson, P.J., 2002, Quantitation and utility of monosialylated α-fetoprotein in the diagnosis of hepatocellular carcinoma with nondiagnostic serum total α-fetoprotein. *Clin. Chemistry*, **48**: 1021 – 1027.

Sahin, U., Tureci, O., Schmitt,H., Cochlovius, B., Johannes, T., Schmits, R., Stenner, F., Luo, G., Schobert, I., and Pfreundschuh., 1995, Human neoplasms elicit multiple specific immune responses in the autologous host. *Proc. Natl. Acad. Sci.*, **92**: 1180 -1183.

Saitoh, S., Ikeda, K., Koida, I., Tsubota, A., Arase, Y., Chayama, K. and Kumada, H., 1994, Serum des-gamma-carboxy prothrombin concentration determination by avidin-biotin complex in small hepatocellular carcinomas. *Cancer*, **74**: 2918 – 2923.

Schafer, D.F., and Sorrell, M.F., 1999, Hepatocellular carcinoma. *Lancet*, **353**: 1253 – 1257.

Selinger, B., and Kellner, R., 2002, Design of proteome–based studies in combination with serology for the identification of biomarkers and novel targets. *Proteomics*, **2**: 1641 – 1651.

Seow, T.K., Ong, S.-E., Liang, R.C.M.Y., Ren, E.-C., Chan, L., Ou, K., and Chung, M.C.M., 2000, Two-dimensional electrophoresis map of the human hepatocellular carcinoma cell libe, HCC-M, and identification of the separated proteins by mass spectrometry. *Electrophoresis*, **21**: 1787 – 1813.

Seow, T.K., Liang, R.C.M.Y., Leow, C.K., and Chung, M.C.M., Hepatocellular carcinoma: from bedside to proteomics. *Proteomics*, **1**: 1249 – 1263.

Shirota, Y., Kaneko, S., Honda, M., Kawai, H.F., and Kobayashi, K.,2001, Identification of differentially expressed genes in hepatocellular carcinoma with cDNA microarrays. *Hepatology*, **33:** 832 – 840.

Simpson, R.J., and Dorow, D.S., 2001, Cancer proteomics: from signalling networks to tumour markers. *Trends in Biotechnol*, **19:** S40 –S48.

Steel, L.F., Mattu, T.S., Mehta, A., Hebestreit, H., Dwek, R., Evans, A.A., London, W.L., and Block, T., 2001, A proteomic approach for the discovery of early detection markers of hepatocellular carcinoma. *Dis. Markers*, **17:** 179 – 189.

Stenner-Liewen, F., Luo, G., Sahin, U., Tureci, O., Koslovski, I., Kautz, I., Liewen, H., and Pfreundschuh, M., 2000, Definition of tumour-associated antigens in hepatocellular carcinoma. *Cancer Epidemiol. Biomarkers Prev.*, **9:** 285 – 290.

Tackels-Horne, D., Goodman, M.D., Williams, A.J., Wilson, D.J., Eskandari, T., Vogt, L.M., Boland, J.F., Scherf, U., and Vockley, J.G., 2001, Identification of differentially expressed genes in hepatocellular carcinoma and metastatic liver tumors by oligonucleotide expression profiling. *Cancer*, **92:** 395 – 405.

Thorgeirsson S.S. and Grisham, J.W. , 2002, Molecular pathogenesis of human hepatocellular carcinoma. *Nature Genetics*, **31:** 339 – 346.

Tonge, R., Shaw, J., Middleton, B., Rowlinson, R., Rayner, S., Young, J., Pognan, F., Hawkins, E., Currie, I., and Davison, M. 2001, Validation and development of fluorescent two-dimensional differential gel electrophoresis proteomics technology. *Proteomics*, **1:** 377 – 396.

Unlu, M., Morgan, M.E., and Minden, J.S., 1997, Difference gel electrophoresis : a single gel method for detecting changes in protein extracts. *Electrophoresis*, **18:** 2071 – 2077.

Wang, Y., Han, K.-J., Pang, X.-W., Vaughan, H.A., Qu, W., Dong, X.-Y., Peng, J.-R., Zhao, H.-T., Rui, J.-A., Leng, X.-S., Cebon, J., Burgess, A.W., and Chen, W.-F., 2002, Large scale identification of human hepatocellular carcinoma-associated antigens by autoantibodies. *J. Immunol.*, **169:** 1102 – 1109.

Watanabe, T., Morizane, T., Tsuchimoto, K., Inagaki, Y., Munakata, Y., Nakamura, T., Kumagai, N., and Tsuchiya, M., 1983, Establishment of a cell line (HCC-M) from a human hepatocellular carcinoma. *Int.J. Cancer*, **32:** 141 – 146.

Wirth, P.J., Hoang, T.N., and Benjamin, T., 1995, Micropreparative immobilized pH gradient two-dimensional electrophoresis in combination with protein microsequencing for the analysis of human liver proteins. *Electrophoresis*, **16:** 1946 – 1960.

Xu, X.-R., Huang, J., Xu, Z.-G., Qian, B.-Z., Zhu, Z.-D., Yan, Q., Cai, T., Zhang, X., Xiao, H.-S., Qu, J., Liu, F., Huang, Q.-H., Cheng, Z.-H., Li, N.-G., Du, J.-J., Hu, W., Shen, K.-T., Lu, G., Zhong, M., Xu, S.-H., Gu, W.-Y., Huang, W., Zhao, X.T., HU, G.-X., Gu, J.-R., Chen, Z., and Han, Z.-G., 2001, Insight into hepatocellular carcinogenesis at transcriptome level by comparing gene expression profiles of hepatocellular carcinoma with those of corresponding noncancerous liver. *Proc. Natl. Acad. Sci.*, **98:** 15089 – 15094.

Yano, N., Habib, N.A., Fadden, K.J., Yamashita, H., Mitry, R., Jauregui, H., Kane, A., Endoh, M., and Rifai, A., 2001, Profiling of the human adult liver transcriptome : analysis by cDNA array hybridization. *J. Hepatology*, **35:** 178 – 186.

Yu, L.-R., Zeng, R., Shao, X.-X., Wang, N., Xu, Y.-H., Xia, Q.-C., 2000, Identification of differentially expressed proteins between human hepatoma and normal liver cell lines by two-dimensional electrophoresis and liquid chromatography-ion trap mass spectrometry. *Electrophoresis*, **21:** 3058-68.

Yu, L.-R., Shao, X.-X., Jiang, W.-L., Xu, D., Chang, Y.-C., Xu, Y.-H., and Xia, Q.-C., 2001, Proteome alterations in human hepatoma cells transfected with antisense epidermal growth factor receptor sequence. *Electrophoresis*, **22:** 3001 – 3008.

Chapter 8

Proteomics and Peptidomics of Gestational Disease

NIGEL M. PAGE

School of Animal and Microbial Sciences, The University of Reading, Reading, RG6 6AJ, UK

1.   INTRODUCTION

Prenatal screening has become an important and routine part of modern obstetric care and the necessity for developing novel diagnostic markers remains a challenging one. The demand for prenatal screening is continually growing and it is estimated that approximately two thirds of pregnant women in the United States are having placental marker serum screening for conditions such as Down's syndrome (DS) (Palomaki *et al.*, 1997). Indeed, the first test a women will encounter to confirm her pregnancy is based on the detection of the α and β polypeptide subunits of human chorionic gonadotrophin (hCG) in her urine, hCG being the most established of the placental peptide markers. The advantages of this test are that not only is hCG secreted by the placenta and specific to pregnancy (except in some forms of cancer such as choriocarcinoma and testicular cancer) but levels of hCG rise exponentially in the first few weeks. Unfortunately, such clear-cut diagnosis is not at present always possible for gestational disease, as placental markers are not necessarily exclusive to the condition being screened for; rather they simply enable an estimation of risk.

This chapter introduces the major gestational diseases in respect to the development of prenatal screening describing the developments from the preproteomics era to current research and to the future role of proteomics in the discovery of novel placental peptide markers. Here, the additional proteomic challenges that are needed to accommodate the study of peptides, being that, by their nature they are smaller and consequently have different

183

H. Hondermarck (ed.), Proteomics: Biomedical and Pharmaceutical Applications, 183–199.

physical chemistry properties from those of proteins are addressed. Hence, the developing field of peptidomics is introduced. It is thus hoped that the application of peptidomics will lead to the design of robust, rapid and clinic friendly diagnostic assays.

2.   THE PLACENTA AND GESTATIONAL DISEASE

The development of the foetus within the maternal uterus is linked anatomically, endocrinologically and metabolically with that of the placenta. This temporary structure with its forty week life span is unique to pregnancy taking on the role of a multiple organ system. It is derived from the outer trophectoderm cells of the blastocyst, which establish physical contact with the maternal uterine endometrium at implantation giving rise to the trophoblast. It is the proliferating cells of the trophoblast that invade and destroy adjacent decidua cells of the endometrium. A process that leads to the general anatomical layout of the placenta being complete within the first three to four weeks of pregnancy. A full blood flow is established at around 10 to 12 weeks - a period that represents a change in the placental environment from relative hypoxia to an increase in oxygen tension (Jaffe *et al.* 1997). Optimal placental perfusion requires the controlled invasion of the trophoblast cells deep into the myometrial spiral arteries; whereby the narrow walls of the spiral arteries are replaced with the trophoblasts, rendering them flaccid and distended, and resulting in an increase of blood flow to the placenta. In the mature placenta, fingers of villi derived from the trophoblast dip into maternal blood spaces providing a large surface area for exchange. This provides nourishment and oxygen for the developing foetus and placenta and allows an ideal place for proteins/peptides to be secreted across the foetal-maternal boundary. Pathologies of the placenta and factors affecting its development play a key role in establishing gestational disease. We, therefore, take a look at the pathological diversities within the major gestational diseases.

1.1   Foetal growth retardation

Foetal growth retardation (FGR) is defined as a predicted foetal weight at term that is under the 10th percentile or below 2.5 kg. It is linked to significantly elevated foetal morbidity and mortality (McCormick, 1985), with the most common causes cited as chronic hypertension, pre-eclampsia, smoking, alcohol, stress and intrauterine infections. It is believed that these conditions cause abnormalities in placental structure and function (Regnault *et al.*, 2002) interfering with implantation, placentation, placental metabolism and transport (Pardi *et al.*, 2002). A plethora of growth factors and hormones such as insulin-like growth factors (IGF), fibroblast growth factors (FGF), epidermal growth factors (EGF), transforming growth factors (TGF) and platelet derived growth factors (PDGF) have all been implicated. Many of which exhibit altered expression within the placenta of FGR pregnancies.

2.1 Pre-term labour

Pre-term birth is a major contributor to perinatal mortality and morbidity and in the USA approximately 11% of births occur before the 37th week (Andrews *et al.*, 2000). Moreover, the actual pre-term birth rate has increased over the last 30 years (Mauldin & Newman, 2001), although advances in neonatal care have resulted in a significant increase in successful outcomes. The regulatory pathways leading to parturition in humans are not well defined and appear to be multifactorial. These include the interleukins, endothelins, oxytocin, urocortin and various steroid hormones. Corticotrophin releasing factor (CRF) has been postulated to be the peptide that sets in motion the positive feedback loops that results in parturition (McLean & Smith, 1999), being released by the placenta in exponentially increasing amounts during pregnancy (Campbell *et al.*, 1987). Leading theories for pre-term birth include infections, inflammatory and ischaemic damage to the placenta (Mauldin & Newman, 2001). Aside from the major economic costs associated with caring for pre-term neonates, epidemiological evidence suggests that there is an increased risk of disease in later life including heart disease, type-2 diabetes and hypertension (Barker, 1989).

2.2 Pre-eclampsia

Pre-eclampsia (PE) affecting 3-10% of pregnancies, is a principal cause of maternal morbidity and mortality accounting for almost 15% of pregnancy-associated deaths and is a major cause of iatrogenic prematurity among new borns. Mild PE includes increases in maternal blood pressure and proteinuria (NIH, 2000) and can develop unpredictably into severe PE over a matter of days or weeks. There is a vast diversity of symptoms associated with severe PE including cerebral oedema, neurological manifestations, liver capsule distension, renal failure, pulmonary oedema and thrombocytopenia (Page, 2002). A defective trophoblastic invasion of the placental bed is thought to result in hypoperfusion and an ischaemic placenta, with the release of unknown factors into the maternal circulation. Markers observed in PE are the powerful vasoconstrictors (*e.g.* endothelins and angiotensins) and those that compensate for these hypertensive effects (*e.g.* atrial natriuretic factor, vasoactive intestinal polypeptide, adrenomedullin). There are also a number of non-vasoactive peptides implicated (*e.g.* leptin, β-hCG, inhibin-A).

2.3 Chromosomal Disorders

Genomic changes, such as aneuploidy, deletions and other chromosomal rearrangements, have long been associated with pregnancy loss and congenital abnormalities. Typically, these are errors in recombination, and therefore cannot often be predicted prior to fertilisation, though certain risk factors, such as increasing maternal age, indicate a greater risk. The most common chromosomal disorder is DS (trisomy 21), a congenital syndrome with a median birth incidence of 1 per 1000 births with rates increased to 33 per 1000 at the age of 45 years (Hook, 1981). The syndrome includes hypotonia, mental and growth retardation, heart defects and an increased incidence of leukemia and Alzheimers disease with associated high rates of infant mortality (Mikkelsen *et al.*, 1990). Other examples of chromosomal disorders include Patau syndrome (trisomy 13), Edwards syndrome (trisomy 18), Turner syndrome (X0), Klinefelter syndrome (XXY), Cri-du-chat syndrome (deleted 5p) and Prader-Willi syndrome (lack of the paternal copy of chromosome 15q). Abnormal serum levels of α-fetoprotein, hCG, inhibin-A and unconjugated oestriol are all associated with DS (Wald *et al.*, 1996).

2.4 Gestational Diabetes

Diabetes mellitus is a common complication of pregnancy affecting 2-15% of pregnancies, of which 90% are classified as gestational diabetes mellitus (Tamas & Kerenyi, 2001). Unlike women with type 1 diabetes, women with gestational diabetes have plenty of insulin. However, the effect of their insulin is partially blocked by a variety of hormones secreted by the placenta, such as oestrogen, cortisol and human placental lactogen. Insulin resistance usually begins about week 20 of pregnancy and increases with placental development. A major problem affiliated with gestational diabetes is macrosomia, a considerably larger than normal baby. This occurs when the transport of nutrients such as glucose across the *utero*-placental unit becomes unregulated resulting in high glucose levels in the foetal and maternal plasma. Leptin is also elevated in infants with type 1 diabetic and gestational diabetic mothers (Persson *et al.,* 1999), while IGF-I, IGF-II and other growth factors including the FGFs are increasingly expressed in such placental tissue (Arnay and Hill, 1998, Hill *et al.*, 1998).

2.5 Trophoblastic Disease

Trophoblastic disease includes gestational trophoblastic disease, a spectrum of rare neoplastic conditions, and gestational trophoblastic tumours (GTTs), the abnormal proliferation of different types of trophoblasts. These diseases vary from partial hydatidiform mole to choriocarcinoma. GTTs are always histologically choriocarcinoma and secrete the β-hCG more abundantly than normal. The serum or urinary level of this subunit is proportional to the tumour volume and represents a fundamental basis for follow-up of these placental tumours (Elegbe *et al.*, 1984). EGF expression is also found to be higher in molar placenta of all gestational ages, linking its role to the proliferative and differentiating activity of the trophoblast (John *et al.*, 1997).

3. OUTCOMES FROM THE PREPROTEOMICS ERA

The first characterisation of a peptide in pregnancy and its role in diagnosis was for hCG by Ascheim (1927). This utilised a bioassay that was designed to observe the enlargement and luteinisation of the corpus luteum of the mouse following injections of urine from pregnant women. Later bioassays applying agglutination methods helped to improve and simplify hCG testing, however, sensitivities in pregnancy diagnosis were not seen till the 1960s when immunoassays were developed (Yagami & Ito, 1965) and with the advent of monoclonal antibodies (Wahlstrom *et al.*, 1981). However, no test, actually identified a gestational abnormality using a protein or peptide until 1972, when it was first reported that maternal serum α-fetoprotein (AFP) levels were raised in foetuses affected by neural tube defect (Brock & Sutcliffe, 1972), a defect that can lead to spina bifida. AFP, a uniquely foetal protein produced by the liver was found to enter the maternal circulation with a linear relationship to week 20 of pregnancy, where higher than normal levels were found to indicate leaking protein from an open neural tube. A year later, raised maternal serum levels of AFP were found to be associated with anencephaly (Brock *et al.*, 1973) and in 1977 the first prenatal screening programme for birth defects was established (Wald *et al.*, 1977). In 1983, Merkatz *et al.* conducted a study of 53 pregnancies affected by chromosomal abnormalities. This followed the observation that a mother who gave birth to a child with trisomy 18 was found to have consistently low levels of AFP throughout pregnancy. They discovered that 43 of the affected cases had serum AFP levels below the mean value for unaffected pregnancies. A wider study taken from pregnancies affected with trisomy 21 found mean AFP values were approximately 25% lower than in unaffected pregnancies (Cuckle *et al.*, 1984). This lead to a method that by also taking into account the maternal age found to be an independent variable, was able to predict 35% of DS pregnancies with a 5% false-positive rate (Cuckle *et al.*, 1987). hCG levels were found to be about twice as high in DS pregnancies (Bogart *et al.*, 1987). This lead to the establishment of multiple screening tests to establish an estimation of risk. In the case of DS screening AFP, hCG and unconjugated oestriol levels (found to be lower in DS pregnancies) became known as the 'triple test' (Wald *et al.*, 1994) and later the 'quadruple test' with the inclusion of inhibin-A (Wald *et al.*, 1996). Much of the history of the preproteomic era has been dominated by similar examples of fortuitous placental peptide/protein isolation, the raising of specific antisera and the controlled matching and screening of placental extracts and maternal serum from both normal pregnancies and those affected by gestational disease. This process more often than not being driven by serendipity and by the availability of patient samples to individual investigators.

4.   PROTEOMICS TO DATE

It is now possible to abandon serendipity and utilise modern proteomic approaches to search for novel markers of pregnancy that may be useful in diagnosis and screening. However, the use of proteomic approaches has yet to see widespread acceptance and incorporation into the field of obstetrics.

To date there are only a handful of studies relating to gestational disease applying proteomics. An extensive PubMed search has revealed only four experimental papers so far published on the proteomics of the placenta of which only two relate to gestational disease and two reviews (Page *et al.*, 2002, Romero *et al.*, 2002). In the first paper Rabilloud *et al.* (2001) report the construction of a human mitochondrial proteome using placenta as the source material. They used two-dimensional (2-D) electrophoresis and peptide mass fingerprinting in an attempt to build a picture of this organelle's proteome. Their approach though not directly aimed at any particular disease could provide interesting reference data. Mitochondria are inherited through the mother and have been shown to play a role in many non-Mendelian inherited diseases (Cummins, 2002). Mitochondria are also an important source of oxidative stress and lipid peroxidation increased activities of which are associated with PE. Not only this, but the number of placental mitochondria are increased during PE (Wang & Walsh, 1998). Hence a fundamental understanding of the mitochondrial proteome could help address this organelle's role in PE and non-Mendelian inherited gestational diseases. In the second paper Ishimura *et al.* (2001) have prepared a database of 150 plasma membrane proteins using 2-D electrophoresis of the rat placenta, which are expressed in a stage specific manner. They have used peptide mapping, amino acid sequence analysis and mass spectrometry to determine stage specific differentiation modifications on G protein subunits in the placenta. In the future animal systems will provide useful proteomics data for the interpretation of human gestational disease, though it is clear from past human and animal studies that there are distinct differences not only between the structures of the placenta but in their endocrinological processes. Some gestational diseases such as PE are also unique to primates. In the third paper, representing the largest functional proteomics study by Hoang *et al.* (2001) the effects of hypoxia on the cytotrophoblast protein repertoire were examined in first trimester human cytotrophoblasts against those maintained under standard tissue culture conditions. It is the cytotrophoblast cells that differentiate and invade the maternal uterus, whereby forming vascular cells in a process determined by the increasingly higher levels of oxygen encountered within the uterine wall (Hoang *et al.*, 2001). The failure of these processes to occur can lead to a shallow uterine invasion that has been associated with both FGR and PE. Following 2-D electrophoresis, forty-three spots were identified for MS which identified differences in the abundance's of enzymes involved in glycolysis and those embracing responses to oxidative stress. Glycolysis enzymes being predicted to rise in response to the increased consumption of glucose during hypoxia. This study provides new information about the generalised mechanisms the cells use to respond to changes in oxygen tension at the maternal foetal interface (Hoang *et al.*, 2001). In the fourth paper, the proteomics of neurokinin B (NKB) was studied by the comparison of normal cytotrophoblast cell preparations with those treated with NKB (Sawicki *et al.*, 2002). Excessive secretion of placental NKB into the circulation during the third trimester of pregnancy is seen in women with PE and has been suggested as a cause for this disease (Page *et al.*, 2000). Their study was performed to determine the possible effects of NKB on normal placenta, their results indicating that this peptide has multiple actions on the trophoblast. These actions are consistent with NKB's role in suppressing normal antioxidant defences and suppressing proteins that block proinflammatory responses (Sawicki *et al.*, 2002). These are facets for some of the placental abnormalities seen in PE.

The development of bioinformatics (the unity of biology and computer science) should also be mentioned for its importance in providing the methods for placental proteome profiling. This area has become one of the fastest growing fields in biological research (reviewed by Luscombe *et al.*, 2001). The momentum of the Human Genome Project has provided the opportunity to gather and store vast arrays of valuable data. We have explored these databases with keywords as simple as 'placenta' or sequence motifs of proteins/peptides to identify partial and uncharacterised cDNA sequences such as those of expressed sequence tags or Tentative Human Consensus sequences (Boguski, 1995). 5' and 3' rapid amplification of cDNA ends of candidate placental genes is then performed to reveal the full-length cDNA sequence of each respective gene which is then translated to reveal and confirm secretory and functional motifs. By these methods we have identified several candidate diagnostic markers (Page *et al.*, 2000, Page *et al.*, 2001). In this manner, bioinformatics can provide a first step alternative to 2-D electrophoresis in the search for novel proteins/peptides. It can also provide a convenient method to establish the prediction of protein variants from encoding donor and acceptor splice sites determined from their genomic DNA regions. This allows proteins/peptides to be predicted that may not be yet annotated in the public databases. Subsequently, antibodies can be raised against such predicted sequences allowing enhanced proteomic analysis. Bioinformatics has therefore become an integral part of proteomics being used to predict novel proteins, for the analysis and interpretation of MS data, the storage, analysis and comparison of gel images and the prediction of interactions between proteins (Vihinen, 2001).

5.   PEPTIDOMICS TO STUDY GESTATIONAL DISEASE

Peptidomics provides many new challenges beyond the application of standard proteomics techniques in the characterisation of peptides (Jurgens & Schrader, 2002). It comes from the recognition that many of the techniques applied to the study of proteins cannot be employed so easily to the analysis of peptides. Whereas 2-D electrophoresis in combination with MS has become the corner stone of most proteomic research there are only a few approaches to date that describe and tackle the analysis of the peptidome. Here, are addressed some of the issues we are tackling in the development of our own peptidomic strategy for the discovery of novel placental peptide markers.

5.1 Placental peptides

Peptides, oppose to proteins, are typically considered those to be up to and around 200 amino acid residues in length, that is, those less than 20 kDa in molecular mass. Examples of different placental peptide sub-groups are the glycosylated polypeptides such as the non-covalently linked α and β chains of hCG, the single polypeptide chains such as placental lactogen (PL), the polypeptide cytokines such as TNF-α and the small peptides such as CRF and NKB. Surprisingly, the placenta has been found to produce many of the known peptides, which have been found to elicit a wide range of functions. Typically, these peptides are synthesised as precursors, which have to be processed by the cell to release the bioactive peptide or peptides in a highly regulated manner. They generally undergo post-translational modifications including acetylation, amidation, carboxylation, glycosylation, phosphorylation and sulphation. Intramolecular disulphide bonding either along the same chain or to link two or more chains is common. The precursor has a strongly hydrophobic signal peptide sequence at its NH_2 terminus, responsible for translocating it to the endoplasmic reticulum. Cleavage of this signal sequence in the endoplasmic reticulum forms the precursor, which is further cleaved, normally at dibasic amino acid residues (*e.g.* Arg-Arg, Lys-Arg) to yield the smaller peptide sequences. Prohormone convertases are responsible for these final cleavages which usually occur in the Golgi apparatus or in the secretory vesicles. The final secretion of the peptide occurs either by constitutive secretion where peptides are released immediately or by regulated secretion where they may be stored prior to release. Secreted peptides have not only been localised to the foetal side of the placenta, that is the syncytiotrophoblasts and cytiotrophoblasts of the villi, but to the membranes of the chorion and amnion, and to the maternal decidual cells and those of the endothelium. Such peptides are shown to have diverse effects including the maternal recognition and adaptation to pregnancy. Whereby, they play an important role in changing the female's reproductive system from a cyclic pattern to a pregnancy state, controlling trophoblast invasion, angiogenesis, growth, metabolism, immune function and cardiovascular responses.

5.2 Tissue/fluid sampling for gestational disease

Placental peptides secreted into the maternal and foetal circulation, amniotic fluid and mother's urine allow their potential detection from a number of different sites. Consequently, there is the need to establish the most suitable tissue/fluid to analyse and the best method for their extraction. Ultimately, non-invasive methods such as maternal venous blood or urine sampling will be advantageous over many of the existing prenatal screening techniques. These include amniocentesis, chorionic villus sampling, cordocentesis and foetal biopsy, which are all very invasive and convey a significantly higher risk of miscarriage (Scott *et al.*, 2002). However, definitive diagnosis for chromosomal disorders at present is gained only by invasive screening as it relies on the culture and karyotyping of cells of foetal origin. In the genomics/transcriptomics age, the placenta has had to play the key role for the source material for identifying novel placental markers (Page *et al.*, 2000). This will continue in the peptidomic age, where normal placenta can be collected from abortion clinics (legally up to week 24 in the UK) or at term (weeks 37 to 42) following obtaining the appropriate local ethical approval. Placenta from other time periods can only be obtained in the UK, if there is a substantial risk to the woman's life or if there are foetal abnormalities. However, peptidomics extends the repertoire of source material beyond that of placental tissue, as pathological changes in a gestational disease may also be detectable in the previously restricted extracellular fluids of the amniotic fluid, foetal and maternal blood and urine. However, collection of some extracellular fluids *e.g.* amniotic fluid and foetal cord blood is highly invasive. Amniotic fluid is normally only collected between week 15 to 16 of pregnancy with no more than 15 mls being taken. While earlier amniocentesis (before 14 weeks) is associated with significant problems, including increased foetal loss, foetal talipes and a reduced amniocyte culture rate (Nicolaides *et al.*, 1994). In the case of cordocentesis for foetal blood, routinely performed after week 18, there is a significant risk of miscarriage (Tongsong *et al.*, 2000). This makes direct peptidomic analysis of maternal blood along with maternal urine, the most convenient and low risk procedure for obtaining an accurate reflection of secreted placental markers in the maternal body fluids. Such sampling can be performed much earlier in pregnancy with the hope of reducing the amount of psychological anxiety and pathological trauma faced by the prospective parents. Neither could such direct maternal information be inferred accurately from placental tissue alone as it is estimated that such measurements would chiefly reflect the higher content of foetal extracellular fluid/tissue. Lin *et al.* (1976) estimated that a 400 g placenta would occupy a 312 ml volume, and would contain 144 ml (46%) of foetal blood, of which, only 36 ml (11.5%) would be derived from the mother.

5.3 Separation and analysis of peptides

In most proteomic studies, peptides are not well portrayed on 2-D electrophoresis gels and are represented by only a few faint spots with low staining capacity which can appear less focussed owing to their higher gel mobility. Moreover, when they are below 10 kDa in molecular mass even on high percentage polyacrylamide gels small peptides are extremely difficult to keep within the gel. For this reason liquid chromatography (LC) has been adopted as the principal separation technique for peptides with reverse phase and ion exchange being the preferred current adaptations. Samples are prepared normally by an initial peptide extraction. This is typically performed by acidifying the sample and conducting a solid phase extraction (*e.g.* using Sep-pak cartridges). In the case of plasma and sera samples additional steps may be taken to deplete the bulk of the abundant sera proteins such as albumin which can comprise 80% of the sample. Often these depletion processes include immunoaffinity (Kennedy, 2001), ultra-filtration (Schulz-Knappe *et al.*, 1997) or gel filtration chromatography (Schulz-Knappe *et al.*, 2001). Combinations of chromatographic procedures are used to reduce the complexity of the peptide array, separating either by hydrophobic interactions or by charge, or both. Such chromatographic procedures have led to the production of peptide banks containing complete spectra of fractions from human plasma obtained by ultra-filtration (Schulz-Knappe *et al.*, 1997). Ultra-filtration from plasma preserves the presence of naturally occurring peptides, in much the same manner as those collected from urine that are below the kidney cut-off size. Characterisation of these peptide banks is performed using MS such as matrix assisted laser desorption/ionisation time of flight MS (MALDI-TOF-MS) to generate mass databases based on peptide molecular masses (Richter *et al.*, 1999). Richter *et al.*, (1999) recorded approximately 5,000 different peptides with 95% of the detected masses smaller than 15 kDa from one such plasma peptide bank. Subsequently, MS-MS sequencing, with its much greater resolution power, revealed the amino acid sequence of many of these circulating peptides. The combined use of these techniques has been termed peptide trapping (Schulz-Knappe *et al.*, 2001). Multidimensional chromatography is also applicable to peptide separation. Such separation involves high pressure LC in the first dimension, followed by MS in the second dimension resulting in distinct high resolution peptide mass fingerprints. The MS data of all samples is then combined into one 2-D diagram, called a peptide display (Schulz-Knappe *et al.*, 2001). From samples volumes of less than 1 ml, more than 1,000 different peptides have been depicted on each peptide display with regard to their relative quantities, molecular masses and chromatographic elution behaviours (Schulz-Knappe *et al.*, 2001).

However, there remain important considerations in the MS analysis of peptides, as compared to those of proteins. These include the fact that the number of specific internal proteolytic (tryptic) sites are limiting and hence generally not available to produce specific peptide fragments. In this regard, MS analysis of proteins has relied heavily on the DNA sequence databases produced from both the genomic and the bioinformatic efforts to predict digested protein fragments. Peptides may be present in many different forms from their mature precursors, to partially processed precursors, to their fully cleaved peptides. They may also be present as genetic variants, splice variants, post-translationally modified or as degradation products. They may also contain previously unrecognised unique processing sites, that cannot yet be predicted. Much of this information is not annotated in the current databases.

6.   DEVELOPING DIAGNOSTIC TESTS

The criteria for any placental marker is that it must be 1) accurate with a good safety profile 2) can be used at the earliest possible period in gestation allowing for informed choice for pregnancy termination and 3) be rapid so as not to lead to unnecessary parental anxiety. In its development the following questions need to be addressed 1) is the marker unique to any one particular condition and 2) can a consensus be determined from a large longitudinal study into the merits of a particular marker? There are no studies to guide the use of placental markers derived from proteomics in the clinic. Clinical proteomics provides the opportunity to develop and utilise the next generation of placental diagnostic markers with the prospect of tailoring some of these towards therapeutic intervention. MS products identified as candidate diagnostic markers can be the intermediate or end products of several different processes. These include those of metabolomics (Fiehn, 2002), interactomics (Govorun & Archakov, 2002) and degradomics (Lopez-Otin & Overall, 2002). Metabolomics comprises the end products of cellular regulatory processes including those of the processing and the modification of proteins/peptides. Interactomics relates to the way different proteins and peptides interact. For example, IGF-I and IGF-II circulate in association with specific binding proteins (IGF-BPs), and their bio-availability during pregnancy depends on the proteolysis of their specific IGF-BPs. Proteins can also be cross-linked such as pregnancy-associated plasma protein-A (PAPP-A) and the proform of eosinophil major basic protein (Overgaard *et al.*, 2003). Degradomics applies to the identification of the substrates and products of protease interaction. Subsequently, there is a constant flux in the dynamics of protein turnover throughout pregnancy, an important and missing dimension in current proteomics. In the development of any diagnostic assay it is meaningful to assess fully the effects of these processes and whether they may cause interference. Otherwise, such a situation could lead to poor correlation in the detection assays employed. For instance, the occurrence of a variety of different molecular/modified forms of the same protein which are not differentiated in a diagnostic assay, the presence or absence of a masking binding protein or an association with another protein, or the presence or absence of circulating proteases could all have profound effects on the final measurement. Many of these functional proteomic outcomes/interactions can not be predicted from merely 2-D electrophoresis or MS. They can only be assessed during assay development and ultimately from the development of specific antibodies and from the outcomes of large longitudinal studies. Markers can also be temporary *e.g.* PAPP-A is a very good marker for DS in the first trimester (between weeks 10 to 14), but when measured in the second trimester the results are very similar to those measured for normal pregnancies (Berry *et al.*, 1997). The diagnostic window chosen is vital.

Methods based on saturation analysis using antibodies have dominated diagnostic assay development. The original concepts being based on the radioimmunoassay. However, these original procedures are impractical in the modern day clinic, being very time consuming and labour intensive. Modifications have included the use of two-site immunoassays based on non-isotopic labels including enzymic, chemiluminescent and fluorescent labels. The definitive diagnostic test will be one performed using a random access immunoassay system which can provide a high degree of automation and speed. The aim being to provide a service (one stop clinic) whereby a woman can be screened for biochemical markers while she attends her routine ultrasound scan. Her results being presented at the end of this session with direct access to immediate advice and counselling. One such immunoassay system is based on time-resolved-amplified-cryptate-emission (TRACE) which provides automated, precise and reproducible measurements within 30 minutes of obtaining a blood sample (Spencer *et al.,*. 1999). TRACE is based on a non-radioactive transfer of energy, that takes place between two fluorescent tracers: a donor, europium cryptate, and an acceptor, that are each bound to an antibody. This technique has already been developed for AFP, β-hCG and PAPP-A and for a range of fertility hormones including follicle stimulating hormone and luteinising hormone. Proteomics may also see direct application in the clinic, in a similar manner to those used to obtain serum proteomic patterns for the diagnosis of cancer (Petricoin & Liotta, 2002). While, microchips of peptides and peptidomimetic compounds may provide powerful tools in the future for high-throughput routine laboratory operations (Pellois *et al.*, 2002).

7.   CONCLUSIONS

Many of the currently used tests for gestational disease provide only an estimation of risk as many mothers and foetuses never actually have or will ever develop the disease. Hence, it is vital to develop new prenatal screening tests that are more reliable and specific. We believe peptide markers may be able to fill this niche, however, ideally they should be unique to the condition and specific to a stage of the disease. It is clear that while peptides are very promising candidates, there is still much to be learnt. At present, not all markers are unique to any one particular condition and no consensus has yet been reached in any study undertaken. The issues are complex and it is hoped that proteomics/peptidomics will be able to compliment the vast amount of knowledge already gained from genomic and bioinformatic studies. Whereby, we will learn fresh data about the way placental markers are expressed, processed, post-translationally modified, secreted and metabolised in each of the different gestational diseases. It is anticipated that this will pave the way forward for the identification of specific targets for the design of robust, rapid and clinic friendly diagnostic assays. And, finally that some of these markers may see applications in therapeutic intervention.

ACKNOWLEDGEMENTS

I would like to thank Nicola Bell who has helped in the final proof reading of this chapter.

REFERENCES

Andrews, W.W., Hauth, J.C., and Goldenberg, R.L., 2000, Infection and preterm birth. *Am. J. Perinatol.* **17:**357-365.

Arnay, E., and Hill, D.J., 1998, Fibroblast growth factor-2 and fibroblast growth factor receptor–1 mRNA expression and peptide localization in placenta from normal and diabetic pregnancies. *Placenta.* **19:**133-142.

Barker, D.J.P., 1989, Rise and fall of Western diseases. *Nature.* **338:**371-372.

Berry, E., Aitken, D.A., Crossley, J.A., Macri, J.N., and Connor, J.M., 1997, Screening for Down's syndrome: changes in marker levels and detection rates between first and second trimesters. *Br. J. Obstet. Gynaecol.* **104:**811-817.

Bogart, M.H., Pandian, M.R., and Jones, O.W., 1987, Abnormal maternal serum chorionic gonadotropin levels in pregnancies with fetal chromosome abnormalities. *Prenat. Diagn.* **7:**623-30.

Boguski, M.S., 1995, The turning point in genome research. *Trends in Biochemical Sciences.* **20:**295-6.

Brock, D.J., Bolton, A.E., and Monaghan, J.M., 1973, Prenatal diagnosis of anencephaly through maternal serum-alphafetoprotein measurement. *Lancet.* **2(7835):**923-924.

Brock, D.J., and Sutcliffe, R.G., 1972, Alpha-fetoprotein in the antenatal diagnosis of anencephaly and spina bifida. *Lancet.* **2(7770):**197-199.

Campbell, E.A., Linton, E.A., Wolfe, C.D., Scraggs, P.R., Jones, M.T., and Lowry, P.J., 1987, Plasma corticotrophin-releasing hormone concentrations during pregnancy and parturition. *Journal of Clinical Endocrinology and Metabolism.* **64:**1054-1059

Cuckle, H.S., Wald, N.J., and Lindenbaum, R.H., 1984, Maternal serum alpha-fetoprotein measurement: a screening test for Down syndrome. *Lancet.* **1(8383):**926-929.

Cuckle, H.S., Wald, N.J., and Thompson, S.G., 1987, Estimating a woman's risk of having a pregnancy associated with Down's syndrome using her age and serum alpha-fetoprotein level. *Br. J. Obstet. Gynaecol.* **94:**387-402.

Cummins, J.M., 2002, The role of maternal mitochondria during oogenesis, fertilization and embryogenesis. *Reprod. Biomed. Online.* **4:**176-182.

Elegbe, R.A., Pattillo, R.A., Hussa, R.O., Hoffmann, R.G., Damole, I.O., and Finlayson, W.E., 1984, Alpha subunit and human chorionic gonadotropin in normal pregnancy and gestational trophoblastic disease. *Obstetrics and Gynecology.* **63:**335-337.

Fiehn, O., 2002, Metabolomics - the link between genotypes and phenotypes. *Plant Mol. Biol.* **48:**155-171.

Govorun, V.M., Archakov, A.I., 2002, Proteomic technologies in modern biomedical science. *Biochemistry (Mosc).* **67:**1109-1123.

Hill, D.J., Petrik, J., and Arany, E., 1998, Growth factors and the regulation of fetal growth. *Diabetes Care.* **21(Suppl 2):**B60-69.

Hoang, V.M., Foulk, R., Clauser, K., Burlingame, A., Gibson, B.W., and Fisher, S.J., 2001, Functional proteomics: examining the effects of hypoxia on the cytotrophoblast protein repertoire. *Biochemistry.* **40:**4077-4086.

Hook, E.B., 1981, Rates of chromosome abnormalities at different maternal ages. *Obstet. Gynecol.* **58:**282-285.

Ishimura, R., Yoshida, K., Kimura, H., Dohmae, N., Takio, K., Ogawa, T., Tanaka, S., and Shiota, K., 2001, Stage-specific modification of G protein beta subunits in rat placenta. *Mol. Cell. Endocrinol.* **174:**77-89.

Jaffe, R., Jauniaux, E., and Hustin, J., 1997, Maternal circulation in the first-trimester human placenta--myth or reality? *Am. J. Obstet. Gynecol.* **176:**695-705.

John, M., Rajalekshmy, T.N., Nair, M.B., Augustine, J., Schultz, G., Nair, M.K., and Balaram, P., 1997, Expression of epidermal growth factor in gestational trophoblastic disease (GTD). *Journal of Experimental Clinical Cancer Research* **16:**129-134.

Jurgens, M., and Schrader, M., 2002, Peptidomic approaches in proteomic research. *Curr. Opin. Mol. Ther.* **4:**236-241.

Kennedy, S., 2001, Proteomic profiling from human samples: the body fluid alternative. *Toxicol. Lett.* **120:**379-384.

Lin, T.M., Halbert, S.P., and Kiefer, D., 1976, Quantitative analysis of pregnancy-associated plasma proteins in human placenta. *J. Clin. Invest.* **57:**466-472.

Lopez-Otin, C., and Overall, C.M. 2002, Protease degradomics: a new challenge for proteomics. Nat. Rev. *Mol. Cell. Biol.* **3:**509-519.

Luscombe, N.M., Greenbaum, D., and Gerstein, M., 2001, What is bioinformatics? A proposed definition and overview of the field. *Methods of Information in Medicine.* **40:** 346-358.

Mauldin, J.G., and Newman, R.B., 2001, Preterm birth risk assessment. *Semin. Perinatol.* **25:**215-222.

McCormick, M.C., 1985, The contribution of low birth weight to infant mortality and childhood morbidity. *New England Journal of Medicine.* **312:**82-90.

McLean, M., and Smith, R., 1999, Corticotropin-releasing hormone in human pregnancy and parturition *Trends in Endocrinology and Metabolism.* **10:**174-178.

Merkatz, I.R., Nitowsky, H.M., Macri, J.N., and Johnson, W.E., 1984, An association between low maternal serum alpha-fetoprotein and fetal chromosomal abnormalities. *Am. J. Obstet. Gynecol.* **148:**886-894.

Mikkelsen, M., Poulsen, H., and Nielsen, K.G., 1990, Incidence, survival, and mortality in Down syndrome in Denmark. *Am. J. Med. Genet. Suppl.* **7:**75-78.

National Institutes of Health, 2000, Working group report on high blood pressure in pregnancy. NIH publication, No. 00-3029 July.

Nicolaides, K., Brizot Mde, L., Patel, F., and Snijders, R., 1994, Comparison of chorionic villus sampling and amniocentesis for fetal karyotyping at 10-13 weeks' gestation. *Lancet.* **344(8920)**:435-439.

Overgaard, M.T., Sorensen, E.S., Stachowiak, D., Boldt, H.B., Kristensen, L., Sottrup-Jensen, L., and Oxvig, C., 2003, Complex of Pregnancy-associated Plasma Protein-A and the Proform of Eosinophil Major Basic Protein. *J. Biol. Chem.* **278**:2106-2117.

Palomaki, G.E., Knight, G.J., McCarthy, J.E., Haddow, J.E., and donhowe, J.M., 1997, Maternal serum screening for Down syndrome in the United States: a 1995 survey. *Am. J. Obstet. Gynecol.* **176**:1046-1051.

Page, N.M., 2002, The endocrinology of pre-eclampsia. *Clin. Endocrinol. (Oxf).* **57**:413-23.

Page, N.M., Butlin, D.J., Lomthaisong, K., and Lowry, P.J., 2001, The characterization of pregnancy associated plasma protein-E and the identification of an alternative splice variant. *Placenta.* **22**:681-687.

Page, N., Butlin, D., Manyonda, I., and Lowry, P., 2000, The development of a genetic profile of placental gene expression during the first trimester of pregnancy: a potential tool for identifying novel secreted markers. *Fetal Diagn Ther.* **15**:237-245.

Page, N.M., Woods, R.J., Gardiner, S.M., Lomthaisong, K., Gladwell, R.T., Butlin, D.J., Manyonda, I.T., and Lowry, P.J., 2000, Excessive placental secretion of neurokinin B during the third trimester causes pre-eclampsia. *Nature.* **405**:797-800.

Page, N.M., Kemp, C.F., Butlin, D.J., and Lowry, P.J., 2002, Placental peptides as markers of gestational disease. *Reproduction.* **123**:487-495.

Pardi, G., Marconi, A.M., Cetin, I., 2002, Placental-fetal interrelationship in IUGR fetuses - a review. *Placenta.* **23SupplA**:S136-141.

Pellois, J.P., Zhou, X., Srivannavit, O., Zhou, T., Gulari, E., and Gao, X., 2002, Individually addressable parallel peptide synthesis on microchips. *Nat. Biotechnol.* **20**:922-6.

Persson, B., Westgren, M., Celsi, G., Nord, E., and Ortqvist, E., 1999, Leptin concentrations in cord blood in normal newborn infants and offspring of diabetic mothers. *Hormone Metabolism Research* **31**:467-771.

Petricoin, E.F., and Liotta, L.A., 2002, Proteomic analysis at the bedside: early detection of cancer. *Trends Biotechnol.* **20(12 Suppl)**:S30-34.

Rabilloud, T., Kieffer, S., Procaccio, V., Louwagie, M., Courchesne, P.L., Patterson, S.D., Martinez, P., Garin, J., and Lunardi, J., 1998, Two-dimensional electrophoresis of human placental mitochondria and protein identification by mass spectrometry: toward a human mitochondrial proteome. *Electrophoresis.* **19**:1006-1014.

Regnault, T.R., Galan, H.L., Parker, T.A., and Anthony, R.V., 2002, Placental development in normal and compromised pregnancies - a review. *Placenta.* **23SupplA**:S119-129.

Richter, R., Schulz-Knappe, P., Schrader, M., Standker, L., Jurgens, M., Tammen, H., and Forssmann, W.G., 1999, Composition of the peptide fraction in human blood plasma: database of circulating human peptides. *J. Chromatogr. B Biomed. Sci. Appl.* **726**:25-35.

Romero, R., Kuivaniemi, H., and Tromp, G., 2002, Functional genomics and proteomics in term and preterm parturition. *J. Clin. Endocrinol. Metab.* **87**:2431-2434.

Sawicki, G., Dakour, J., and Morrish, D.W., 2002, Proteomics of neurokinin B in the placenta demonstrates novel actions in suppressing placental antioxidant defenses. Proceedings of the Swiss Proteomics Society. In Congress Applied Proteomics. Editors: (Palagi, P.M., Quadroni, M., Rossier, J.S. Sanchez, J.C., and Stocklin, R.), FontisMedia SA, Lausanne, Switzerland, pp.102-104.

Schulz-Knappe, P., Schrader, M., Standker, L., Richter, R., Hess, R., Jurgens, M., and Forssmann, W.G., 1997, Peptide bank generated by large-scale preparation of circulating human peptides. *J. Chromatogr. A.* **776**:125-132.

Schulz-Knappe, P., Zucht, H.D., Heine, G., Jurgens, M., Hess, R., and Schrader, M., 2001, Peptidomics: the comprehensive analysis of peptides in complex biological mixtures. *Comb. Chem. High. Throughput Screen.* **4**:207-217.

Scott, F., Peters, H., Boogert, T., Robertson, R., Anderson, J., McLennan, A., Kesby, G., and Edelman, D., 2002, The loss rates for invasive prenatal testing in a specialised obstetric ultrasound practice. *Aust. N. Z. J. Obstet. Gynaecol.* **42**:55-58.

Spencer, K., Souter, V., Tul, N., Snijders, R., and Nicolaides, K.H., 1999, A screening program for trisomy 21 at 10-14 weeks using fetal nuchal translucency, maternal serum free beta-human chorionic gonadotropin and pregnancy-associated plasma protein-A. *Ultrasound Obstet. Gynecol.* **13**:231-237.

Tamas, G., and Kerenyi, Z., 2001, Gestational diabetes: current aspects on pathogenesis and treatment. Exp. Clin. Endocrinol. *Diabetes.* **109Suppl2**:S400-411.

Tongsong, T., Wanapirak, C., Kunavikatikul, C., Sirirchotiyakul, S., Piyamongkol, and W., Chanprapaph, P., 2000, Cordocentesis at 16-24 weeks of gestation: experience of 1,320 cases. *Prenat. Diagn.* **20**:224-228.

Vihinen, M., 2001, Bioinformatics in proteomics. *Biomol. Eng.* **18**:241-248.

Wahlstrom, T., Stenman, U.H., Lundqvist, C., Tanner, P., Schroder, J., and Seppala, M., 1981, The use of monoclonal antibodies against human chorionic gonadotropin for immunoperoxidase staining of normal placenta, pituitary gland, and pituitary adenomas. *J. Histochem. Cytochem.* **29**:864-865.

Wald, N.J., Cuckle, H., Brock, J.H., Peto, R., Polani, P.E., and Woodford, F.P., 1977, Maternal serum-alpha-fetoprotein measurement in antenatal screening for anencephaly and spina bifida in early pregnancy. Report of U.K. collaborative study on alpha-fetoprotein in relation to neural-tube defects. *Lancet.* **1(8026)**:1323-1332.

Wald, N.J., Densem, J.W., Smith, D., , G.G., 1994, Four-marker serum screening for Down's syndrome. *Prenat. Diagn.* **14**:707-716.

Wald, N.J., Densem, J.W., George, L., Muttukrishna, K., and Knight, P.G., 1996, Prenatal screening for Down's Syndrome using inhibin-A as a serum marker. *Prenatal Diagnosis.* **16**:143-153

Wang, Y., and Walsh, S.W., 1998, Placental mitochondria as a source of oxidative stress in pre-eclampsia. *Placenta.* **19**:581-586.

Yagami, Y., and Ito, Y., 1965, Immunological assay of human serum chorionic gonadotrophin in normal and abnormal pregnancy. *J. Jpn. Obstet. Gynecol. Soc.* **12**:82-88.

Chapter 9

Applications of Proteomics Technologies for Drug Discovery

KEWAL K. JAIN
Jain PharmaBiotech, Blaesiring 7, CH-4057 Basel, Switzerland

1. INTRODUCTION

Historically, pharmaceutical products have been developed primarily through the random testing of thousands of synthetic compounds and natural products. The traditional approach to drug discovery is based on generation of a hypothesis based on biochemistry and pharmacological approach to a disease. Targets are defined on the basis of this hypothesis and lead discovery is a matter of chance. The classical drug-discovery effort also involves isolating and characterizing natural products with some biological activity. These compounds are then 'refined' by redesigning their molecular structure to yield new entities with higher biological activity and lower toxicity/side effects. The main limitation of such a process is that the discovery of natural products with defined biological activity is essentially a hit-or-miss approach and therefore lacks a rational basis.

The first major advance in drug discovery was combinatorial chemistry, which involves using high-throughput technologies for preparing large numbers of compounds for use in screening against a variety of biological targets. In an effort to overcome some of the difficulties associated with traditional drug discovery, genomics was used as a means to improve our understanding of disease. It was hoped that a comprehensive knowledge of an organism's genetic makeup would lead to more efficient drug discovery. Although useful, DNA sequence analysis alone does not lead efficiently to new target identification, because one cannot easily infer the functions of gene products, or proteins, and protein pathways from DNA sequence. Major stages in drug discovery are listed in Table 1.

H. Hondermarck (ed.), Proteomics: Biomedical and Pharmaceutical Applications, 201–227.
© 2004 *Kluwer Academic Publishers. Printed in the Netherlands.*

Table 1. Major stages in preclinical drug discovery and informatics components

Identification of novel drug targets from protein sequence databases
Characterization of biological functions of target proteins
Evaluation of the therapeutic relevance of target protein functions
Identification of ligands by searching of protein–ligand interaction
Development of in vitro assays
Screening for active compounds (hits)
Transformation of hits into leads
Optimization of leads by target 3D structure- and/or lead-based drug design
Assessment of optimized leads in models
Selection of preclinical candidates

Drug discovery is a lengthy and expensive process with shortage of promising drug leads. Functional genomics and proteomics have provided a huge amount of new drug targets. High-throughput screening and compound libraries produced by combinatorial chemistry have increased the number of new lead compounds. The challenge now is to increase the efficiency of testing lead efficacy and toxicity. In practice, this is not easy because an infinite number of genes, proteins and other molecules interact with each other in signaling pathways to direct cell function. With advances in proteomic technologies, there is an increasing interest in the application of these to improve the drug discovery process. Because most of the drugs act on proteins, it is important to focus drug discovery efforts at this level.

2.   PROTEINS AND DRUG ACTION

Many of the pharmaceutically important regulation systems operate through proteins (i.e., post-translationally). Major drugs act by binding to proteins. For example a "protease inhibitor" drug is designed to disable the protease enzyme (which is a protein) that allows a particular virus to reproduce. A drug with the right shape can latch onto the surface of the protease protein, and keep it from doing its job. If the protease is disabled, the virus can't reproduce itself, so the damage it can inflict is limited. To find useful compounds like a protease inhibitor for a particular virus, scientists need to be able to understand the shape and the function of both the compound itself and the protein it affects.

The majority of drug targets are proteins that are encoded by genes expressed within tissues affected by a disease. It is estimated that there are approximately 10,000 different enzymes, 2,000 different G-protein-coupled receptors, 200 different ion channels, and 100 different nuclear hormone receptors encoded in the human genome. These proteins are key components of the pathways involved in disease and, therefore, are likely to be a rich source of new drug targets. Proven drug targets share certain other characteristics which can only be identified by understanding their expression levels in cells and cannot be determined by their gene sequences alone. Drug targets are (1) often expressed primarily in specific tissues, allowing for selectivity of pharmacological action and reducing the potential for adverse side effects and (2) generally expressed at low abundance in the cells of the relevant organ. An effective target discovery system would therefore enable the detection of genes that encode for proteins expressed in specific tissues at low abundance, thereby permitting the rapid identification of proteins, which are likely to be targets for therapeutic and diagnostic development. For the reasons discussed above, source material for the identification of new proteins is shifting from the tissue-culture cells to the discovery of proteins that change in actual human tissues.

3. PROTEOMIC TECHNOLOGIES

Numerous proteomic technologies that are available have been described elsewhere (Jain, 2003). Protein purification and expression profiling, which facilitates parallel analyses of expressed proteins, dynamic descriptions of protein regulation and detailed biochemical characterization of protein function, all provide important information on novel targets for drug discovery. Technologies that are useful for drug discovery are listed in Table 2.

Table 2. Proteomic technologies that are useful for drug discovery

2-D polyacrylamide gel electrophoresis (2-D PAGE) for protein separation
Mass spectrometry (MS) for identification of ligands
Phage antibody libraries for target discovery
Yeast 2-hybrid system for protein-protein interaction studies
Multiplexed high performance liquid chromatography coupled to MS: HPLC/MS
Matrix-Assisted Laser Desorption Ionization Mass Spectrometry (MALDI-MS) and MALDI-TOF (time of flight)-MS
Isotope-coded affinity tag peptide labeling
Protein biochips/microarrays
Cellular proteomics
Subcellular proteomics
Peptide mass fingerprinting
Biosensors for detection of small molecule-protein interactions
3-D structural proteomics for drug design
Chemical probes for quantification of changes in protein activities
Target inactivation technologies
Laser capture microdissection for obtaining pure cells from tissues under direct vision
Display technologies: fluorescence and chemiluminescence
Bioinformatics

3.1   2-D polyacrylamide gel electrophoresis

2-D gel electrophoresis is still the workhorse for obtaining protein expression patterns in cells. In high-format mode, it can produce gels containing up to 10,000 distinct proteins and peptide spots. 2-D polyacrylamide gel electrophoresis has traditionally been the gold standard discovery-based tool for proteomics (Gorg *et al*, 2000). Quantitative study of global changes in protein expression in tissues, cells or body fluids can be conducted using 2-D gels and image analysis. This method has the advantages of direct determination of protein abundance and detection of post-translational modifications such as glycosylation or phosphorylation, which result in a shift in mobility. Mass spectrometry may be used for the subsequent characterization of proteins of interest. Because thousands of proteins are imaged in one experiment, a picture of the protein profile of the sample at a given point in time is obtained, enabling comparative proteome analysis. Protein expression changes may give clues to the role of certain proteins in disease and some of the identified proteins map to known genetic loci of a disease. The major problem with this technique is only a small percentage of the entire proteome and be visualized and newer technologies are being developed that can access other regions of the proteome.

3.2   Mass spectrometry

Mass spectrometry (MS) has played an increasingly important role in drug discovery including analytical characterization of potential drug molecules and metabolic identification. A mass spectrophotometer consists of three components: (1) an ionization source, (2) a mass analyzer, and (3) a detector. Mass spectral analysis requires that the analyte be introduced into the mass spectrometer as a gaseous ion.

New ionization methods and mass analyzers have extended the applicability and overall sensitivity of to mass spectrometry to macromolecular targets. Mass spectrometry can be used to directly study the covalent and noncovalent interactions of drug molecules and biomolecule targets. Inhibitors that bind irreversibly or covalently to a target are easily studied by use of MS because the covalent complex survives intact in the gas phase. MS is combined with other proteomic technologies as well for use in drug discovery: *e.g.*, matrix-assisted laser desorption ionization MS (MALDI-MS) and electrospray ionization (ESI).

3.3 Electrospray ionization

The utility of electrospray ionization (ESI) lies in its ability to produce singly or multiply charged gaseous ions directly from an aqueous or aqueous/organic solvent system by creating a fine spray of highly charged droplets in the presence of a strong electric field. The sample solution is typically sprayed from the tip of a metal nozzle maintained at approximately 4000 V. Dry gas, heat, or both are applied to the highly charged droplets, causing the solvent to evaporate. Evaporation causes the droplet size to decrease, while the surface charge density increases. Ions are transferred to the gas phase as a result of their expulsion from the droplet and are then directed into a mass analyzer through a series of lenses. ESI-MS is useful for probing a wide range of biological problems as a detector for HPLC and capillary zone electrophoresis, in the study of noncovalent complexes, and for obtaining structural information. ESI does have limitations in that it is not very tolerant of the presence of salts (>1.0 mM), nor is it practical for the analysis of multicomponent samples. Fortunately, in several aspects of mass analysis where ESI is not very useful, Matrix-Assisted Laser Desorption Mass Spectrometry (MALDI-MS) has proven to be very effective.

3.4 Matrix-Assisted Laser Desorption/Ionization Mass Spectrometry

Matrix-Assisted Laser Desorption/Ionization Mass Spectrometry (MALDI-MS), was one of the major breakthroughs in mass spectrometric methods for analysis of proteins. MALDI has become a widely used method for determination of biomolecules including peptides, proteins, carbohydrates, glycolipids and oligonucleotides. A modification of it, MALDI-TOF (time of flight) was introduced in 1994 and commercialized in 1995. The principle of this involves bombardment of a mixed matrix of solutes by ultraviolet laser pulse in an electric field to desorb and ionize cocrystalized sample/ matrix (e.g., 2,5-dihy-droxybenzoic acid) from a metal surface. The matrix serves to minimize sample damage from the laser beam by absorbing the incident laser energy, resulting in the sample and matrix molecules being ejected into the gas phase and ionized. Once ions are formed in the gas phase, they can be electrostatically directed to a mass analyzer. Velocity of distribution of absorbed ions has a broad range. Fast ions arrive in a lower mass but slow and fast ions arrive at the same time. The time a molecule takes to move from the point of ionization to the detector is a function of the mass-to-charge ratio (m/z) of a particle and is termed TOF. Resolution of the technique has improved since its initial introduction and MALDI-MS has emerged as an effective bioanalytical tool having unique capabilities in handling complex mixtures (such as proteolytic digests) and in high-sensitivity (femtomole or even subfemtomole) measurements.

3.5   Isotope-coded affinity tag peptide labeling

Isotope-coded affinity tag (ICAT) peptide labeling is an approach that combines accurate quantification and concurrent sequence identification of individual proteins in complex mixtures (Gygi, *et al* 1999). This method is based on a newly synthesized class of chemical reagents used in combination with tandem mass spectrometry. The method consists of four steps:

1. The reduced protein mixtures representing two-cell state are treated with two different versions of ICAT reagent - one light and one heavy.
2. The labeled sampled are combined and proteolytically digested to produce peptide fragments.
3. The tagged cysteine-containing fragments are isolated by avidin affinity chromatography.
4. The isolated tagged peptides are separated and analyzed by microcapillary tandem MS which provides both identification of peptides by fragmentation in MS-mode and relative quantitation of labeled pairs by comparing signal intensities in MS mode.

The advantages of ICAT over 2-D gel electrophoresis that has the potential for full automation, and thus for high-throughput proteomic experiments. There is no need to run time-consuming experiments and because it is based on stable isotope labeling of the protein, there is no need for metabolic labeling or no radioactivity is involved. ICAT can be used for the analysis of several classes of proteins such as membrane proteins and low copy number proteins that are poorly tractable by 2-D gels. Most importantly, it provides accurate relative quantification of each peptide identified. The limitations of this technique are that the proteins must contain cysteine and the large size of the tag compared to some small peptides and may interfere with peptide ionization. These, however, can be overcome by designing different reagents with specificities for other peptide chains and using a smaller tag group.

ICAT is an emerging technique for differential expression proteomics, and its full potential remains to be fully evaluated. Advances in sample fractionation at the protein level, sample fractionation at the peptide level, and improved data acquisition schemes, will all be required for the full potential of ICAT to be realized. New separation systems, such as ultra-high pressure nanoscale capillary LC will improve the peak capacity for ICAT experiments, leading to improved proteome coverage. New MS technologies, such as the high sensitivity, high-throughput MALDI–TOF instrument, can be expected to have a very significant impact in ICAT proteomics (Moseley, 2001).

3.6   Phage antibody libraries for target discovery

Cambridge Antibody Technology (CAT) has developed combinatorial phage antibody libraries. Antibodies are a key tool for investigating the activities of proteins. CAT bioinformatics analyses gene sequences, and identifies those, which make a protein of potential interest. A small portion of each protein is made synthetically (the peptide). An antibody that binds the peptide is isolated. The antibody is then used to probe for protein expression in human tissues.

CAT's antibodies can be used directly to analyze the presence or absence of a protein in diseased and normal tissue to give evidence of 'guilt by association'. CAT can test whether proteins implicated by an association with disease have a direct role in causing that disease by using antibodies to directly neutralize or possibly mimic the effects of the protein in both in vitro and in vivo models of the disease. This gives data about the involvement of the protein as a causative agent in the disease.

CAT has developed ProxiMol as a new target discovery technology. It is so-called because it uses antibodies to identify molecules in close proximity to an original target. This can be used to investigate proteins in their natural context and provide important information relevant to disease progression leading to the identification of further targets. This technology has been used extensively in its target discovery programs.

3.7   Drug discovery through protein-protein interaction studies

A key to the understanding of the function of any protein target is the knowledge of its cellular partners, the other proteins that interacts with that target in the course of performing its biological role. This has been made possible by technological progress in purification of protein complexes and interacting proteins couples, the identification of the interacting partners with sensitivity and specificity, the management of data on protein interaction pathways, and efforts to put that predictive information to work in reconstructing pathway functions and predicting sites of intervention to disrupt or modulate that pathway's outcome. Understanding the role of a specific target in the context of a biochemical pathway allows an accelerated choice of optimal therapeutic targets.

The extent of protein dual tasking may require a rethinking of the drug-discovery process. Many drugs aim to block a single function of a protein. It is now clear that each protein is performing many roles, which could all be affected. Future therapies might use multiple drugs in combination to manipulate a pathway.

The rationale for protein-protein interaction studies is that proteins that bind together are more likely than not to be functionally related. Protein-protein interactions are important because they help in delineating a protein that's involved in a disease process and also likely to be a potential drug target.

3.8   Two-hybrid protein interaction technology for target identification

Two-hybrid protein interaction technology enables isolation and characterization of potential drug targets by identifying specific proteins that bind with other peptides or proteins that are known to be part of a signaling pathway. This protein interaction technology is directed at:

a) Mapping an entire protein-protein intracellular functional pathway in disease relevant cells.
b) Finding new proteins interacting with other new and known proteins.
c) To screen identified protein targets against a library of peptides in order to identify each active protein-peptide interaction site on the target.
d) Eliminating potential targets rapidly because they interact with multiple signaling pathways.

Using this technology, a protein that gives a detectable signal (reporter protein), such as fluorescence, is split into two inactive parts. One part of the reporter protein is fused with a specific protein known to be involved in a signaling disease-relevant pathway (bait protein). Multiple copies of the other part of the reporter protein are fused one by one with all the proteins known to be present in the cell type being studied (library protein). When the bait protein binds to a specific library protein, the two parts of the reporter protein reunite and become active again, thereby generating a detectable signal. An improved version of the two-hybrid protein interaction method in yeast cells is used.

3.9 Biosensors for detection of small molecule-protein interactions

Biosensors have the ability to detect not only protein-protein interactions, but also small molecule-protein interactions that play a role in preclinical evaluation of drug candidates. The current technology enables the analysis of these interactions on a microchip format that is highly sensitive and utilizes very little sample. Biosensors are utilized in three main discovery areas: (1) protein-protein interactions, (2) ligand identification, and (3) small molecule interactions. The sensitivity of biosensors has progressed to allow for the detection of small molecules binding to proteins and lipids.

3.10 Liquid chromatography/mass spectrometry

In liquid chromatography (LC), proteins are kept in solution so that a higher percentage of the sample is analyzed using affinity chromatography and having access to the entirely accurately assembled genome of the organism under study. Protein mixture may be digested to analyze at peptide level. Both single dimensional high-pressure liquid chromatography (LC) and multidimensional LC (LC/LC) can be directly interfaced with the mass spectrometer to allow for automated collection of tremendous quantities of data. While there is no single technique that addresses all proteomic challenges, the shotgun approaches, especially LC/LC-MS/MS-based techniques such as MudPIT (multidimensional protein identification technology), show advantages over gel-based techniques in speed, sensitivity, scope of analysis, and dynamic range (McDonald and Yates 2002). Advances in the ability to quantitate differences between samples and to detect for an array of post-translational modifications allow for the discovery of classes of protein biomarkers that were previously undetectable. Multiplexed LC systems are coupled to MS to analyze affinity-tagged cellular lysates and protein mixtures and someday might replace 2-D gel-based systems.

3.11 Protein biochip technology

Biochip is a broad term indicating the use of microchip technology in molecular biology and can be defined as arrays of selected biomolecules immobilized on a surface. Most of the biochips use nucleic acids as information molecules but protein biochips are also proving to be useful. The basics of protein biochip construction are similar to those of DNA chip as the glass or the plastic surface is dotted with an array of molecules (these can be DNA or antibodies) designed to capture proteins. Fluorescent markers or other methods of detection reveal the spots that have captured proteins and light up.

The value of protein biochip technology in drug discovery and development is being recognized within the pharmaceutical industry. Current research and development in protein biochip technology indicates excellent future prospects for protein biochip technology in the drug development process. Profiling proteins will be invaluable, for example, in distinguishing the proteins of normal cells from early-stage cancer cells, and from malignant, metastatic cancer cells that are the real killers. In comparison with the DNA microarrays, the protein microarrays, or protein biochips, offer the distinct possibility of developing a rapid global analysis of the entire proteome. Microarrays of an entire eukaryotic proteome can be prepared and screened for diverse biochemical activities. The microarrays can also be used to screen protein-drug interactions and to detect posttranslational modifications. Thus, the concept of comparing proteomic maps of healthy and diseased cells may allow us to understand cell signaling and metabolic pathways and will form a novel base for pharmaceutical companies to develop future therapeutics much more rapidly. Various applications of protein biochip technology in relation to drug discovery are shown in Table 3.

Table3. Applications of protein biochip technology in relation to drug discovery

Analysis of the proteome
Comparison between genomic and phenomic data
Elucidation of mechanism of drug action
Elucidation of pathophysiology of disease
Protein profiling
Drug discovery: target identification and validation
Toxicoproteomics
Study of protein-protein interactions and protein-DNA interactions
Disease biomarker discovery
Combination of protein biochip with laser capture microdissection
Cancer biomarker discovery
Detection of beta amyloid biomarkers in Alzheimer's disease
Protein markers for identification of microorganisms and their responses to drugs

3.11.1   ProteinChip with Surface Enhanced Neat Desorption

Ciphergen has introduced a new generation prototype of ProteinChip Arrays, which are expected to have particular utility for drug discovery applications. These new polymeric surfaces further extend Ciphergen's capabilities in Surface Enhanced Neat Desorption (SEND), a patented technology that Ciphergen has pioneered and continues to advance.

Ciphergen's new ProteinChip SEND surfaces are modified with homogenous polymeric coatings that interact with bursts of focused laser energy to allow the direct creation of ions, without introducing unacceptable levels of chemical noise, for analysis via SELDI detection using Ciphergen's ProteinChip Systems. The elimination of the need to add chemical "matrix" solutions as a separate step, common to all MALD-MS methods, not only facilitates and simplifies research but also allows for better quantitation. Further, these SEND surfaces enable the analysis of small, low molecular weight molecules by reducing the chemical noise background otherwise created by matrix desorption. Polymeric blends containing both laser energy absorbing and functional binding groups are being developed to allow both on-chip chromatography as well as affinity immobilization of target molecules, such as receptors, which can then be screened against low molecular weight combinatorial drug candidates using this novel protein biochip approach. In addition to offering immediate advantages over conventional LC-MS and MALDI-TOF-MS systems, these new SEND ProteinChip Arrays mark an important step toward creating a novel chemical genomics-based drug discovery platform which could be used for on-chip small molecule screening.

3.12   Cellular proteomic approaches

Cellular proteomic systems are being used increasingly to assess the activity of proteins. Protein activities corresponding to spotted cDNAs are evaluated by various phenotypic criteria applied at each discrete locus. Microarrays whose features are clusters of live cells that express a defined cDNA at each location have been used as an alternative to protein microarrays for the identification of drug targets, and as an expression cloning system for the discovery of gene products that alter cellular physiology (Ziauddin and Sabatini, 2001). Proteins involved in tyrosine kinase signaling, apoptosis and cell adhesion, and with distinct subcellular distributions were identified by this approach.

Cellular proteomic approaches such as this promise to provide unparalleled advances towards annotation of gene function and the identification of novel drug targets. Genes can be systematically evaluated for their participation in signaling pathways as well as cellular processes and physiological phenotypes through overexpression. Assays can be done in arrayed format and across discrete loci, thereby expanding the range of possible cell-based functional screens. Furthermore, parallel analysis affords the identification of genes with less potent activities (i.e. slightly above background), and does not necessarily require a robust reporter system for success.

3.12.1   Topological proteomics

Topological proteomics means visualization of the abnormal cellular protein networks, cell by cell. Conventional proteomics is performed in three steps: separation of complex protein mixtures, characterization of the separated proteins, and database searching to identify the composition of the complex. The employed techniques such as 2-D electrophoresis and mass spectrometry use tissue homogenates. Therefore, only quantitative changes of the most abundant proteins with particular biochemical properties can be detected. Any topological information, information on the cellular and sub-cellular distribution of proteins, that determines protein-interactions in networks, is lost.

MelTec GmbH (Magdeburg, Germany) has developed the Multi-Epitope-Ligand-Kartographie (MELK), a robotic whole cell imaging technology for topological proteomics. This technology enables deciphering of protein networks, their protein components, their function and localization, and visualizes any protein independent of its biochemical properties, expression level or localization. Due to MELK's ability to screen in situ at the single cell level, it is especially suited for the study of diseases in which invasive cells play a crucial role such as immune-mediated diseases, cancer and arteriosclerosis. By using MelTec's Neuronal Cell Detection System (NCDS) algorithm, the biomathematical basis of MELK, the researchers were able to automatically monitor in tissue sections the number of fluorescent marked migratory cells, the positions of these cells and the phenotype of these cells, thus identifying and localizing particular subsets of cells, e.g. identifying through unique protein networks the individual immune-cells that invade a tissue or organ (Nattkemper, *et al* 2001). The NCDS system enabled researchers to conduct high-throughput, reproducible and valid statistical analysis of protein networks involved in disease pathways. To learn about cellular function and mechanisms, the groups of proteins that define a particular cellular fingerprint reflecting topological information, were considered, rather than just looking at individual proteins and their molecular information. Applications of topological proteomics relevant to drug discovery are in target identification and prioritization as well as lead optimization.

3.13   Subcellular proteomics

Knowledge about the subcellular localization of a protein may provide a hint as to the function of the protein. The combination of classic biochemical fractionation techniques for the enrichment of particular subcellular structures with the large-scale identification of proteins by mass spectrometry and bioinformatics provides a powerful strategy that interfaces cell biology and proteomics, and thus is termed 'subcellular proteomics'.

Proteomics can no longer be equated with a single 2-D electrophoresis gel. Greater information can be obtained using targeted biological approaches based upon sample prefractionation into specific cellular compartments to determine protein location, while novel immobilized pH gradients spanning single pH units can be used to display poorly abundant proteins due to their increased resolving power and loading capacity. Subcellular proteomics, in addition to its exceptional power for the identification of previously unknown gene products, is the basis for monitoring important aspects of dynamic changes in the proteome such as protein transloction. Furthermore, this new approach can direct biological focus towards molecules of specific interest within complex cells and thus simplify efforts in proteome-based drug discovery.

3.14   Peptide mass fingerprinting

Peptide mass fingerprinting (PMF) is the primary tool for mass spectrometry identification of proteins in proteomic studies. The basis of PMF is that the set of peptide masses obtained by mass spectroscopic analysis of a digested protein provides a characteristic profile or "fingerprint" of that protein. This mass profile is then compared with possible peptide masses calculated from theoretical in silico digestion of known proteins. Several algorithms have been used to facilitate this process and a number of programs are available on the Internet for protein identification by peptide mass fingerprinting. Protein identification by PMF is a widely recognized technique for proteins from 2-D gels and can give excellent results. Using PMF or a combination of mass and sequence, i.e. peptide-sequence tags, current protein or EST data-bases can now be searched in less than a minute, enabling real-time feedback to direct further mass spectrometry for rapid confirmation of a predicted hit. The PMF approach is most suited to completely sequenced genomes such as yeast but its drawback is the limited ability to deal with protein mixtures, particularly if the components are of vary widely in the degree of abundance. At very low levels protein digests invariably contain keratins, which are often introduced inadvertently. In addition, there may be peptides from IgG used for immunoprecipitation. Thus the protein of interest may be completely hidden by other more abundant proteins.

3.15   3-D structural proteomics for drug design

Structural information is a key component of the processes of hypothesis, synthesis, testing, and redesign that constitutes structure-based drug design. Knowing the structure of a protein of interest enables researchers to map the detailed binding-site energetics between the protein target and a small molecule that is a potential drug candidate. When joined with targeted combinatorial libraries and medicinal chemistry considerations, protein structure information can drive a relatively fast and directed lead optimization process.

3.15.1   Role of 3-D structure of protein in drug discovery

The primary sequence of a protein, i.e. the linear string of amino acids, determines its 3-D structure, which in turn determines its function. By determining the 3-D structure of "novel" genes identified by genome sequencing, one gains an insight into the gene function and identifies gene products that are potential therapeutic targets. Availability of 3-D structure will facilitate discovery of drugs specifically targeted to the membrane receptor. Novel lead compounds can be designed by an approach based on 3-D structure of the target protein by X-Ray crystallography. Lead compounds can be optimized for pharmacokinetic parameters such as improved bioavailability by modifying the key sites of binding regions, thus reducing the attrition rate of the compounds in development.

3.15.2   Automated 3-D protein modeling

Computational methods are expected to play an increasingly important role in functional analysis of the proteins discovered in fully sequenced genomes. Automated protein modeling is already used for functional analysis of proteins and 3-D protein modeling. The availability of local similarity search algorithms enables the identification of suitable template structures even at very low sequence similarity levels. Inaccuracies of protein models can result from automated alignment procedures which introduce errors by placing insertions and deletions incorrectly and these can be remedied by identification and subsequent multiple alignments of all members of the protein family. Over 16,000 3-D macromolecular structures - nearly 14,000 of them classed as proteins and peptides - have been deposited in the Protein Data Bank, a publicly accessible repository that's operated by the Research Collaboration for Structural Bioinformatics.

Several of the structures in the database are the same basic protein bound to different ligands. The unique protein structures, therefore may number only 1,500.

3.16   Chemical probes

Chemical probes can enable the quantitation of changes in protein activities in any cell type and tissue over a range of normal and pathological conditions. These probes work by specifically binding to a common structural element shared by all members of a protein family, consequently avoiding proteins that are either bound to an inhibitor or are in an inactive state. The technology can be used to monitor the activity of both secreted and membrane-bound proteins including a large number of pharmaceutically important protein classes such as serine hydrolases, cysteine hydrolases, kinases and phosphatases, believed to be rich in druggable targets.

3.17   Peptide probes

Individual probes are designed to elicit a desired phenotypic change in the cell that can be separated from the vast majority of irrelevant ones and characterized further. For example, probes that selectively kill cancer cells can be retrieved and used subsequently to enable drug development. Upon introduction into a cell with retroviral vectors, the peptidic probes act in a fashion similar to small molecule drugs (i.e., they have specific binding or inhibitory properties). Specific probes that induce a desired response can be recovered using high-throughput screening systems (e.g., fluorescence-activated cell sorting). The probes are then used to recover the protein drug target. This cell-based "function-first" approach establishes a link between a selected protein and a disease state. Ultimately, this process can generate a validated novel intracellular target that is suitable for high-throughput small molecule drug screening. Potential advantages of peptide probes include their ability to probe the intracellular proteome for novel targets with no prior knowledge of biochemical pathways required

3.18   Target inactivation technologies

3.18.1   Target inactivation by antisense approaches

Antisense molecules are synthetic segments of DNA or RNA, designed to mirror specific mRNA sequences and block protein production. The use of antisense drugs to block abnormal disease-related proteins is referred to as antisense therapy. Synthetic short segments of DNA or RNA are referred to as oligonucleotides. Target mRNA and protein expression can be inhibited using antisense oligonucleotides to facilitate this process and determine genetic pathways. Antisense technology is considered to be a viable option for high-throughput determination of gene function and drug target validation. An example of application of this approach is in vitro and in vivo inhibition of interleukin (IL)-5-mediated eosinopoiesis by murine IL-5R alpha antisense oligonucleotide (Lach-Trifilieff, *et al* 2001). Advantages of antisense for gene fictionalization and target validation are speed, selectivity and flexibility.

Conventional methods for validating protein targets are usually indirect, involving modification (dominant negative mutants), deletion (genetic knockout), or suppression (anti-sense or ribozymes) of the corresponding gene or mRNA. Most of these approaches have limitations. Results of gene knockout approaches can be inconclusive due to embryonic lethality or compensatory effects caused by genetic redundancy. Inducible, time- and space-resolved gene knockouts try to address this problem with varying success. Moreover, only certain animal systems are amenable to this kind of genetic manipulation. mRNA suppression by antisense or ribozyme strategies creates retarded effects since protein disappearance lags behind the effects on mRNA levels. This approach is also not free of functional compensation problems. Moreover, specificity problems can arise from dealing with the still unknown and vast complexity of the whole transcriptome of a cell. For functional inactivation at the protein level, specific inhibitory drugs or function-blocking antibodies can be used but they are often very difficult to obtain. New techniques are being developed to remedy some of these limitations.

3.18.2   Chromophore-assisted laser inactivation

Chromophore-assisted laser inactivation (CALI) is a powerful approach to inactivate protein function by targeted induction of photochemical modifications. CALI is capable of knocking out one subunit/domain at a time or all together and creates an acute loss of protein function at a given point in time. Protein inactivation is transient in living cells since new protein synthesis can replace the damaged protein fraction. This transient knockdown mimics the dose-dependent effects of a drug, which makes CALI especially suitable for identification, and validation of targets.

After a protein target is functionally validated by CALI, antibodies can be generated using phage display and screened for antibodies that are neutralizing, thereby circumventing structural proteomic approaches. This represents a new approach of rapidly developing therapeutic antibody leads from the simultaneous identification and validation of targets, and can be extended to small molecule drugs. CALI is a highly versatile tool for validating disease relevant targets at the protein level (Rubenwolf, *et al* 2002). This approach also takes into account post-translational modifications like phosphorylation, glycosylation or acylation, thereby enlarging its applicability for many different types of targets.

3.19   Laser capture microdissection

Laser capture microdissection (LCM) provides an ideal method for extraction of cells from specimens in which the exact morphologies of both the captured cells and the surrounding tissue are preserved. The following proteomics technologies, which are relevant to drug discovery, can be combined with LCM:
a) 2-D polyacrylamide gel electrophoresis (PAGE)
b) Mass spectrometry
c) Matrix-Assisted Laser Desorption Mass Spectrometry
d) High performance liquid chromatography
e) ProteinChip: Surface-Enhanced Laser Desorption/Ionization

Arrays using LCM-procured cancer epithelial cells can test the functional status of the pathways of interest and may be used for rapid identification of targets for pharmacological intervention, as well as to assessment of the therapy in correcting the deranged pathways. The impact of proteomics on human cancer and diseases will not be limited to the identification of new biomarkers for early detection and new targets. These tools will be used to design rational drugs tailored according to the molecular profile of the protein circuitry of the diseased cell (Jain, 2002a).

3.20   Fluorescent detection of proteins

Protein expression encompasses an enormous dynamic range. Since rare proteins cannot be amplified by any type of PCR method, sensitive detection is critical to proteome projects. Fluorescence methods deliver streamlined detection protocols, superior detection sensitivity, broad linear dynamic range and excellent compatibility with modern microchemical identification methods such as mass spectrometry. Two general approaches to fluorescence detection of proteins are the covalent derivatization of proteins with fluorophores or noncovalent interaction of fluorophores through direct electrostatic interaction with proteins. One approach for quantifying fluorescence is to use a photomultiplier tube detector combined with a laser light scanner. In addition, fluorescence imaging is performed using a charge-coupled device camera combined with an ultraviolet light or xenon arc source. Fluorescent dyes with bimodal excitation spectra may be broadly implemented on a wide range of analytical imaging devices, permitting their widespread application to proteomics studies and incorporation into semiautomated analysis environments.

3.20.1   Fluorescent labeling of proteins in living cells

Recombinant proteins have been fluorescently labeled in living cells by extracellular administration of fluorescein. The small ligand thus designed is permeates membranes and does not fluoresce until it binds with high affinity and specificity to the tetracysteine domain. Such in situ labeling adds much less mass than does green fluorescent protein and offers greater versatility in attachment sites as well as potential spectroscopic and chemical properties. This system provides a recipe for slightly modifying a target protein so that it can be singled out from the many other proteins inside live cells and fluorescently stained by small nonfluorescent dye molecules added from outside the cells.

3.21   Bioinformatics

The use of large-scale technologies for the analysis of gene expression and protein levels will further add to both the volume and complexity of the data. Conventional means of storing, analyzing and comparing related data are already overburdened; new ways of analyzing data are needed that employ sophisticated statistical algorithms and powerful computers. High-throughput analysis has become increasingly common in proteome projects and requires automated analysis of the mass spectrometric data. An important part of automation is quality control and therefore development of methods to determine the quality of search results has become a focus. A few examples of application of bioinformatics in proteomics-based drug discovery are described here briefly.

3.21.1   In silico search of drug targets by Biopendium

The detection of similarities between amino acid sequences is fundamental to pharmaceutical research *in silico*. Conventional techniques can only detect relationships between sequences when at least 25% of their residues match. Inpharmatica' proprietary Biopendium enables one to intelligently interrogate such data via comparison with proprietary information on protein relationships. The Biopendium is a wholly integrated resource that currently contains pre-calculated analyses on over 600,000 sequences revealing approximately 600 million sequence/structure/function relationships. By carrying out extensive protein sequence comparisons, the Biopendium is able to reveal key relationships between individual proteins and their predicted functions. Applications of the Biopendium include the following:

Identification of novel distant homologues of pharmaceutically relevant protein families

- Identification and prioritization of drug targets
- Identification a new starting point for lead identification
- Identification of other proteins that may interact with the ligand of interest, pointing to potential specificity problems and side effects

3.21.2   Protein structural database approach to drug design

This approach consists of web-based technologies that integrate genomics data and tools for rational drug design, enabling the analysis of gene sequences, protein structures, and combinatorial libraries. These web-based tools provide a seamless integration of pharmaceutically valuable gene/protein sequences with structure-based drug design tools, and can be used, for example, to take sequences to 3-D templates to virtual screens in order to find non-peptide leads, or to find drug leads and functional probes for novel disease genes rapidly. The web interface acts as a front-end to a host of public domain databases and bioinformatics sites. The efficiency of this gene-sequence-derived computational approach effectively lowers the threshold for producing chemical probes of gene function. In addition, comparative structural data are expected to provide new tools for designing target-specific drugs.

4.   SIGNALING PATHWAYS AND PROTEOMICS

About one-fifth of all human genes encode proteins involved in signal transduction. Large-scale proteomics projects now in progress seek to better define critical components of signal transduction networks, to enable more intelligent design of therapeutic agents that can specifically correct disease-specific signaling alterations by targeting individual proteins. There are instances in which protein interaction technologies have been specifically adapted to identify small molecule agents that regulate protein response in physiologically desirable ways that are relevant to future drug discovery efforts.

Signal transduction can be analyzed by currently available proteomic technologies because signaling pathways contain enzymes, which modify high-abundance proteins other than those of the pathway. Thus, modulation of the signaling through a pathway will produce a "footprint" in the proteome that is characteristic of a specific cell phenotype (Resing, 2002).

The reversible phosphorylation of tyrosine residues is an important mechanism for modulating biological processes such as cellular signaling, differentiation, and growth, and if deregulated, can result in various types of cancer. Therefore, an understanding of these dynamic cellular processes at the molecular level requires the ability to assess changes in the sites of tyrosine phosphorylation across numerous proteins simultaneously as well as over time. A sensitive approach based on multidimensional liquid chromatography/mass spectrometry enables the rapid identification of numerous sites of tyrosine phosphorylation on a number of different proteins from human whole cell lysates (Solomon *et al*, 2003). This methodology should enable the rapid generation of new insights into signaling pathways as they occur in states of health and disease.

4.1   Identification of protein kinases

Protein kinases are coded by more than 2,000 genes and thus constitute the largest single enzyme family in the human genome. Most cellular processes are in fact regulated by the reversible phosphorylation of proteins on serine, threonine, and tyrosine residues. At least 30% of all proteins are thought to contain covalently bound phosphate. A novel method can determine if drugs and drug targets are effective in combating disease by identifying the key regulatory protein "switches" (phosphorylation sites) inside human cells (Ficarro, *et al* 2002). This method can be used to identify novel targets in disease, to compare the effects of different drug candidates, and to develop assays that can be used throughout pre-clinical and clinical development.

4.2   G-protein coupled receptors as drug targets

G-protein coupled receptors (GPCRs) are an important class of drug targets that exist as proteins on the surface membranes of all cells and are also referred to as 7-TM or serpentine receptors as they cross the membrane seven times. The GPCRs are a superfamily of proteins accounting for approximately 1% of the human genome and are associated with a wide range of therapeutic categories, including asthma, inflammation, obesity, cancer, cardiovascular, metabolic, gastrointestinal and central nervous system diseases. Purified multiple GPCRs in a functional form can be used for the identification of tight binding ligands. There are estimated to be about 2,000 GPCRs within the human body with potential availability as drug discovery targets. GPCRs have historically been valuable drug targets, but to date there are only approximately 100 well-characterized GPCRs with known ligands, of which only about half are currently targets of commercial drugs. Approximately 60% of all currently available prescription drugs interact with these receptors.

Nearly all molecules known to signal cells via G proteins have been assigned a cloned G-protein-coupled-receptor (GPCR) gene. Functions have been elucidated for about 160 of these receptors, and about 140 receptors are left without a function or a ligand the so-called "orphan" receptors. Several of these novel receptor systems have been identified by use of NeoGene's proprietary orphan receptor strategy (Civelli, *et al* 2001). Completion of the Human Genome Project suggests that there are over 1,000 orphan GPCRs, which hold great promise to refill the pharma pipeline. The discovery of all the GPCR genes in the genome and the identification of the unsolved receptor-transmitter systems, by determining the endogenous ligands ("deorphaning"), represents one of the most important tasks in modern pharmacology.

5.   CHEMICAL GENOMICS APPROACH TO DRUG DISCOVERY

There is no standard definition of chemical genomics but it implies study of how small molecules interact with cells. This would include, for example, experiments in which drug treatment of cells has been studied using large-scale expression analysis, or large-scale protein analysis. Chemical ligands are used in genomics approaches to understand the global functions of proteins. Chemical genomics is thus relevant to proteomics and the term chemoproteomics may be used when proteomic approaches are used. Methods have been described for preparing microarrays of functionally active proteins on solid supports, which may also prove useful for evaluating small-molecule specificity - in the short term with relatively small arrays of related proteins and in the longer term with comprehensive 'proteome arrays' (MacBeath, 2001).

A cell-permeable, target-specific chemical ligand that perturbs posttranslational modification of a downstream target protein can be studied by protein profiling technologies, such as 2-D gel electrophoresis. A change in modification of a protein can be detected by comparison of the protein migration patterns before and after treatment with the chemical ligand. The identity of the protein can then be revealed by mass spectrometry. Chemical ligands can also be combined with several existing genomic tools to devise chemical genomics approaches for gene expression, protein profiling, and global analysis of genetic interactions to produce complete biochemical and genetic profiles of the drug target-proteins (Zheng and Chan, 2002).

Chemical genomics is applied to the discovery and description of all possible drug compounds (all of the chemical possibilities) directed at all possible drug targets (the 100,000 or more proteins coded by the human genome). This approach has the potential to cover a broad range of therapeutic areas, because while gene families code for structurally similar proteins, each protein in a gene family can have a very different biological function. Different targets within a gene family may be implicated in widely different diseases.

As a drug discovery strategy in the post-genomic era, chemical genomics is also called chemogenomics and unites medicinal chemistry with molecular biology using low-resolution sequence homology to identify genomic targets of interest. The corresponding proteins can then be screened using affinity selection to identify small-molecule ligands with high affinity to targets of known or unknown function. Ligands identified by this method can be used in secondary biological assays to check for biological activity. In case of targets of unknown function, low-throughput secondary assays can be used to determine the therapeutic relevance of a novel target (Fig. 1). In practice, the chemogenomics approach is accelerating the drug discovery and will increase the flow of new drug candidates into development.

Figure 1. Chemical genomic approach to drug discovery

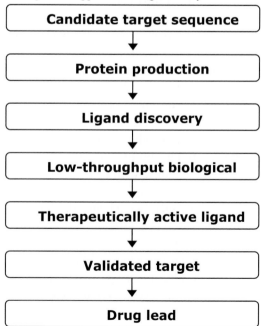

A chemical genomics approach has been used to explore the genetic interaction between TOR (target of rapamycin protein) and other yeast genes on a genomic scale. In one study, the rapamycin sensitivity of individual deletion mutants generated by the Saccharomyces Genome Deletion Project was systematically measured (Chan, *et al* 2000). In contrast to conventional genetic analysis, this approach offers a simple and thorough analysis of genetic interaction on a genomic scale and measures genetic interaction at different possible levels. It can be used to study the functions of other drug targets and to identify novel protein components of a conserved core biological process such as DNA damage checkpoint/repair that is interfered with by a cell-permeable chemical compound.

6. COMBINED APPROACHES

6.1 Genomics and proteomics

Proteomic technologies are now being integrated into the drug discovery process as complimentary to genomic approaches. This offers the scientists the ability to integrate information from the genome, expressed mRNAs and their respective proteins as well a subcellular localization. By focusing on protein activity levels, or expression levels, researchers are able to learn more about the role proteins play in causing and treating disease. Proteomics also aids in deciphering the mechanisms of disease and increasing both the opportunity to develop drugs with reduced side effects and an increased probability of clinical trial success. Proteomics has the potential to increase substantially the number of drug targets and thereby the number of new drugs. Automation of proteomics on a scale similar to that used for genome sequencing may be needed and this is feasible by adapting the many tools already developed for genomics for application to proteomic technologies.

6.2 Combination of RNA and protein profiling

Both mRNA and protein expression data may be used in case of tissue samples for an insight into disease mechanisms and molecular effects of drugs. mRNA expression has the advantage for genes expressed at low level as PCR (polymerase chain reaction)-mediated target amplification can be used. However, mRNA isolated from body fluids is of poor quality due to degradation and proteins are the most relevant surrogate markers for monitoring biological disturbances resulting from disease and drug treatment. There is poor correlation between mRNA and protein expression.

mRNA degradation, post-translational modification and post-transcriptional regulation of gene expression make it difficult to extrapolate from mRNA to protein profiles. Analysis of mRNA alone does not provide an accurate picture of proteins present in the cell. mRNA profiling, therefore, should be combined with appropriate protein measurements. RNA-protein complexes have emerged as a novel class of targets for drug discovery.

6.3   Metabolomics and proteomics

Metabolomics is the study of small molecules and their interactions within an organism, which is critical to the drug-discovery process. The importance of metabolomic studies is indicated by the finding that a large proportion of the 6,000 genes present in the genome of *Saccharomyces cerevisiae*, and of those sequenced in other organisms, encode proteins of unknown function. Many of these genes are "silent", i.e. they show no overt phenotype, in terms of growth rate or other fluxes, when they are deleted from the genome. One study has demonstrated how the intracellular concentrations of metabolites can reveal phenotypes for proteins active in metabolic regulation (Raamsdonk, *et al* 2001). Quantification of the change of several metabolite concentrations relative to the concentration change of one selected metabolite can reveal the site of action, in the metabolic network, of a silent gene. In the same way, comprehensive analyses of metabolite concentrations in mutants, providing "metabolic snapshots", can reveal functions when snapshots from strains deleted for unstudied genes are compared to those deleted for known genes.

An integrated approach to build, test, and refine a model of a cellular pathway includes analysis of perturbations in critical pathway components using DNA microarrays, quantitative proteomics, and databases of known physical interactions (Ideker, *et al* 2001). An approach combining proteomics with metabolomics is pursued by a collaboration of Thermo Electron Corporation and Paradigm Genetics Inc (Research Triangle Park, NC). The companies will jointly design and develop the next generation of chromatography/mass spectrometry systems to create a new platform for identifying and validating metabolite biomarkers important to the development of safe and effective drugs. The unique combination of gene expression profiling (determining the level of activity of genes in an organism at a specific time), metabolic profiling (determining the identity and quantities of chemicals in an organism at a specific time), and phenotypic profiling (measuring the physical and chemical characteristics of an organism at a specific time), with data from all systems being managed and analyzed in Paradigm Genetics' FunctionFinder bioinformatics system, creates a new paradigm for industrializing functional genomics.

7. PROTEOMICS FOR DRUG DISCOVERY IN VARIOUS THERAPEUTIC AREAS

Application of proteomics in various therapeutic areas is described in various chapters of this book. Two important areas of drug discovery where proteomics plays an important role are cancer and central nervous system (CNS). These are mentioned here briefly.

7.1 Proteomics for cancer drug discovery

Proteomic technologies are now being integrated with genomic approaches for cancer drug discovery and target validation (Jain, 2002b). Among the large number of proteomic technologies available for this purpose, the most important ones are 3-D protein structure determination, protein microarrays, laser capture microdissection and study of protein-protein and protein-drug interactions. Cancer biomarkers and several cell pathways are important drug targets. Several technologies are in use both in the academic as well as the industrial sectors and the results in cancer drug discovery are encouraging.

7.2 Proteomics for CNS drug discovery

Proteomic technologies are suitable for study of the brain tissues and can be applied for discovery of targets for drug to treat neurological disorders (Jain, 2002c). Particularly suited for this approach are diseases with protein pathology such as Alzheimer's disease. Important receptors for CNS drugs include proteins such as G-protein-coupled receptors, N-methyl-D-aspartate receptors and protein kinases. Molecular diagnostics can be based on proteins detected in the cerebrospinal fluid and the same proteins can serve as drug targets. Proteomics will complement pharmacogenomics and facilitate the development of personalized medicines for neurological disorders.

8. CONCLUSIONS

Introduction of proteomic technologies has added a useful dimension to drug discovery. Proteomics technologies are useful in two important therapeutic areas: cancer and central nervous system. By helping to elucidate the pathomechanism of diseases, proteomics will help the discovery of rational medications that will fit in with the future concept of personalized medicines.

Proteomics technologies are not used alone but rather they are integrated with genomic and chemical approaches. Finally, bioinformatics plays an important role in drug discovery.

REFERENCES

Civelli, O., Saito, Y., Lin, S., *et al*, 2001,The orphan receptor strategy and the discovery of novel neuropeptides. *Trends in Neurosciences* **24**: 230-237.

Ficarro, S.,B., McCleland, M.,L., Stukenberg, P.,T., *et al*, 2002, Phosphoproteome analysis by mass spectrometry and its application to Saccharomyces cerevisiae. *Nature Biotechnology* **20**: 301-305.

Gorg A., Obermaier C., Boguth G., *et al*, 2000, The current state of two-dimensional electrophoresis with immobilized pH gradients. *Electrophoresis*. **21**: 1037-53.

Gygi, S.,P., Rist, B., Gerber, S.A., *et al*, 1999, Quantitative analysis of protein mixtures using isotope coded affinity tags. *Nature Biotechnology* **17**: 994-999.

Ideker, T., Thorsson, V., Ranish, J.A., *et al*, 2001, Integrated genomic and proteomic analyses of a systematically perturbed metabolic network. *Science* **292**: 929-34.

Jain, K. K., 2003, *Proteomics: Technologies, Markets and Companies*. Jain PharmaBiotech Publications, Basel.

Jain, K. K., 2002a, Application of Laser Capture Microdissection to Proteomics. *Methods in Enzymology*. **356**: 157-167.

Jain, K. K., 2002b, Proteomics-based anticancer drug discovery and development. *Techn Cancer Res & Dev* **1**: 231-236.

Jain, K. K., 2002c, Role of Neuroproteomics in CNS Drug Discovery. *Targets* **1**: 95-101.

Lach-Trifilieff, E., McKay, R.,A., Monia, B.,P., *et al*, 2001, In vitro and in vivo inhibition of interleukin (IL)-5-mediated eosinopoiesis by murine IL-5R alpha antisense oligonucleotide. *Am J Respir Cell Mol Biol* **24**: 116-22.

McBeath, G., 2001, Chemical genomics: what will it take and who gets to play? *Genome Biology* **2**: 2005.1-2005.6.

McDonald, W.H. and Yates, J.R. 3rd, 2002, Shotgun proteomics and biomarker discovery. *Dis Markers*.**18**: 99-105.

Moseley, M.,A., 2001, Current trends in differential expression proteomics: isotopically coded tags. *Trends in Biotechnology* **19** Suppl:S10-S16.

Nattkemper, T.,W., Ritter, H.J., Schubert, W., 2001, A neural classifier enabling high-throughput topological analysis of lymphocytes in tissue sections. *IEEE Trans Inf Technol Biomed.* **5**: 138-49.

Raamsdonk, L.M,, Teusink,, B., Broadhurst, D., *et al*, 2001,A functional genomics strategy that uses metabolome data to reveal the phenotype of silent mutations. *Nature Biotechnol* **19**: 45-50.

Resing, K., A., 2002, Analysis of signaling pathways using functional proteomics. *Ann N Y Acad Sci* **971**: 608-14.

Rubenwolf,S., Niewohner, J., Meyer, E., *et al,* 2002, Functional proteomics using chromophore-assisted laser inactivation. *Proteomics*. **2**: 241-6.

Salomon, A.,R., Ficarro, S.,B., Brill, L.,M., *et al*, 2003, Profiling of tyrosine phosphorylation pathways in human cells using mass spectrometry. *PNAS USA* **100:** 443-8.

Zheng, X.,F.,S., Chan., T.,F., 2002, Chemical genomics in the global study of protein functions. Drug Discovery Today **7:** 197-205.

Ziauddin, J. and Sabatini, D.M., 2001, Microarrays of cells expressing defined cDNAs. *Nature.* **411:** 107-10.

Chapter 10

Proteomics and Adverse Drug Reactions

MARTIN R. WILKINS
Experimental Medicine and Toxicology, Imperial College London, Hammersmith Hospital, Du Cane Road, London W12 0NN, UK

1. INTRODUCTION – THE SCALE AND NATURE OF THE PROBLEM

"Primum non nocere" is the physician's guiding maxim. Drugs are closely scrutinised during development, both for evidence of toxic effects and to define appropriate dose regimens that maximise benefit-risk ratios. Indeed, a cautious approach to drug development is one of the main reasons for the high rate of attrition and the relatively small number of drugs that make it to the market place from the new chemical entities identified. Nonetheless, despite this and even with the greatest care in therapeutic prescribing, adverse reactions to drugs occur in clinical practice – and all too frequently.

An adverse drug reaction (ADR) has recently been defined as "An appreciably harmful or unpleasant reaction, resulting from an intervention related to the use of a medicinal product, which predicts hazard from future administration and warrants prevention or specific treatment, or alteration of the dosage regimen, or withdrawal of the product" (Edwards and Aronson, 2000). Accurate data on their incidence are limited but there is general agreement that they are common and costly. In 2001, 2,168,248 toxic events affecting 8% of the population were reported in the USA (Litovitz et al, 2000). Of these, 748,094 were drug related and all classes of therapeutic agent are represented. The problem is estimated to be responsible for 4.7% of hospital admissions, rank between 4th and 6th in the leading causes of

229

H. Hondermarck (ed.), Proteomics: Biomedical and Pharmaceutical Applications, 229–242.
© 2004 *Kluwer Academic Publishers. Printed in the Netherlands.*

death in the US (Lazarou et al, 1998) and cost around $136 billion per year (Johnson & Bootman, 1995).

ADRs can take various forms. They may be clinically silent for long periods and only apparent on clinical investigation. Among the most common presentations are skin rash, liver dysfunction, Q-T prolongation on the ECG, neurological disturbance and renal impairment. Some ADRs are predictable and preventable but many are not. From a clinical perspective, ADRs can be broadly divided into type A and type B. Type A reactions are more common and may be predicted from the known properties of the drug. Arguably the more dangerous are Type B reactions. Typically the latter are unrelated to the known pharmacology of the drug and are not dose-dependent. They are often regarded as idiosyncratic and may depend heavily on host factors. Some have an immunological basis but we know very little about the underlying mechanisms. Type B reactions are thought to account for up to 20% of ADRs and they carry a high mortality rate.

In an ideal world it would be possible to identify individuals at risk of an adverse response to a given drug and an alternative agent would be used.

Box 1: Unmet need in the management of ADRs for an investigation

- To identify susceptible individuals
- To detect early tissue damage
- That can be performed on accessible body tissues
- That identifies the responsible drug
- That avoids the need for rechallenge

Failing this, the diagnosis would be made early and the responsible agent identified with confidence. In reality neither is rarely possible. The frequency of ADRs testifies to the fact that they often catch physicians, pharmacists and patients unawares and this is a particular concern where tissue damage is involved. The clinical appearance of skin rashes rarely discriminates between drugs and a rise in liver enzymes or the appearance of protein in the urine are generic responses to tissue injury and usually indicate the organ affected rather than nominate the agent responsible. There is an unmet need for a clinical investigation that can identify individuals that are susceptible to an adverse reaction from a drug they may be prescribed and assist in the diagnosis of an ADR, identifying the culprit and avoid the need to rechallenge patients (see Box 1).

2. CURRENT DIAGNOSTIC OPTIONS

Currently the diagnosis of ADRs requires clinical skill rather than scientific proficiency. It needs a low threshold of suspicion and benefits from clinical experience, and where possible, pattern recognition, supported by standard haematological, biochemical and histological services.

The panel of biochemical analyses available in most routine clinical laboratories measure liver and renal function and a limited number of protein markers of tissue damage. They identify the affected organ and can be used to monitor progress but lack aetiological specificity. Thus, liver function tests measure enzymes that leak into the circulation as a result of hepatic damage but point rather vaguely to the cause. In a jaundiced patient, changes in the circulating levels of aspartate transaminase and alkaline phosphatase are useful in distinguishing between an obstructive cause, for example due to gall stones, and hepatocellular damage from drugs but do not discriminate between hepatic toxins. Similarly, the measurement of plasma levels of "cardiac enzymes", troponins and the myocardial isoform of creatine kinase can be used to identify and follow myocardial damage while natriuretic peptide levels are used to diagnose and monitor treatment in cardiac failure - but do not define the cause. These markers are all useful pointers but there is considerable room for improvement.

Where a patient is on a single agent and the clinical presentation fits with well-documented descriptions of a causative association with that drug, the diagnosis is straightforward. A more usual scenario is that the patient is on several medicines and the reaction could be due to one of a number of drugs. The possible culprits may be narrowed by identifying a temporal association with the introduction of a new agent or an increase in dose or by observing the effect of selective drug withdrawal. However, when the patient is very ill and needs continuous treatment, withdrawing therapy may not be easy and it has to be undertaken with great care.

Among the measures that might assist diagnosis and attribute causality in ADRs are nomograms (Lanctot and Naranjo, 1995), rechallenging the patient systemically with the drug and skin testing (DeLeo, 1998). Rechallenging the patient to establish causality may be scientifically correct but carries obvious risks to the patient and the likelihood of conducting such a challenge is inversely related to the severity of the ADR. Skin testing has been conducted for contact allergy for many years. Patch testing, prick testing and intradermal drug injection can be employed to examine for an adverse response to drugs with reduced risk of a life-threatening reaction but experience with this approach is limited (Barbaud et al., 1998). There are concerns about the sensitivity and specificity of this type of drug challenge and, as with contact allergy, distinction has to be made between irritation and true allergy.

Box 2: Assignment of causality of adverse drug reaction

- Certain – Consistent temporal association, including clinical course following withdrawal of drug and, where appropriate, rechallenge.
- Probable – Consistent temporal association but not confirmed by rechallenge
- Possible – Likely association but could be explained by another disease or drug.
- Unlikely – Temporal relationship to drug administration not consistent with causality
- Conditional – Lack of data necessary for proper assessment but more data being examined
- Unassessable - Lack of data necessary for proper assessment

---

The management of severe ADRs may well entail multiple changes in therapy and the use of corticosteroids as a 'blanket' measure to reduce inflammation and immunological responses. The patient may recover but the physician is left with assigning the probability of a causative association (ranging from certain to unassessable, Box 2) and the patient may be deprived of a useful agent on the grounds of an opinion rather than hard data.

3.   PRACTICAL PROTEOMICS IN TOXICOLOGY

3.1   The tissue

The application of protoemics to the management of ADRs requires the measurement of protein levels in an appropriate, accessible tissue. Plasma constitutes the largest accessible version of the human proteome and it comprehensively samples the human phenotype, giving a picture of the state of the body at a particular point in time. Nearly 300 proteins have been identified in plasma to date and many more are there to be discovered (Anderson and Anderson, 2000). These have been classified into 8 categories (Box 3). Other fluids and tissues can be obtained from patients for analysis (for example, saliva, tears, urine, hair, skin, liver etc). The disadvantage of these samples is that the protein complement is either a small subset of the 'plasma proteome' or provides only a restricted view of local cellular activity. This may be very relevant for some ADRs, for example, adverse skin reactions (Boxman et al., 2002). But in practical terms, efforts focusing on plasma and the methodologies that can utilise this material are most likely to succeed in bringing protein profiling to the bedside.

Box 3: The constituents of the plasma proteome[1]

- Proteins secreted by solid tissues that act in plasma - classical plasma proteins for liver and gut e.g. albumin
- Immunoglobulins - some 10 million antibodies
- "Long distance" receptor ligands - classical peptide and protein hormones
- Local receptor ligands - short-distance mediators of cellular response e.g. cytokines
- Temporary passengers - non-hormone proteins travelling to their site of action e.g.lysosomal proteins
- Tissue leakage products – as a result of tissue damage e.g.cardiac troponins
- Aberrant secretions – e.g. as a result of tumour development
- Foreign proteins – from infectious organisms

1. Modified from Anderson and Anderson, 2002

3.2   The methodology

Traditionally, markers of cellular response to disease or drugs have evolved from the pursuit of candidate molecules. Protein profiling surveys a broad range of proteins simultaneously without preconception of the molecules that might be involved. This approach is more efficient, improves statistical power and offers the possibility of identifying protein signatures of drug activity. The protein 'barcode' could then be used in diagnosis and monitoring without further resolution of the specific proteins captured or the proteins could be identified and specific immunoassays developed.

Practical strategies for protein profiling are the subject of much research. Two-dimensional electrophoresis gels require relatively large sample volumes and are not amenable to high-throughput analyses. The potential of mass spectrometry to yield comprehensive profiles of peptides and proteins in biological fluids without the need to first carry out protein separations is more attractive. An example of this approach is known as surface-enhanced laser desorption/ionization (Petricoin et al., 2002). Microlitre quantities of serum from many samples are applied to the surface of a protein-binding plate, with properties to bind a class of proteins. The bound proteins are treated and analysed by matrix laser desorption ionisation (MALDI). The mass spectra patterns obtained for different samples reflect the protein and peptide contents of these samples. It has been applied to the analysis of serum from cancer patients. Patterns that distinguish between cancer patients and normal subjects with remarkable accuracy have been reported for several types of cancer (Petricoin et al., 2002). The main drawbacks of direct analysis of tissues or biological fluids by MALDI are the preferential detection of proteins with a lower molecular mass and the difficulty in determining the identity of proteins owing to post-translational modifications obscuring the correspondence of measured and predicted masses. Nonetheless the technology has significant potential for exploring markers of drug toxicity in biological fluids from patients.

There is substantial interest in the development of microarrays or biochips for the systematic analysis of thousands of proteins (Fields et al., 2002). New classes of capture agents include aptamers (SomaLogic, http://www.somalogic.com/), ribozymes (Archemix, http://www.archemix.com/), partial-molecule imprints (Aspira Biosystems, http://www.aspirabio.com) and modified binding proteins (Phylos, http://www.phylos.com). Protein arrays have been used in the detection and characterisation of antibodies in immune disorders, such as rheumatoid arthritis, and could be employed to characterise immune responses to drugs (Robinson et al., 2001). Among the main challenges in making biochips for the global analysis of protein expression is the current lack of comprehensive sets of genome-scale capture agents such as antibodies and that proteins undergo numerous post-translational modifications that may be crucial to their functions. These modifications are generally not captured using either recombinant proteins or antibodies that do not distinctly recognize specific forms of a protein. This may be circumnavigated by arraying proteins isolated directly from cells and tissues following protein fractionation schemes (Madoz-Gurpide et al., 2002).

3.3 Proteomics in drug development and clinical practice

There is considerable interest in the use of proteomics in drug development to identify biomarkers that correlate with therapeutic response or signal a toxic reaction to chemical entities (Steiner and Witzmann, 2000). The range of protein changes that occur following exposure of a tissue to a drug may provide reassurance that the drug targets selectively an appropriate biochemical pathway or they can elucidate new drug targets. Applied to toxicology, the changes detected by proteomics may precede macroscopic tissue damage and serve as a sensitive indicator of potential toxicity. They may also provide insight into biochemical pathways of toxicity.

Toxicity studies must often be performed in animal models and concerns over the extrapolation of results to humans probably represent the greatest single hurdle in modern toxicological practice. There is an expectation that proteomics will provide "bridging biomarkers" – protein signatures detected in preclinical investigations that can be employed to alert investigators of adverse events in clinical studies. At present various safety factors are used to establish a safe dose for humans in preclinical evaluation of new chemicals (Smith, 2001). To allow for interindividual differences, a safety factor of 10 is applied. This is multiplied by another factor of 10 to allow for the possibility that humans may be more susceptible than the animals tested. A further factor of 10 is applied if there are particular concerns or the drug is to be used in children. This can result in an extremely large safety margin and the over cautious termination of some drugs in development. Understanding the mode of action of a toxic event in an animal model can help interpret the relevance (or lack of) to humans; changes can be species-specific and the recognition of this can be useful in decisions on whether to proceed with the development of a particular drug. A characteristic proteomic signature of toxicity would allow a detailed examination of the dose-response relationship in preclinical experiments, provide the opportunity to compare chemicals and find the most suitable for further development and offer a biomarker(s) for the early detection of adverse reactions in clinical trials.

As proof of principle, there are several reports of studies in rodents where the object has been to characterise the toxic effects of specific chemicals on the liver or kidneys. Distinct protein patterns have been associated with exposure to peroxisome proliferators activated receptor (PPAR) agonists and non-steroidal anti-inflammatory drugs that can be used to screen new chemical entities for activity (Anderson et al., 1996; Eberini et al., 1999). Proteomics identified the relationship between changes in the expression of a calcium-binding protein, calbindin-D 28kDa, and nephrotoxicity from ciclosporin A in renal biopsies (Aicher et al., 1998). Similar studies in rats have identified a novel protein in serum that may be a marker of renal toxicity from gentamycin (Kennedy, 2001). Changes in the pattern of proteins excreted in the urine of patients receiving radiocontrast agents have been documented that might represent markers of impending nephropathy (Hampel et al., 2001).

Not all the changes observed in a proteomic experiment will be related to toxicity; some will be adaptive. While more difficult to conduct, it is important to recognise that more useful information may be gained from clinical research and every effort should be made to facilitate this. Such studies may take the form of collecting samples (blood, urine) from all subjects in a clinical trial for retrospective analysis when the outcomes of exposure are known. Alternatively it would be possible to establish tissue banks (of blood, urine, skin or liver biopsies) from patients with clinically well characterised ADRs.

Central to the success of proteomics applied to toxicology will proteome reference libraries. These datasets will permit comparisons between drugs, clarify species-dependent effects, allow predictions to be made about the potential for toxicity and provide a diagnostic resource for consultation.

4.   COMPETING TECHNOLOGIES

4.1   Genetics

The role of genetics is attracting considerable attention. Numerous examples of genetic predisposition to ADRs exist in the literature. The recent sequencing of the human genome has focused attention on the more widespread use of genotype data to reduce the potential of an adverse drug response and permit personalised medicines.

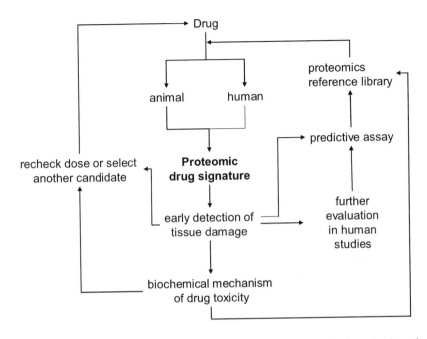

Figure 1. Use of a proteomic drug signature to inform toxicity studies in animals and humans.

Genetic variation can influence both drug kinetics and drug targets. Much of the literature is concerned with the clinical relevance of genetic polymorphisms in drug metabolising enzymes but data are accumulating on the contribution of variations in receptors, ion channels, enzymes and immune response to variation in drug response. To date these studies have examined the association with candidate genes but there is growing interest and speculation about the application of single nucleotide polymorphism (SNP) profiles (Roses, 2000).

SNPs are single base differences in DNA sequence that are distributed throughout the human genome at an average density of 1 every 1.9 kilobases (The International SNP Map Working Group, 2001). Around 60,000 are within coding regions and likely to be functionally active but it is not necessary to examine for a functional link between SNP and any particular drug. Patients could be divided into responders/non-responders or ADR/no-ADR groups according to their SNP haplotype, without preconception of the underlying mechanism. This approach is attractive for its potential predictive value and the avoidance of ADRs. Moreover, at the time of clinical presentation, knowledge of a patient's genotype may add to the probability that drug A rather than drug B is the causative agent. However, it is of limited use in diagnosis. It is also important to realise that this approach remains speculative as there are significant obstacles to the use of SNP profiling, which include the need for low-cost, high-throughput genotyping platforms, the large sample sizes required to demonstrate an association and ethical and social issues.

4.2   Differential gene expression

Recent developments in technology and bioinformatics permit the rapid assay and interpretation of 25,000+ gene transcripts on small solid-state "chips". The technologies for DNA microarray analysis are still evolving. There is a tendency by manufacturers to favour oligonucleotide- over complementary DNA-based microarrays, and progress has been made on adoption of data analysis standards. This technique has been employed to study gene expression in a variety of tissues in response to different perturbations, such as hypoxia, gene knockout and drugs. It has been used to subclassify tumours at the molecular level in an attempt to improve predictions about prognosis and target treatment more effectively. In principle, it could be used to detect and define the characteristic change in expression of several genes following exposure to a drug – that is, detect a gene signature associated with toxicity to a drug and thus be of value in diagnosis.

This approach has the advantage of sensitivity, in that very low levels of transcripts can be measured, but has some significant limitations. Unfortunately, although it can be applied to solid tissue samples, such as skin, liver and renal biopsies, it is not easy to isolate good quality mRNA from biological fluids. This limits the use of gene expression profiling in more accessible samples, such as blood, urine and synovial fluid. In addition, post-translational modification of proteins and concerns over the correlation between mRNA and protein abundance mean that reliance on the measurement of transcript levels will not provide the full story and further information could be gained from protein profiling.

Proteomics has advantages over gene expression profiling. Proteins can be measured reliably in a broader range of biological tissues (eg blood, CSF, synovial fluid) than mRNA transcripts and are the "business molecules". On the negative side, it is more difficult to detect proteins expressed in low abundance. Moreover, for the investigation to enter clinical practice, a rapid assay of protein markers is required. However, once the biomarkers that characterise a drug response have been identified, these proteins could be screened by standard immunoassay. The possibility of measuring several such proteins simultaneously using antibodies immobilised on "chips" (antibody arrays) is very attractive.

4.3 Metabonomics

The global analysis of cellular metabolites in biological fluids and tissues rivals proteomics for the integrated assessment of whole organism wellbeing. The biomarker information that is present in the NMR spectra of fluid and tissue samples is potentially very rich, as hundreds of compounds that represent a variety of metabolic pathways are measured simultaneously. Like proteomics, metabonomics makes no assumptions about the nature of the toxicity and seeks molecular signatures of effect. NMR spectroscopy of biofluids when coupled with pattern recognition analysis can be an efficient method of investigating toxicity profiles of xenobiotics. In addition, the development of high-resolution magic angle spinning (MAS) technology permits very high quality ^{1}H NMR spectra from small samples of whole tissue with no sample pre-treatment.

The use of metabonomics has been explored in the assessment of rodent responses to specific toxins and novel markers have been identified. For example, a combination of changes in urinary levels of trimethylamine-N-oxide, N,N-dimethylglycine, dimethylamine and succinate are associated with renal papillary damage, for which no prior biochemical markers existed (Holmes et al., 1995). A wide range of toxins are now being investigated using a metabonomics approach. A consortium (the Consortium for Metabonomic Technology, COMET) from the pharmaceutical industry and academia is generating databases of ^{1}H NMR spectra of rat and mouse urine and blood serum (Lindon et al., 2003). From this expert systems are being devised for the prediction of liver and kidney toxicity.

Metabonomics may be regarded as being closer to the mechanism of tissue damage than proteomics and can detect lower abundance molecules but has its limitations. Like gene expression and protein profiling it detects adaptive as well as toxic changes and these need to be distinguished in the interpretation of the results. It can be applied to clinically accessible samples but certain pathologies, such as liver fibrosis, are associated with negligible effects on biofluids, as metabolic derangement does not occur until there is significant tissue damage.

5.   CONCLUSIONS - AN INTEGRATED APPROACH TO THE MANAGEMENT OF ADRs

The conventional approach to preventing and managing adverse drug reactions has serious limitations. Proteomics offers an exciting advance but has to be considered in context. Without question, it is best employed together with other emerging platform technologies. Gene expression profiling, proteomics and metabonomics may all be used to screen out toxic drugs early in development with greater confidence and bring to the market agents with a better safety profile, a known mechanism of toxicity (and by implication, a better chance of avoiding an adverse event) or an assay of clinical value in the early detection of an adverse response. These assays, together with clinical genetics, may enable more personalised prescribing, with a view to improving the benefit:risk ratio.

In addition to prevention, there is diagnosis. It is fanciful to expect that the "omic" technologies will prevent all adverse events. Aside from errors of prescribing (not a trivial problem) and drug-drug or drug-environment interactions, unexpected so-called idiosyncratic adverse events will continue to present to physicians. In the clinical world, patients are often exposed to more than one drug and this can complicate the picture. There is no doubt that an investigation that identifies which drug of a combination a patient is taking is responsible for an adverse effect and should therefore be discontinued would be a significant advance. Proteomics (and, indeed, gene expression and metabonomics) is currently being evaluated in the assessment of known toxins for specific signatures or fingerprints of toxicity. It may be necessary to study the effect of drug/toxin combinations to characterise the presence of additivity, synergy or antagonism in biomarker profiles.

There is wide recognition that the success of this approach depends upon collaboration and the generation of well validated databases. The Human Proteome Organisation (HUPO, http://www.hupo.org) was founded to consolidate national and regional proteome organizations into a worldwide organization, encourage the spread of proteomics technologies and assist in the coordination of public proteome initiatives aimed at characterizing specific tissue and cell proteomes. Pilot international initiatives already underway are directed at identifying proteins detectable in normal serum and plasma and their range of variation with age, ethnicity and physiological state, and a liver proteome study to identify proteins expressed in the liver. These initiatives have attracted substantial interest and will be integrated with efforts in protein informatics to achieve data standardization on the one hand, and data curation on the other (Hanash, 2003).

REFERENCES

Aardema, M.J., and MacGregor, J.T., 2002, Toxicology and genetic toxicology in the new era of "toxicogenomics": impact of "-omics" technologies. *Mutation Res.* **499:** 13-25.

Aicher, L., Wahl, D., Arce, A., Grenet, O., and Steiner, S., 1998, New insights into cyclosporine A nephrotoxicity by proteome analysis. *Electrophoresis* **11:** 1998-2003.

Anderson, N.L., Esquer-Blasco, R., Richardson, F., Foxworthy, P., Eacho, P., 1996. The effects of peroxisome proliferators on protein abundance in mouse liver. *Toxicol. Appl. Pharmacol.* 137: 75-89.

Anderson, N.L., and Anderson, N.G., 2002, The human plasma proteome. *Mol. Cell. Proteom.* **1:** 845-867.

Bandara, L.R., and Kennedy, S., 2002, Toxicoproteomics – a new preclinical tool. *DDT* **7:** 411-418.

Barbaud, A., Reichert-Penetrat, S., Tréchot, P., Jacquin-Petit, M-A., Ehlinger, A. et al., 1998, The use of skin testing in the investigation of cutaneous adverse drug reactions. *Br. J. Dermatol.* **139:** 49-58.

Boxman, I.L.A., Hensbergen, P.J., Van der Schors, R.C., Bruynzeel, D.P., Tensen, C.P., et al., 2002, Proteomic analysis of skin irritation reveals the induction of HSP27 by sodium lauryl sulphate in human skin. *Br. J. Dermatol.* **146:** 777-785.

DeLeo, V.A., 1998, Skin testing in systemic cutaneous drug reactions. *Lancet* **352:** 1488-1490

Eberini, I., Miller, I., Zancan, V., Bolego, C., Puglshi, L., et al., 1999. Proteins of rat serum IV. Time course of acute-phase protein expression and its modulation by indomethacine. *Electrophoresis*, **20:** 846-853.

Edwards, I.R., and Aronson, J.K., 2000, Adverse drug reactioins: definitions, diagnosis and management. *Lancet* **356:** 1255-1259.

Gale, E.A.M., 2001, Lessons from the glitazones: a story of drug development. *Lancet* **357:** 1870-1875.

Hampel, D.J., Sansome, C., Sha, M., Brodsky, S., Lawson, W.E. et al., 2001, Toward proteomics in uroscopy: urinary protein profiles a radiocontrast medium administration. *J. Am. Soc. Nephrol.* **12:** 1026-1035.

Hanash, S., 2003, Disease proteomics *Nature* **422:** 226-232.

Holmes, E., Bonner, F and Nicholson, J.K., 1995, Comparative studies on the nephrotoxicity of 2 bromoethanamine hydrobromide in the Fisher 344 rat and he multimammate desert mouse (Mastomys natalensis) *Arch Toxicol* **70:** 89-95.

Jain, K.K., 2001, Proteomics: delivering new routes to drug discovery – Part 2. *DDT* **6:** 829-832.

Johnson, J.A., and Bootman, J.L., 1995, Drug-induced morbidity and mortality: a cost of illness model. *Arch. Intern. Med.* **155:** 1949-1956.

Kaufman, D.W., and Shapiro, S., 2000, Epidemiological assessment of drug-induced disease. *Lancet* **356:** 1339-1343.

Kennedy, S., 2001, Proteomic profiling from human samples: the body fluid alternative. *Toxicol. Lett.* **120:** 379-384.

Lazarou, J., Pomeranz, B., Corey, P. 1998. Incidence of adverse drug reactions in hospitalised patients: a meta-analysis of prospective studies. JAMA, 279, 1200-1205.

Lindon, J.C., Nicholson, J.K., Holmes, E., Antti, H., Bollard, M.E., et al., 2003, Contemporary issues in toxicology. The role of metabonomics in toxicology and its evaluation by the COMET project. *Toxicol. Appl. Pharmacol.* **187:** 137-146.

Litovitz, T.L., Klein-Schwartz, W., White, S., et al., 2001, 2000 Annual report of the American Association of poison control and toxic exposure surveillence system. *Am. J. Emerg. Med.* **19:** 337-395.

Madoz-Gurpide, J., Wang, H., Misek, D. E., Brichory, F. & Hanash, S. M., 2001, Protein based microarrays: a tool for probing the proteome of cancer cells and tissues. *Proteomics* **1:** 1279–1287.

Man, W.J., White, I.R., Bryand, D., Bugelski, P., Kamilleri, P., et al., 2002, Protein expression analysis of drug-medicated hepatotoxicity in the Sprague-Dawley rat. *Proteomics* **2:** 1577-1585.

Naaby-Hansen, S., Waterfield, M.D., and Cramer, R., 2000, Proteomics – post-genomic cartography to understand gene function. *TRENDS Pharmacol. Sci.* **22:** 376-384.

Nicholson, J.K., Connelly, J., Lindon, J.C., and Holmes, E., 2002, Metabonomics: a platform for studying drug toxicity and gene function. *Nature Rev. Drug Discov* **1:** 153-161.

Petricoin III, E.F., Hackett, J.L., Lesko, L.J., Puri, R.K., Gutman, S.I., et al., 2002, Medical applications of microarray technologies: a regulatory science perspective. *Nature Genet. Suppl.* **32:** 474-479.

Petricoin, E.F. Zoon, C.K., Kohn, E.C., Barrett, J.C., and Liotta, L.A., 2002, Clinical proteomics: translating benchside promise into bedside reality. *Nature Rev* **1:** 683-695

Pirmohamed, M., and Park, B.K., 2001, Genetic susceptibility to adverse drug reactions. *TRENDS Pharmacol. Sci.* **22:** 298-305.

Robinson, W. H. *et al.* , 2002, Autoantigen microarrays for multiplex characterization of autoantibody responses. *Nature Med.* **8:** 295–301.

Roses, A.D., 2000, Pharmacogenetics and the practice of medicine. *Nature* **405:** 857-865.

Smith, L.L., 2001, Key challenges for toxicologists in the 21st century. *TRENDS Pharmacol. Sci.* **22:** 281-285.

Steiner, S., Witzmann, F.A., 2000. Proteomics: Applications and opportunities in preclinical drug development. *Electrophoresis* 21: 2099-2104.

The International SNP Map Working Group., 2001, A map of the human genome sequence variation containing 1.42 million single nucleotide polymorphisms. *Nature,* **409:** 928-933.

Chapter 11

Proteomics for Development of Immunotherapies

JEAN-FRANÇOIS HAEUW and ALAIN BECK
*Centre d'Immunologie Pierre Fabre, 5 Avenue Napoléon III, 74160 Saint-Julien en Genevois,
France (www.cipf.com)*

1.   INTRODUCTION TO BIOPHARMACEUTICALS

Recombinant DNA technology has come of age. The first true application of this technology was the manufacture of therapeutic proteins. The first such biopharmaceutical product to come on the market was recombinant human insulin. This product, produced in *Escherichia coli* cells, was granted a marketing licence by the Food and Drug Administration (FDA) in October 1982. Until recently, all biopharmaceutical products were protein-based. However, since the 1990s, nucleic acid-based biopharmaceuticals have also come to prominence, being employed in gene therapy and anti-sense technology. The first anti-sense drug Vitravene®, for the treatment of cytomegalovirus retinitis in AIDS patients, was approved by the FDA in 1998. Today the definition of biopharmaceuticals refers to pharmaceutical preparations involving a protein produced by recombinant DNA technology or a nucleic acid-based compound as drug substance or active ingredient. Most notable among these biotechnology products are those for human therapeutic use, including hormones, growth factors, cytokines, monoclonal antibodies and vaccines. These products are used in a large number of therapeutic segments for the treatment of widespread pathologies, such as cancer, rheumatoid arthritis or asthma, as well as rare genetic disorders like Fabry or Gaucher diseases.

We will focus this chapter on therapeutic proteins, since their development, analytical characterization and commercialization are more advanced than that of nucleic acids.

243

H. Hondermarck (ed.), Proteomics: Biomedical and Pharmaceutical Applications, 243–278.
© 2004 *Kluwer Academic Publishers. Printed in the Netherlands.*

1.1   The biopharmaceutical market

Since the beginning of the 1990s approvals of new biotechnology drugs and vaccines are rapidly increasing, with 22, 32 and 24 new biotech drug approvals and new indications for already approved drugs in 1999, 2000 and 2001, respectively. Biopharmaceuticals accounted for approximately 30% of all new molecular entities approved in the USA by the FDA in 2001 and 2002. In Europe, the situation is similar. Two hundred and sixty-five new pharmaceutical products have been granted marketing licences by the European Commission since 1995. Ninety-five of these were of biotechnology, representing 36% of all new drugs approved within this time-frame (Walsh, 2003). There are now more than 100 biopharmaceuticals already approved and available to patients in some world regions at least, with 88 having received approval within the European Union (Walsh, 2003). Among these 88 protein-based products, hormones and cytokines represent the largest categories, with 23 and 18 products respectively. Additional product categories include recombinant blood factors and related products, monoclonal antibodies and a range of subunit vaccines.

It is estimated that 250 million people have already benefited from medicines and vaccines developed through biotechnology, saving and improving their life. There are around 2000 biotechnology companies worldwide, and more than 180,000 employees working in this growing sector. Total biopharmaceuticals sales reached approximately 22 billion US Dollars in 2001, corresponding to 6 % of the total pharmaceutical market, and are expected to reach 45-50 billion US Dollars by 2006. In 2001, the leading product category was constituted by erythropoietins, representing 26% of total sales. Immunotherapy products accounted for around 17% of total biopharmaceuticals sales, with 14% for monoclonal antibodies and 3% for vaccines. Other products accounting for significant market values were: interferons (16% of total sales), insulins (9% of total sales) and growth hormones (8% of total sales). Examples of a few highly successful biopharmaceuticals and vaccines are: Amgen's erythropoietins for anemia and dialysis ('Epogen®/aranesp' 2001 sales: US$ 2.150 billion); Schering-Plough's alpha interferon product for treating hepatitis C and various cancers ('Intron®' 2001 sales: US$ 1.447 billion); Eli Lilly's product line for diabetics ('Humulin®' 2001 sales: US$ 1.061 billion) and GlaxoSmithKline's hepatitis B vaccine ('Engerix B®' 2001 sales: US$ 750 million). Some antibody products launched in the past few years have almost reached blockbuster status, such as Johnson & Johnson/Schering Plough's Remicade® for Crohn's disease and rheumatoid arthritis, and Rituxan® for non-Hodgkin's lymphoma from IDEC/Genentech, with sales of more than US$ 750 million in 2001.

Biopharmaceuticals are among the most expensive of all pharmaceutical drugs. For example, in the 1980s the estimated cost per patient for a breast cancer treatment with a chemotherapy drug was around 150 Euros for 6 months. Today, this cost has increased more than 100 fold with the use of the humanized monoclonal antibody Herceptin®/trastuzumab, which was launched in 1998. The annual cost per patient for this treatment is indeed estimated between 30.000 and 50.000 Euros. These high costs of treatment are mainly due to a long, difficult and expensive process required to discover and develop safe and effective new medicines. Indeed it takes on average 10 to 15 years to bring a drug to market from inception. Before a company can administer a drug to human people, it requires an average of 6.5 years of basic discovery work and preclinical testing in animals. Clinical trials start with Phase I toxicity testing. The drug is administered to 20 to 100 healthy volunteers for about a year and a half. The tests study a safety profile of the drug (determination of safe dosage range, detection of most frequent side effects) and determine how it is absorbed, distributed, metabolised and excreted. In phase II, the drug is given to 100 to 500 volunteer patients (people with the disease) to determine its effectiveness. This phase takes about 2 years. Finally, Phase III involves 1,000 to 5,000 patients to confirm efficacy and identify adverse events. The development cost is now estimated between US$ 600 and 800 billion. Moreover only one out of 5,000 compounds that enter preclinical testing is approved for launching.

The financial potential partly explains the 500 new biopharmaceuticals in various stages of development worldwide. It can be expected that an average of 10 new biopharmaceuticals will reach the market every year during the coming 5 to 7 years, which will contribute to the continued growth of this segment. As recently estimated by the American Association of Pharmaceutical Researchers and Manufacturers (PhRMA), 371 biotechnology medicines are in development in the USA (http://www.phrma.org). Among these, 178 medicines are developed for cancer, and 68 others for infectious diseases including 21 drugs for HIV infection. When these medicines are classified by product category, immunotherapy products represent the most important class with 98 recombinant vaccines and 75 monoclonal/engineered antibodies.

1.2   Production of biopharmaceuticals

Among the host systems used for recombinant protein production, microbial and mammalian cells are the most widely used. Microbial systems combine outstanding advantages such as easy development of production, high yield and low costs of development and production. However they often require complex refolding and purification operations. Since they do not allow post-translational modifications such as glycosylation, their use is limited to more "simple" biopharmaceuticals. Among microbial systems, *Escherichia coli* is the most widely used. Heterologous expression in *E. coli* often leads to the accumulation of unfolded and biologically inactive protein in the cytoplasm, deposited as inclusion bodies. When compared to extracellular expression, this insoluble intracellular expression complicates downstream processing since multiple steps are required before chromatographic purification: cell disruption, separation of inclusion bodies from cell debris and soluble proteins, protein solubilisation with strong chaotropic agents, like guanidinium hydrochloride or urea, and reducing agents, and last refolding by dilution with folding buffer. Correctly folded and active protein is further purified by chromatographic methods.

Mammalian systems are considered to be more costly to develop because of complex nutritional requirement, slow growth and susceptibility to physical damage. However they are regarded as systems of choice for more complex biopharmaceuticals, like antibodies and fragments, involving multiple peptidic chains and requiring substantial post-translational modifications. In general, animal cell culture-derived biopharmaceuticals are produced as extracellular proteins and thus secreted into the medium. Further to the success of engineered antibodies, the share of mammalian cell systems has now reached around 60% of all systems used to produce protein-based pharmaceuticals.

Yeast is the third expression system used to produce biopharmaceuticals. As mammalian systems, they possess the ability to carry out post-translational modifications of proteins, although the glycosylation pattern usually varies somewhat from the patterns observed on the native protein or on the protein expressed in mammalian cells. Two recombinant proteins expressed in *Saccharomyces cerevisiae* are now approved for general medical use: hepatitis B surface antigen vaccine and the anticoagulant Hirudin®. Alternative promising production systems, in particular transgenic animal and plant systems, are still in development but these systems have to prove that they are technically and economically attractive.

Independently of the expression system used, a production process requires many stages to obtain a purified product; typically 10 to 20 steps compose upstream and downstream processing. Downstream processing usually consists of two to four chromatographic steps.

1.3   Purity analysis of recombinant pharmaceuticals

For biopharmaceuticals, absolute purity is impossible to achieve. Thus, purity is a relative term, depending mainly on the methods used to evaluate presence or absence of impurities and contaminants. Biopharmaceuticals are heterogeneous as a result of complex expression and purification processes. In addition, depending on certain manufacturing and storage conditions, physico-chemical changes can be observed, which can affect product purity and efficacy.

A wide range of analytical techniques is used in the characterization of recombinant proteins, including proteomics tools. By definition, proteomics is the simultaneous analysis of complex protein mixtures like tissue extracts, cell lysates, subcellular fractions or biological fluids at a given time and under precisely defined conditions. Therefore, proteomics tools can also be applied to quantify and identify complex mixtures of proteins in a purified biopharmaceutical. Purity analysis occurs at three stages of the production process: during the production (in-process controls), at the end of the purification process on the bulk material, and after the final formulation on the finished product.

Potential impurities can come from host organism, production process or raw materials. In fact, impurities can be divided in two categories: process-related and product-related impurities. Process-related impurities are components derived from the manufacturing process, including fermentation ingredients and host cell components. Product-related impurities are variants of the protein product, which can result from fragmentation, chemical modification, glycosylation variation, aggregation, denaturation or unfolding. Table 1 summarizes major potential impurities and some methods for their characterization. Other detectable materials are considered to be contaminants, which can be defined as all adventitiously introduced entities that are not part of the production process. These include mycoplasma, bacterial, pyrogenic and viral contaminants. These contaminants will not be discussed in this chapter.

Levels of process and product-related impurities must be within safety limits established by international regulatory agencies FDA (http://www.fda.gov), EMEA (http://www.eudra.org) and WHO (http://www.who.int), which thus require the development of sensitive and validated quantitative methods from manufacturers. Major impurities are often product-related impurities while minor impurities are generally process-related impurities.

While the product itself or its related impurities are likely to be non immunogenic (except for recombinant vaccines against infectious diseases for which a high immunogenicity is required), repeated administration of host cell derived impurities can induce potential deleterious effects such as oncogenicity, specific toxicity or unwanted immunological responses creating allergic or anaphylactic reactions. A detection range of 1-100 parts per million (ppm) of residual host cell proteins has been quoted as a regulatory benchmark (FDA, 1985) since host cell proteins may cause an immune response in patients at levels as low as 100 ppm. Therefore it might be important to quantify and identify minor impurities to assure greater safety of the product. Modified forms of the protein of interest must also be removed during the purification process and identified in the final product, since they may compromise the product by reducing its potency.

Table 1. Types of potential impurities in recombinant pharmaceuticals and principal methods of detection and analysis

Impurity	Methods of analysis
Process-related impurities	
Upstream processing derived	
Media proteins (albumin, insulin…)	Immunoassay
Amino acids	RP-HPLC
Antibiotics	HPLC, immunoassay
Downstream processing derived	
Denaturant	HPLC
Reducing agent	HPLC
Solvent	GC
Trace metals	Atomic absorption spectroscopy
Enzymes	Immunoassay
Column leachables	Immunoassay
Host cell derived	
Proteins	SDS-PAGE, immunoassay
DNA (genomic and plasmidic)	Hybridization assay
Endotoxins	LAL, rabbit test
Product-related impurities	
Aggregates	SDS-PAGE, SEC
Denatured forms	SEC
Deamidations	IEF, IEC, chromatofocusing
Oxidations	IEC, peptide mapping
Misfolded conformers	Peptide mapping
N- or C-terminal variants	RP-HPLC, IEC, peptide mapping, IEF
Fragmented products	SDS-PAGE, SEC, MS

RP-HPLC, reverse phase HPLC; GC, gas chromatography; LAL, *Limulus* amoebocyte lysate; SEC, size-exclusion chromatography; IEF, isoelectric focusing; IEC, ion exchange chromatography; MS, mass spectrometry.

The range of chromatographic methods now available allows the standard production of protein products which are more than 95-98% pure. The challenge resides now more in the development and use of sensitive techniques to detect and quantify any impurities that may be present into the final product.

On the basis of two case studies of immunotherapy biopharmaceuticals developed in our Centre, this chapter will present and discuss different methodologies to characterize therapeutic proteins derived from recombinant DNA, together with different techniques used to analyse and quantify product- and process-related impurities. The first case study is a recombinant subunit vaccine, designated BBG2Na, to protect humans against respiratory syncytial virus (RSV). The second case study is a recombinant bacterial protein, called rP40, which we develop as an antigen carrier and immunological adjuvant for human vaccines.

2. BBG2NA, A RECOMBINANT SUBUNIT VACCINE AGAINST HUMAN RESPIRATORY SYNCYTIAL VIRUS

Respiratory syncytial virus is a member of the *Pneumovirus* genus and the *Paramyxoviridae* family. It is the main etiologic agent of serious respiratory disease in infants and young children (Collins *et al.*, 1996), and also causes serious disease in immunocompromised individuals and the elderly (Fouillard *et al.*, 1992; Mlinaric-Galinovic *et al.*, 1996). Approximately two-thirds of infants are infected with RSV during the first year of life, and nearly 100 % have been infected by the age of 2. Annual epidemics result in more than 100,000 hospital admissions per year in the United States (Shay et al., 1999). Despite its medical importance, no vaccine is available more than 40 years after the discovery of the virus. Potential candidates for RSV vaccines include the two major surface glycoproteins : the fusion (F) and the attachment (G) proteins (Piedra, 2003). Two subgroups, namely RSV-A and B, have been identified, primarily based on G protein genetic variability. Both proteins induce neutralizing antibodies and protective immunity in animal models. However, native G protein induces subgroup-specific protection, while F protein protects against both subtypes. In contrast, we have recently described a recombinant chimeric protein that protects mice and rats against both RSV-A and B challenge. This recombinant vaccine, called BBG2Na, was shown to elicit protective immunity against RSV in rodents after parenteral (Plotnicky-Gilquin *et al*, 2000, Klinguer *et al* , 2002) as well as after mucosal administration (Klinguer *et al*, 2001). Furthermore, no lung immunopathology following RSV challenge in mice immunised with BBG2Na was observed, contrary to the case of formalin-inactivated RSV vaccine (Plotnicky *et al*, 1999). Promising results were obtained in phase I (Power *et al*, 2001) and phase II

(Plotnicky *et al*, 2002) clinical trials. A phase III clinical trial has been initiated.

2.1   Production of BBG2Na

BBG2Na is a well defined chimeric protein consisting of the central conserved region, including residues 130-230, of human Long strain RSV-A G protein (G2Na) genetically fused to the albumin-binding domain of streptococcal protein G (BB) (Libon *et al*, 1999; Goetsch *et al*, 2003). BBG2Na was produced in *E. coli* with high yield and clinical-grade quality. A biologically inactive protein was produced intracellularly in inclusion bodies. After solubilisation of inclusion bodies with strong denaturant and reducing agents, protein was refolded by dilution and overnight incubation at room temperature. The correctly folded, and biologically active, protein was further purified to homogeneity. The industrial purification process involved five high-resolution chromatographic steps: cation exchange chromatography, hydrophobic interaction chromatography, size exclusion chromatography, anion exchange chromatography and final desalting size exclusion chromatography. The process did not include any biological product of animal origin. Thus, contaminations by viruses or non-conventional transmissible agents were unlikely to occur. In-process controls were performed to assess the reproducibility of manufacturing process and to ensure the elimination of impurities during purification. BBG2Na was extensively characterized, both from an immuno-chemical and a physico-chemical point of view, according to FDA and European Medicines Evaluation Agency guidelines for recombinant proteins and vaccines for human use.

2.2   Characterization of BBG2Na

2.2.1   Structural characterization and confirmation

BBG2Na contains 349 amino acids. Its molecular weight was determined by mass spectrometry. The mass of purified BBG2Na measured by electrospray ionisation mass spectrometry (ESI-MS) was 38672.48 Da $\pm$ 2.65. It was in accordance with the calculated mass ($C_{1708}H_{2760}N_{460}O_{541}S_6$ = 38672 Da) deduced from the gene sequence. The primary structure was assessed by peptide mapping and N- and C-terminal sequencing. N-terminal sequence analysis showed that a single sequence was detected, MKAIFVLNAQ, which corresponds exactly to the first 10 amino acids at the N terminus of BBG2Na as predicted from the DNA sequence. Reverse phase HPLC analysis of a tryptic digestion of BBG2Na was used for identification and primary structure confirmation (see example in Fig. 1). Trypsin cleaves proteins at the C-terminal side of the basic residues lysine and arginine. Twenty-one peaks were resolved. Most of the peptides were characterized by mass spectrometry, allowing the confirmation of 88% of the primary sequence. Due to the structure of the BB part, several peptides are repeated 2 or 3 times but were only identified in one peak. A second overlapping map was generated using staphylococcal protease V8 which cleaves specifically glutamyl peptide bonds. A combination of the two maps allowed a 100% confirmation of the primary amino acid sequence deduced from the nucleic acid sequence.

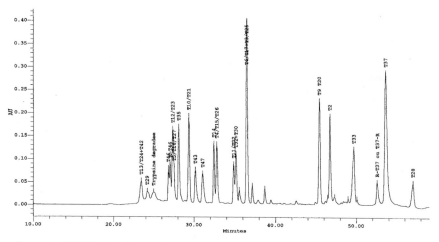

Figure 1. BBG2Na tryptic peptide map.

Two peptides of the tryptic peptide map were of particular interest for primary structure confirmation. First, peptide T47, identified as the C-terminal peptide (calculated mass = 985.52 Da; measured mass = 985.06 Da), was isolated and used to confirm C-terminal integrity by Edman degradation. Second, peptide T37 was used for disulfide bridge localisation because it contains the 4 cysteine residues of the molecule. This 36 amino acid peptide was identified on the basis of its mass (calculated mass = 4150.72 Da; measured mass = 4150.00 Da) and purified for further disulfide bond assignment studies (Fig. 2A). It was submitted to thermolysin sub-digestion. This enzyme non specifically cleaves peptide bonds involving the amino groups of hydrophobic residues. The different peptides generated were analysed by LC/ESI-MS (Fig. 2B). The cysteine pairings deduced from the measured masses were in agreement with a [C1-C4/C2-C3] native type of disulfide linkages connecting Cys[173] (C1) to Cys[186] (C4) and Cys[176] (C2) to Cys[183] (C3) (Fig. 2C).

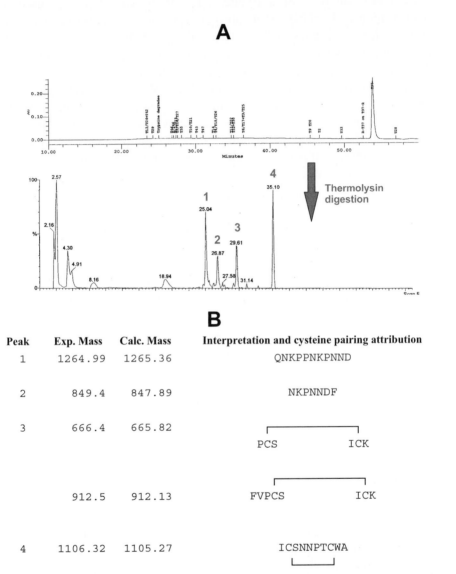

Figure 2. BBG2Na disulfide bridge assignment. A, thermolysin digestion analysis of peptide T37; B, LC-MS data of peptides released after thermolysin digestion of peptide T37.

The native pairing of cysteine residues in purified BBG2Na was also confirmed by using synthetic isomers of peptide T37, with different cysteine pairings as a result of different oxidation conditions, as standards for HPLC analyses and thermolysin peptide mapping (Beck *et al*, 2000). Moreover, circular dichroism and nuclear magnetic resonance studies on BBG2Na and peptides deriving from G2Na part demonstrated that the immunodominant central conserved region of the G protein of human RSV A including the 4 cysteine residues was the most structured domain. The three-dimensional structure of this "cysteine nose" was similar to the structure determined for the homologous protein of bovine RSV (Doreleijers *et al.*, 1996).

2.2.2 Physico-chemical properties

Chromatographic methods, such as reverse phase HPLC (RP-HPLC) and size exclusion chromatography, were used for identity and purity control, and also for quantitative measurement. The major impurity, which can be detected by RP-HPLC after the two first purification steps (Fig. 3), was isolated and analysed by mass spectrometry. The mass measured by Matrix Assisted Laser Desorption Ionisation Time Of Flight (MALDI-TOF) mass spectrometry for this impurity was around twice the mass of the monomer (78265 Da), and was interpreted to be a dimer of BBG2Na.

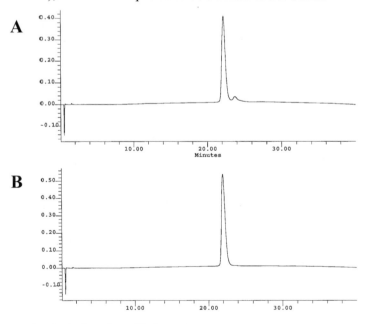

Figure 3. Analysis of purified BBG2Na by RP-HPLC. A, BBG2Na after two chromatographic steps; B, clinical grade BBG2Na.

Dimeric and oligomeric forms of BBG2Na were also detected by SDS-PAGE, but only for the protein obtained before the first gel filtration step (Fig. 4). These product-related impurities were identified by western blotting with monoclonal antibodies directed at the G2Na or BB parts. These impurities were absent from clinical grade batches purified with five chromatographic steps (Fig. 3 and 4).

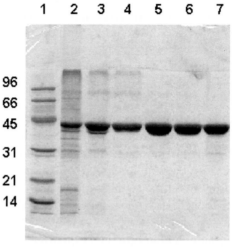

Figure 4. Analysis of in-process and purified BBG2Na by SDS-PAGE electrophoresis. Proteins (10μg) were separated on a 12.5% acrylamide gel before staining with Coomassie blue. Lane 1, molecular mass markers (kDa); lane 2, BBG2Na raw extract; lane 3, cation exchange chromatography step; lane 4, hydrophobic interaction chromatography step; lane 5, size exclusion chromatography step; lane 6, anion exchange chromatography step; lane 7, desalting chromatography step.

BBG2Na was characterized by an isoelectric point (pI) higher than 9.3 by using isoelectric focusing and capillary electrophoresis. As expected for basic proteins, BBG2Na was strongly adsorbed at a neutral pH by aluminium phosphate adjuvant, which exhibits an acidic point of zero charge (pZC) of 5.0 (Dagouassat *et al*, 2001b). Surprisingly, it was also adsorbed by aluminium hydroxide adjuvant (pZC = 11.1), which normally binds molecules with acidic isoelectric points. These observations may be explained by the bipolar two-domain structure of the BBG2Na chimera which is not reflected by the global basic pI of the whole protein. The BB domain has an acidic pI (5.5) and is responsible for adsorption to aluminium hydroxide adjuvant, whereas the G2Na domain with a highly basic pI (> 10) mediates binding to aluminium phosphate adjuvant (Dagouassat *et al*, 2001b). This unique property of BBG2Na makes it eminently suitable for combination to mono- and multivalent aluminium- or phosphate-containing vaccines already on the market or in development.

2.3   Detection and quantitation of impurities

2.3.1   Product-related impurities

Since the BBG2Na purification process involves five chromatographic steps, the final purified bulk product is highly pure. The different methods mentioned above were used to detect product-related impurities. Neither truncated and aggregated forms nor other modified forms (deamidated, isomerised, mismatched S-S linked or oxidized forms) were identified. Nevertheless, these methods allowed the demonstration of the clearance of different product-related impurities throughout the purification process. For example, elimination of BBG2Na aggregates after the third chromatography step, i.e. size exclusion chromatography step, was assessed by RP-HPLC (Fig. 3), SDS-PAGE (Fig. 4), western blot and gel-filtration analyses. Similarly, SDS-PAGE and western blot analyses were used to demonstrate the elimination of a truncated form of BBG2Na which was identified as one of the major impurities in the initial extract (Fig. 4). This impurity, characterized by an apparent molecular weight very close to the molecular weight of BBG2Na (Fig. 4), was separated from BBG2Na only by SDS-PAGE. The resolution of numerous analytical chromatographic methods was insufficient to separate it from BBG2Na. The second chromatography step of the purification process, i.e. hydrophobic interaction chromatography step, allowed the specific elimination of this impurity (Fig. 4, lane 4).

2.3.2   Process-related impurities

This part will mainly focus on host cell proteins (HCPs) quantification. Process-specific HCP assays are in general targeted to be in place prior to the initiation of phase III clinical trials. Immunoassays are the most specific and sensitive techniques available for detecting and quantifying protein impurities. There are two methods commonly employed to quantify protein impurities in biopharmaceuticals: enzyme-linked immunosorbent assays (ELISA) and immunoligand assays (ILA). Both methods are able to detect very low ppm level of impurities. ELISA have been developed to measure host protein impurities in a number of recombinant proteins including human growth hormone (Anicetti *et al.*, 1986), insulin (Baker *et al.*, 1981) and staphylokinase (Wan *et al.*, 2002). ILA assays have been used to detect protein impurities in recombinant bovine somatotropin (Whitmire and Eaton, 1997) and human erythropoietin (Ghobrial *et al.*, 1997). The major advantage of ILA is that antigens and antibodies can bind in liquid phase, thus strongly favouring the preservation of their native conformation.

We have developed and validated a process-specific ILA assay based on the Threshold system of the company Molecular Devices for strain-specific host cell-derived *E. coli* protein contaminants in in-process and purified lots of BBG2Na (Dagouassat *et al.*, 2001a). This assay is based upon use of rabbit anti-HCP polyclonal antibodies derived from null cells. A manufacturing scale preparation of HCP was made starting with cell paste of *E. coli* RV308 containing a plasmid vector which has been cured of the expression gene and consequently did not express the BBG2Na protein. The normal extraction and purification procedure was further applied, and the final HCP preparation for immunization was obtained after the first chromatography step. Since it is essential that the null cell HCP immunogen preparation not be contaminated with the biopharmaceutical product of interest, exhaustively cleaned and dedicated equipments were used to prepare HCP immunogen. One of the challenges is the generation of a polyclonal antibody which is highly specific and sensitive for each of the proteins in the complex mixture used as immunogen. HCP immunoreagent was produced by a cascade immunization procedure as described by Thalhamer and Freund (1984). This method allowed elimination, after each boosting dose, of the most immunogenic proteins on immobilized purified antibodies, in order to further immunize animals with poorly immunogenic proteins and also obtain a response against these proteins. Ideally, all of the proteins detected by sensitive silver stain on a SDS-PAGE should also be detected in an immunoblot by the progressively enhanced anti-HCP immune IgG. In practice, some poorly antigenic proteins which are detectable by silver stain may never elicit a detectable immune response. In our case, after three rounds of cascade immunizations the serum demonstrated specificity for more than 80% of the proteins present in the HCP mixture, including weaker antigens, as shown by 1-D and 2-D electrophoresis and western blot (Fig. 5). Most of the proteins under 30 kDa detected by western blot were highly basic proteins, characterized by a pI higher than 9 by 2-D electrophoresis (Fig. 5D). We were satisfied that most of the HCP were detectable by the immunoreagent at an acceptable level of sensitivity. Assuming that further immunizations would not have shown additional reactivities, the rabbits were sacrificed and the anti-HCP serum was collected and purified by affinity on immobilized Protein G. This purified HCP antiserum had no cross-reaction to BBG2Na and related proteins in western blot and ELISA. It allowed the detection of HCP by western blot into the BBG2Na extract and after the 2 first chromatography steps. No HCP was detected by western blot at the end of the purification process, thus confirming the high purity of the final bulk of BBG2Na obtained after 5 chromatography steps. Nevertheless, the purified serum was used to quantify

HCP throughout the process of purification of BBG2Na and allowed the detection of very low levels of HCP into the final purified bulk.

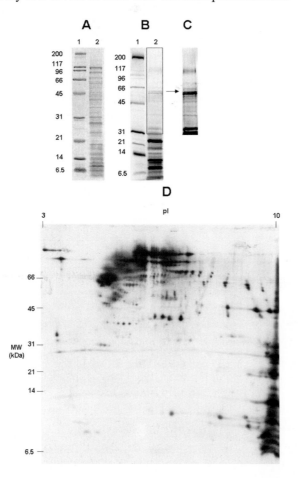

Figure 5. Specificity of the anti-HCP immune serum. A: SDS-PAGE analysis of HCP (16%, Coomassie blue staining). Lane 1, molecular mass markers (kDa); lane 2, HCP (10µg). B and C: immunoblotting with a rabbit polyclonal anti-HCP serum. After electrophoresis, proteins were transferred to a PVDF membrane. Immunolabelled proteins were visualized using ECL reagents. Lane 1, molecular mass markers (kDa); lane 2, HCP (2µg); lane C, over-exposition of the higher part of lane B2. D: 2-D PAGE western blot analysis of HCP. One hundred µg of HCP were loaded onto a 3-10 IPG strip by in-gel rehydration. After focusing, proteins were separated in the second dimension on a 15% homogeneous gel and further transferred to a PVDF membrane. Proteins were immunolabelled with the rabbit anti-HCP serum and further with a HRP-labelled mouse anti-rabbit IgG antibody before visualization using ECL reagents.

The double-sandwich ILA utilizes the purified anti-HCP antibodies separately labelled with fluorescein and biotin (Fig. 6) (Briggs *et al.*, 1990). The first stage of the assay consists in the formation of a tripartite complex between HCPs and the two labelled antibodies. After addition of streptavidin, these immune complexes are then captured on a biotinylated nitrocellulose membrane. Detection stage starts with incubation of the membrane with an anti-fluorescein-urease conjugate. The urease hydrolyses added urea into ammonia that alters the pH of the solution and consequently changes the surface potential on the membrane, which is then measured by a silicon sensor. The general detection range reported by Molecular Devices for HCP ILA assays is 2-160 ng/ml for *E. coli* HCPs and 2.5-200 ng/ml for Chinese Hamster Ovary cells HCPs.

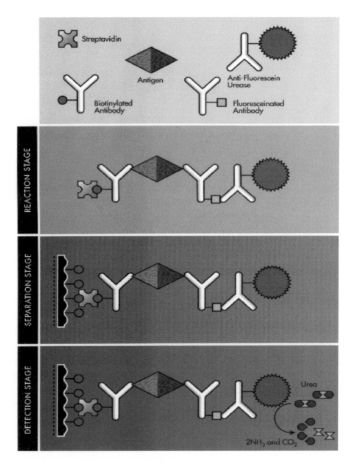

Figure 6. Immunoligand assay reaction scheme. (reprinted with permission from Molecular Devices).

The HCP-immunoligand assay had detection and quantification limits of 0.5 and 1.2 ng/ml of HCP respectively, equivalent to 10 and 30 ppm respectively. In order to calculate the amount of HCP contained in the samples, the standard curve obtained with HCP was plotted using the quadratic regression [$y=a+bx+cx^2$] (Fig. 7). This assay allowed the measurement of HCP in the BBG2Na final bulk since values in the range of 195 +/- 30 ppm were determined. We were also able to observe the clearance of impurities throughout the purification process (Table 2). Before the first chromatographic step, HCP concentration was 1.8%, equivalent to 18,000 ppm. The level of impurities decreased after each purification step to around 200 ppm after the last step.

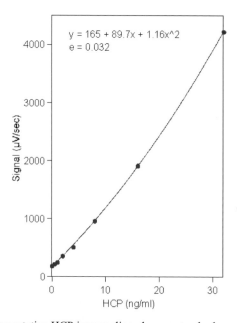

Figure 7. Representative HCP immunoligand assay standard curve.

Table 2. Removal of protein impurities through the purification process of BBG2Na.

Chromatography step	Content of HCP	HCP removal
	ppm[a]	%
1	18,000	-
2	3,000	83.3
3	800	95.5
4	700	96.1
5	200	98.9

[a] ppm values were calculated as the HCP amount (ng/ml) corrected for total protein content.

Another ILA assay has also been developed for the quantification of residual DNA in clinical BBG2Na batches (Lokteff *et al.*, 2001). As for HCP, this assay demonstrated the clearance of residual DNA during the downstream process. The amount of residual DNA found in the final bulks was below 20 pg per 300 µg of BBG2Na, which was the highest dose of antigen tested in clinical trials. This was confirmed with sensitive hybridization assays based on specific probes for plasmid and genomic DNA (Lokteff *et al.*, 2001).

A number of tests were also developed and further used routinely to detect other process-related impurities in clinical batches, such as a *Limulus* amoebocyte lysate test to quantify bacterial endotoxins (< 1 I.U./mg of BBG2Na) and a specific RP-HPLC assay to detect tetracycline (no trace detected) since this antibiotic was used for selection by drug pressure during the fermentation step.

3.   P40, A PROMISING NEW CARRIER AND ADJUVANT FOR HUMAN VACCINES

Most conventional vaccines consist of killed organisms or purified proteins derived from these organisms. Safer vaccines use subunit antigens corresponding to immunogenic epitopes, such as peptides or poly- and oligosaccharides. However, such epitopes are usually not immunogenic by themselves. Coupling such antigens to carrier proteins and associating them to adjuvants has been reported to increase their immunogenicity, converting them into T-dependent antigens. Among the molecules able to boost the immune system, pathogen-associated components represents the largest class. Diphteria and tetanus toxoids, deriving from the toxins of *Corynebacterium diphteriae* and *Clostridium tetani* respectively, are carrier proteins successfully used in humans in conjugated *Haemophilus influenzae* type b, pneumococcal and meningococcal vaccines. Bacterial integral membrane proteins, such as *E. coli* TraT and the group B meningococcal outer membrane protein, have been described to be excellent carriers for chemically conjugated antigens. OMPC, the outer membrane protein complex of *Neisseria meningitidis*, is also used as a carrier for *Haemophilus influenzae* type b and meningococcal capsular polysaccharides in human vaccines.

Other pathogen-associated molecules called PAMPs, for pathogen-associated molecular patterns, have been recently described (Jeannin *et al.*, 2002). These highly conserved structures are recognized by innate cells. Recent evidence shows that this recognition can mainly be attributed to non-clonally distributed receptors called pattern-recognition receptors, or PRRs, such as scavenger and Toll-like receptors. Binding of PAMPs to PRR induces the production of reactive oxygen and nitrogen intermediates, pro-inflammatory cytokines, and up-regulates expression of co-stimulatory molecules on antigen presenting cells, subsequently stimulating T cells and initiating the adaptive immunity. These properties of PAMPs are now exploited to develop new approaches in immunotherapy and vaccine strategies.

We have recently described *Klebsiella pneumoniae* outer membrane protein A (OmpA), called P40, to be part of this class of molecules since it targets dendritic cells, induces their maturation and favors antigen cross-presentation (Jeannin *et al.*, 2000). OmpA is one of the major proteins in the outer membrane of gram-negative bacteria, present at 10^5 copies per cell. It is believed to occur in a monomeric form in its native state. OmpA is highly conserved among gram-negative bacteria and is thought to consist of two domains. The three-dimensional structure of the 19 kDa transmembrane domain has been determined in solution using NMR (Arora *et al.*, 2001; Fernandez *et al.*, 2001) and after crystallisation by X-ray diffraction (Pautsch and Schulz, 1998; 2000). This N-terminal part of the protein, including amino acid residues 1 to 177, forms a membrane-spanning domain and crosses the outer membrane eight times in antiparallel β-strands, leading to a typical amphiphilic β-barrel. The C-terminal moiety is thought to be periplasmic. In addition to non-physiological functions, OmpA is required for the maintenance of structural cell integrity and the generation of normal cell shape (Sonntag *et al.*, 1978). The capacity to form pores in the bacterial membrane (Sugawara and Nikaido, 1992; Saint *et al.*, 1993; Bond *et al.*, 2002) and confer serum resistance and pathogenicity to *E. coli* K1 (Prasadarao *et al.*, 1996) has been demonstrated.

P40, which contains at least one T helper epitope (Haeuw *et al.*, 1998), was demonstrated to be as potent as the reference carrier tetanus toxoid in inducing strong antibody responses against B cell epitopes such as peptides (Haeuw *et al.*, 1998; Rauly *et al.*, 1999) and bacterial polysaccharides (Rauly *et al.*, 1999; Haeuw *et al.*, 2000; Libon *et al.*, 2002) coupled to it. Moreover the protein also works as a potent immunological adjuvant for specific CTL response induction. This was demonstrated with peptides from two melanoma associated differentiation antigens: Melan-A and tyrosinase-related protein 2. We found that peptides Melan-A$_{26-35}$ A27L and TRP-2$_{181-188}$ were able to generate a specific CTL response when coupled or mixed with P40 (Beck *et al.*, 2001a and b; Miconnet *et al.*, 2001). The CTL response generated against the TRP-2 peptide in the presence of P40 was associated with tumour protection in experimental models (Miconnet *et al.*, 2001). Together with the fact that P40 is able to activate antigen presenting cells and favor internalisation and cross-presentation of antigens (Jeannin *et al.*,

2000), these results make the use of P40 a particularly practical immunotherapy approach. P40 is currently evaluated in clinical phase I in human in association with a CTL peptide epitope in melanoma.

3.1   Production of P40

The P40 expression vector, called pVALP40, was designed such that it encodes, under the control of the tryptophan promoter, the *K. pneumoniae* OmpA protein with a 9 amino acid sequence derived from the trp operon leader sequence fused to its N-terminal end. This N-terminal extension was necessary for high-yield expression of *K. pneumoniae* OmpA in *E. coli*. P40 was over-expressed as an intracellular protein in inclusion bodies. The inclusion bodies were solubilized in 7 M urea in the presence of 10 mM dithiothreitol within 2 h at 37°C, and renaturation was achieved by diluting the mixture 13-fold in the presence of a detergent (Haeuw *et al.*, 1998). Since P40 is a membrane protein, the presence of a detergent was essential for complete renaturation. The zwitterionic detergent N-tetradecyl-N,N-dimethyl-3-ammonio-1-propanesulfonate, or Zwittergent 3-14, was chosen from 15 other detergents tested for its higher capacity to solubilize and renature P40. It was used at the final concentration of 0.1 % (mass/vol.) for renaturation, and also at the same concentration in the different buffers used throughout the purification process. The refolded protein was further purified by a two-step ion exchange chromatography process, including anion exchange and cation exchange chromatography steps (Haeuw *et al.*, 1998). No biological product of animal origin was included in the process. The purified P40 was further extensively characterized.

3.2   Characterization of P40

3.2.1   Structural characterization and confirmation

P40 contains 344 amino acid residues. For ES-MS analysis, since the detergent used for renaturation and purification of P40 is incompatible with subsequent spectrometric analyses, it was eliminated. The protein was precipitated by addition of cold ethanol. Following the removal of detergent, mass measurement of the purified P40 was performed by ES-MS after solubilization in formic acid. The measured mass, 37.061 +/- 2 Da, was in accordance with the theoretical mass (37.059 Da) deduced from the DNA sequence (Fig. 8).

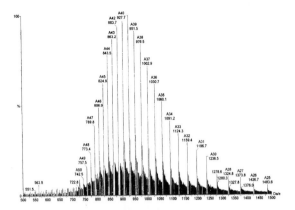

Figure 8. ES-MS analysis of purified P40.

The primary structure was assessed by peptide mapping and N- and C-terminal sequencing. N-terminal sequence analysis showed that a single sequence was detected, MKAIFVLNAA, which corresponds exactly to the first 10 amino acids at the N terminus of P40 as predicted from the DNA sequence. Reverse phase HPLC analysis of a digestion of P40 with a lysine-specific endopeptidase, *i.e.* endoproteinase Lys-C, was used for identification and primary structure confirmation (Fig. 9). Endoproteinase Lys-C hydolyzes peptide bonds at the carboxylic side of lysine residues. The seventeen peaks resolved were characterized by mass spectrometry, allowing the confirmation of 99 % of the primary sequence.

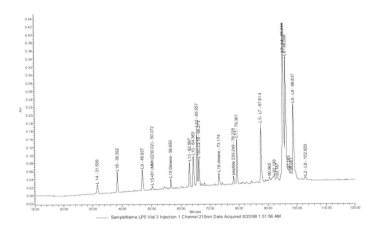

Figure 9. Endoproteinase Lys-C peptide map of P40.

The C-terminal sequence of P40 was confirmed by sequencing the C-terminal peptide identified from the peptide map: the peak annotated 'L18' on the chromatogram corresponds to the C-terminus amino acids 337-344 (measured mass: 870,61 Da; theoretical mass: 870,96 Da). The sequence deduced from the sequencing analysis, EVVTQPQA, was in accordance with the theoretical amino acid sequence.

When P40 was digested with endoproteinase Lys-C in the presence of the detergent used to refold and purify the protein, several peptides identified by LC-MS analysis contained 3 to 4 theoretical fragments. These peptides were localized in the membrane domain of P40 and resulted from an incomplete digestion of this part of the protein. This was mostly due to the non accessibility of some lysine residues in the β-barrel formed in the presence of the detergent. On the basis of the two-dimensional model of the β-barrel domain of *E. coli* OmpA (Vogel and Jähnig, 1986; Pautsch and Schulz, 1998), most of the lysine residues of P40 can be predicted to be resistant to proteolysis because of their location into β strands, with their side chains pointing toward the internal channel, or into the short extracellular or periplasmic loops which are inaccessible to proteases. According to this model, only two lysine residues of P40 would be accessible to proteolysis: lysine 78 and 121. On the opposite, a P40 protein produced without any detergent was more susceptible to digestion with endoproteinase Lys-C. The membrane domain of this protein was thought to be non correctly refolded, thus rendering accessible all the lysine residues which were inaccessible in the β-barrel. In this case, the digestion with endoproteinase Lys-C was complete and all the peptides of the theoretical map were identified by LC-MS. Together with data coming from SDS-PAGE and circular dichroism analyses, these data confirmed the β-barrel structure of the N-terminal domain of P40.

3.2.2   Physico-chemical properties

SDS-PAGE electrophoresis was used to analyse P40 during its purification and monitor its refolding. P40 exhibited different electrophoretic mobilities when it was boiled or not in the presence of SDS (Fig. 10).

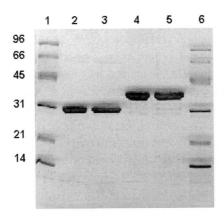

Figure 10. Analysis of purified P40 by SDS-PAGE electrophoresis. Proteins (10μg) were separated on a 12.5% acrylamide gel before staining with Coomassie blue. Lanes 1 and 6, molecular mass markers (kDa); lane 2, P40 non reduced and non heated; lane 3, P40 reduced and non heated; lane 4, P40 non reduced and heated at 100°C for 10 min in sample buffer; lane 5, P40 reduced and heated at 100°C for 10 min in sample buffer.

The heat-denatured P40 migrated with an apparent molecular weight of 36 kDa, which agrees with the molecular mass determined by mass spectrometry and deduced from amino acid sequence, whereas the non denatured form of the protein migrated with a lower apparent molecular mass of approximately 31 kDa. This phenomenon of heat-modifiability has been well described for *E. coli* OmpA (Heller, 1978; Dornmair *et al.*, 1990) and is a common property to bacterial outer membrane proteins with a β-barrel membrane domain. Boiling such proteins in the presence of SDS extends their β structure to α helices and thereby causes reduction of SDS binding and electrophoretic mobility.

Two chromatographic methods were also developed for identity and purity control: reverse phase HPLC and gel filtration (Haeuw *et al.*, 1998). These methods were able to detect and quantify P40 dimers. They were also used for quantitative measurement, together with classical colorimetric assays and an ELISA based on a P40 specific monoclonal antibody.

3.3   Detection and quantitation of impurities

3.3.1   *Escherichia coli* OmpA

A unique process-related impurity was co-purifying with the protein of interest. This impurity is *E. coli* OmpA. It is highly homologous to P40, with more than 80% of similarity. Because of their high homology, the behaviour of these proteins on various chromatographic supports is thought to be similar. Therefore a complete elimination of *E. coli* OmpA is practically impossible to achieve with conventional purification techniques. Considering that they would also share similar activities, we assumed that moderate levels of *E. coli* OmpA in P40 lots would not be considered as a problem by the authorities. However, such an impurity must be detected and quantified apart from the others. A few examples have been published where quantification methods are based on specific antibodies. For example, a major HCP contaminant of recombinant human acidic fibroblast growth factor, S3 ribosomal protein, was quantitated by western blot using specific monoclonal antibodies (O'Keefe *et al.*, 1993). Since P40 and *E. coli* OmpA have very close molecular weights and structures, the resolution of electrophoretic and chromatographic methods was insufficient to specifically detect *E. coli* OmpA impurity in P40 batches. To develop specific immunoassays, monoclonal antibodies were prepared. A comparison of *E. coli* OmpA and P40 primary sequences revealed that the major differences were located in the membrane domain of both proteins (Fig. 11). Besides the 9 amino acid sequence fused to the N-terminal end of P40, the sequence of *E. coli* OmpA presents two deletions of 5 amino acids when compared to the equivalent P40 sequence (Fig. 11). On the basis of this observation, monoclonal antibodies directed at peptides TGFINNNG and TKSNVYGK, including one deletion each, were generated.

```
Ec OmpA  ---------APKDNTWYTGAKLGWSQYHD TGFI-----NNNGP THENQLGAGAFGGYQV
rP40     MKAIFVLNAAPKDNTWYAGGKLGWSQYHDTGF YGNGF QNNNGP TRNDQLGAGAFGGYQV

Ec OmpA  NPYVGFEMGYDWLGRMP YKGSVENGAYKAQGVQL TAKLGYP ITDDLD IYTRLGGMVWRA
rP40     NPYLGFEMGYDWLGRMA YKGSVDNGAFKAQGVQL TAKLGYP ITDDLD IYTRLGGMVWRA

Ec OmpA  DTKSN-----VYGKNHDTGVSPVF AGGVEY AITPE IATRLEYQWTNNIGDAHTIGTRPD
rP40     DSKGNYASTGVSRSEHDTGVSPVF AGGVEW AVTRD IATRLEYQWVNNIGDAGTVGTRPD
```

Figure 11. Comparison of the amino acid sequences of the membrane domain of E. coli OmpA and P40.

One of the numerous monoclonal antibody specific for the octapeptide TKSNVYGK was able to specifically detect *E. coli* OmpA in P40 lots by western blot (Fig. 12). A semi-quantitative 1D western blot assay was developed with this monoclonal antibody. The level of *E. coli* OmpA impurity in several lots was established at less than 1%.

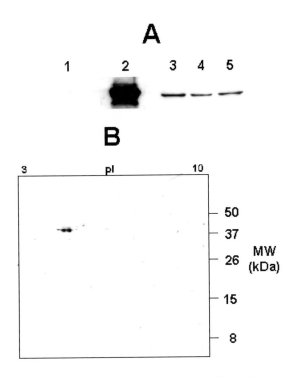

Figure 12. Detection of Escherichia coli OmpA in purified P40 by western blot. A. 1-D PAGE western blot. Proteins were separated on a 12% acrylamide gel. Lane 1, 15μg of HCP; lane 2, 2.5μg of purified E. coli OmpA; lanes 3-5, 15μg of 3 different batches of purified P40. B. 2-D PAGE western blot. Fifty μg of P40 were loaded onto a 3-10 IPG strip by in-gel rehydration. After focusing, proteins were separated in the second dimension on a 15% homogeneous gel. Further to 1 and 2-D electrophoresis, proteins were transferred to a PVDF membrane. Proteins were immunolabelled with the anti-TKSNVYGK monoclonal antibody, and further with a HRP-labelled goat anti-mouse IgG antibody before visualization using ECL reagents.

The purified anti-TKSNVYGK monoclonal antibody was used to quantify *E. coli* OmpA throughout the process of purification of P40. A sequential double-sandwich ILA assay, based on the use of two specific anti-*E. coli* OmpA antibodies, was developed. A biotin-labelled anti-TKSNVYGK monoclonal antibody was first incubated with the P40 solution. After addition of streptavidin, the complex formed between the antibody and residual *E. coli* OmpA was captured by filtration through a biotinylated nitrocellulose membrane. The membrane was further incubated with a rabbit fluorescein-labelled polyclonal anti-OmpA serum, obtained by immunizing rabbits with purified *E. coli* OmpA. The detection stage was then performed as described above in paragraph 2.3.2. The *E. coli* OmpA-ILA assay had detection and quantification limits of 5 and 10 ng/ml respectively. A standard curve was generated by plotting the signal obtained for different concentrations of purified *E. coli* OmpA (Fig. 13).

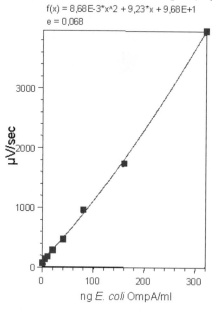

Figure 13. Representative E. coli OmpA immunoligand assay standard curve.

The sensitivity range of the ILA assay was comprised between 10 and 320 ng/ml. A quadratic regression allowed to calculate the amount of *E. coli* OmpA contained in the tested samples. Values of less than 1 % were determined for different bulks of P40 (Table 3).

Table 3. Residual content of *E. coli* OmpA and host cell proteins in pilot batches of P40.

P40 bulks	*E. coli* OmpA content[a]	HCP content[b]
	%	%
1	0.9	1.0
2	0.8	0.8
3	0.9	3.5

[a] g of *E. coli* OmpA / 100 g of P40.

[b] g of HCP / 100 g of P40.

3.3.2 Other host cell proteins

As for BBG2Na, we have developed a process-specific ILA assay for strain-specific host cell-derived proteins based on the Molecular Devices Threshold system. This assay is also based on the use of rabbit anti-HCP polyclonal antibodies. HCP were prepared from a *E. coli* strain which did not express the P40 protein. HCP preparation was stopped at the extraction level. Since the absence of anti-OmpA antibodies in the anti-HCP serum was essential, we were also obliged to delete the OmpA gene by site-directed mutagenesis in order to ensure that there would be no expression of OmpA in the *E. coli* strain used for immunization. The absence of OmpA expression was assessed by western blot with the anti-TKSNVYGK monoclonal antibody described above. Cleaned and dedicated equipments were used to prepare HCP immunogen. After four immunizations with HCP, the rabbits were sacrificed and the collected anti-HCP serum was purified by affinity chromatography on immobilized protein G. The purified HCP antiserum did not cross-react with P40 and related proteins in western blot (Fig. 14A). It allowed to detect HCP by western blot into the final bulk of P40 (Fig. 14A). More than 80% of the HCP stained by silver nitrate on a 2-D gel were detected by western blot with the purified antiserum (Fig. 14B). A higher detection level could certainly be obtained with a cascade immunization protocol (Thalhamer and Freund, 1984) as previously described for the BBG2Na RSV vaccine in paragraph 2. There was no need for such a highly sensitive antiserum at this stage of development. All the proteins detected on a large pI range (i.e. 3-10) 2-D gel were comprised between pI 5 and 7.5. Thus, the two-dimensional analyses of P40 were all realized with pI 5-8 IPG strips.

P40 monomer and dimer spots were assigned on the silver nitrate stained 2-D gel (Fig. 14B) on the basis of a 2-D western blot labelled with a P40 specific monoclonal antibody. Different charged isoforms of P40 were observed between pI 5 and 6.5. This phenomenon has been described for many other bacterial outer membrane proteins which resolve into at least two-charged isoforms when analysed by 2-D electrophoresis, and specially for the *E. coli* OmpA by Molloy *et al.* (1998; 2000). The exact nature of this charge variance has not been established but we suppose that deamidations could largely contribute to the protein heterogeneity. Indeed, P40 contains three Asn-Gly sites located in the membrane domain (Fig. 11) which are known to be typical deamidation sites. Multiple isoforms of *E. coli* OmpA spots were also identified on the 2-D gel (Fig. 14B) on the basis of a 2-D western blot analysis with the anti-TKSNVYGK monoclonal antibody as shown in figure 12. Several other proteins, called HCP, were identified by a classical proteomics approach (Fig. 14B and Table 4).

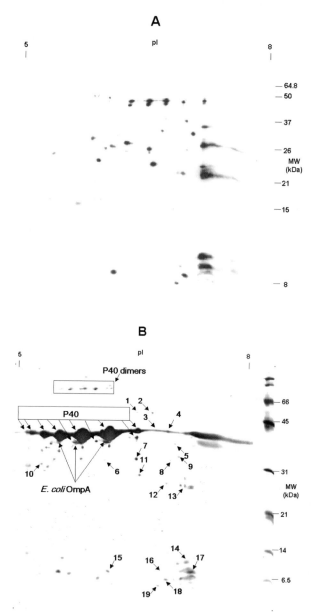

Figure 14. 2-D PAGE analysis of P40. One hundred and fifty µg of P40 (batche #3) were loaded onto a 5-8 IPG strip by in-gel rehydration. After focusing, proteins were separated in the second dimension on a 15% homogeneous gel before (A) transfer to a PVDF membrane and immunoblotting with the rabbit anti-HCP serum, or (B) silver nitrate staining.

For identification, protein spots were cut from colloidal Coomassie Brilliant Blue-stained gels equivalent to the gel shown in Fig. 14B. Excised spots were destained and digested with trypsin. Peptides were extracted from the pieces of gel with acetonitrile and further analysed by MALDI-TOF mass spectrometry (Autoflex mass spectrometer, Bruker-Daltonics). Proteins were identified by searching in NCBI non redundant database using Mascot software. All the proteins detected with the HCP antiserum (Fig. 14A) were identified by using this approach. Some of the identified proteins were found in different spots: for example spots 6, 7 and 8, which were separated by 0.2 pI unit, were identified to be periplasmic D-ribose-binding protein (Table 4). Among the proteins identified were two particular sets of proteins: first a set of periplasmic transporters and second a set of ribosomal subunits. For the ribosomal proteins identified, the calculated pI and molecular weight were not in accordance with the theoretical values. These proteins were fragments of the native proteins with lower pI and molecular weight. A degradation, which can occur during production and extraction stages, could explain the presence of these fragments of ribosomal proteins. Similar results were obtained with the two other batches of P40 evaluated.

Table 4. List of the HCP identified from a 2-D gel of P40.

Spot #	Protein name	SWISS-PROT Accession no.	Theoretical M_r [a] / pI	Sequence coverage %
1 / 2	Survival protein	P21202	47.2 / 6.48	45 / 47
3 / 4	Sulfate-binding protein precursor	P06997	36.6 / 6.35	36 / 47
5	Thiosulfate binding protein	P16700	37.6 / 7.78	60
6 - 8	D-ribose periplasmic binding protein	P02925	30.9 / 6.85	33 - 55
9	FKBP-type peptidyl-prolyl cis-trans isomerase	P45523	28.8 / 8.39	42
10	Histidine-binding periplasmic protein	P39182	28.5 / 5.47	62
11	Periplasmic glutamine-binding protein	P10344	27.1 / 8.44	41
12 / 13	Superoxide dismutase manganese	P00448	23 / 6.45	30 / 51
14	50S ribosomal subunit protein L1	P02384	24.7 / 9.64	44
15 / 18	Cold shock protein	P36997	7.4 / 8.09	86 / 86
16	30S ribosomal subunit protein S10	P02364	11.7 / 9.68	77
17	50S ribosomal subunit protein L10	P02408	17.7 / 9.04	37
19	30S ribosomal subunit protein S7	P02359	17,5 / 10.3	40

[a] Molecular weight in kDa.

The HCP antiserum was also used to quantify HCP by ILA assay throughout the process of purification of P40. The reaction steps were the same as those described above for BBG2Na impurities (Fig. 6). The sensitivity range of the P40 HCP-ILA assay was comprised between 5 and 100 ng/ml (Fig. 15), and detection and quantification limits were 5 and 10 ng/ml respectively.

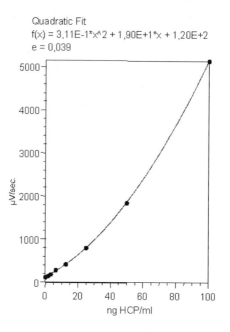

Figure 15. Representative HCP immunoligand assay standard curve.

HCP content was determined for the same lots of P40 which were also analysed for residual *E. coli* OmpA (table 3). Even if these 3 lots were produced with the same process, they were shown to contain different quantities of residual HCP. Two batches had quite similar HCP content of about 1%, whereas the HCP percentage determined for the third batch was more than 3-fold higher. Even if this type of assay is not required by the authorities for clinical phase I, we used it as a tool to evaluate our purification process. This result demonstrates that the purification process will need to be improved for further clinical phases, and particularly for phase III, in order to reduce the content of HCP into the final bulks of purified P40, with a special emphasis on the reproducibility of the contamination level.

4.   CONCLUSIONS

A comprehensive understanding of the chemical structure, impurity profile, biological properties and degradation pathways is an absolute requirement prior to the development and marketing of any drug substance. In contrast to small drugs, biopharmaceuticals are large entities ranging in size and complexity from small synthetic peptides to conjugated antibodies. They are in general heterogeneous as a consequence of both the biosynthetic processes used and the subsequent manufacturing and storage of the products. The therapeutic performance of biopharmaceuticals strongly depends on their secondary and tertiary structures. Changes in structure may lead to loss of therapeutic effect and safety concerns, such as the introduction of unwanted immune reactions against the biopharmaceutical. Microheterogeneities, co- and post-translational modifications, together with process associated modifications must be detected, identified and monitored. Currently, no battery of tests can be described which would be applicable to all types of products. A general guide to the requirements for setting specifications to support new biopharmaceutical marketing applications is given in a document of the International Conference on Harmonization entitled "ICH Topic Q6B, Specifications: Test Procedures and Acceptance Criteria for Biotechnological/Biological Products" (for complete reference, see the References section). This document does not recommend specific procedures but lists examples of technical approaches which might be considered for structural characterization.

Until now there is only one recombinant vaccine commercialized worldwide: the hepatitis B vaccine is produced in yeast (*Saccharomyces cerevisiae*) or mammalian cells such as CHO cells. As for other molecules produced in such type of cells, a special attention must be paid to the most complex post-translational modifications which can occur, namely N- and O-glycosylations. This type of modifications has not been addressed in this chapter since the described examples were both produced in *E. coli*. Glycoproteins are in general mixtures of glycoforms, that is the same polypeptide backbone with different populations of glycans attached to it. The glycosylation of a biopharmaceutical has been shown to have an influence on its solubility, resistance to proteases, biological activity, pharmacokinetics and immunogenicity. Some degree of glycans heterogeneity are accepted, but this must be consistent, within specifications and between production batches. There are now some robust and rugged methods to carry out the characterization of glycosylation profiles of biopharmaceuticals such as antibodies (Roberts *et al.*, 1995; Raju *et al.*, 2000) or recombinant hormones like erythropoietin (Yuen *et al.*, 2003) or human growth hormone (Kawasaki *et al.*, 2003).

As it has been shown for two examples in this chapter, proteomic techniques, and specially electrophoresis and mass spectrometry, can now be applied routinely to the characterization of biopharmaceuticals. These techniques together with quantitative immunoassays (ILA or ELISA) can be used to detect, identify and quantify all the proteins of a recombinant vaccine and thus to qualitatively and quantitatively determine the specific set of proteins, or proteome, of a vaccine which could also be called "vaccinome". A vaccinome would include not only the active components of the vaccine, such as for example the S, pre-S1 and pre-S2 proteins for the hepatitis B vaccine, and their variants, but also all the identified impurities.

ACKNOWLEDGEMENTS

The authors would like to thank Laurence Zanna, Virginie Robillard, Marie-Claire Bussat, Christine Libon, Nathalie Corvaia, Thien Nguyen, Laurent Chevalet, Jean-Claude Corbière, Véronique Legros, Nathalie Zorn and Alain Van Dorsselaer for their contribution to the work described in this chapter.

REFERENCES

Anicetti, V.R., Feshkins, E.F., Reed, B.R., Chen, A.B., Moore, P., Geier, M.D. and Jones, A.J.S., 1986, Immunoassay for the detection of *E. coli* proteins in recombinant DNA derived human growth hormone. *J. Immunol. Methods* **91**:213-224.

Arora, A., Abildgaard, F., Bushweller, J.H. and Tamm, L.K., 2001, Structure of outer membrane protein A transmembrane domain by NMR spectroscopy. *Nat. Struct. Biol.* 8:334-8.

Baker, R.S., Schmidtke, J.R., Ross, J.W. and Smith, W.C., 1981, Preliminary studies on the immunogenicity and amount of *Escherichia coli* polypeptides in biosynthetic human insulin produced by recombinant DNA technology. *Lancet* **2**:1139-1142.

Beck, A., Zorn, N., Bussat, M.C., Haeuw, J.F., Corvaia, N., Nguyen, T.N., Bonnefoy, J.Y. and Van Dorsselaer, A., 2000, Synthesis and characterization of Respiratory Syncytial Virus protein G related peptides containing two disulfide bridges. *J. Pept. Res.* **55**:24-35.

Beck, A., Bussat, M.C., Klinguer-Hamour, C., Goetsch, L., Aubry, J.P., Champion, T., Julien, E., Haeuw, J.F., Bonnefoy, J.Y. and Corvaia, N., 2001a, Stability and CTL activity of N-terminal glutamic acid containing peptides. *J. Pept. Res.* **57**:528-538.

Beck, A., Goetsch, L., Champion, T., Bussat, M.C., Aubry, J.P., Klinguer-Hamour, C., Haeuw, J.F., Bonnefoy, J.Y. and Corvaia, N., 2001b, Stability and CTL-activity of P40/ELA melanoma vaccine conjugate. *Biologicals* **29**:293-298.

Bond, P.J, Faraldo-Gomez, J.D. and Sansom, M.S. 2002, OmpA: a pore or not a pore? Simulation and modeling studies. *Biophys. J.* **83**:763-775.

Briggs, J., Kung, V.T., Gomez, B., Kasper, K.C., Nagainis, P.A., Masino, R.S., Rice, L.S., Zuk, R.F. and Ghazarossian, V.E., 1990, Sub-femtomole quantitation of proteins with Threshold®, for the biopharmaceutical industry. *BioTechniques* **9**:598-606.

Collins, P.L., McIntosh, K. and Chanock, R.M., 1996, Respiratory syncytial virus. In *Fields - Virology* (B.N. Fields, D.N. Knipe and P.M. Howley, eds), Lippincott-Raven, Philadelphia, pp. 1313-1351.

Dagouassat, N., Haeuw, J.F., Robillard, V., Damien, F., Libon, C., Corvaia, N., Lawny, F., Nguyen, T.N., Bonnefoy, J.Y. and Beck, A., 2001a, Development of a quantitative assay for residual host cell proteins in a recombinant subunit vaccine against human respiratory syncytial virus. *J. Immunol. Methods* **251**:151-159.

Dagouassat, N., Robillard, V., Haeuw, J.F., Plotnicky-Gilquin, H., Power, U.F., Corvaïa, N., Nguyen, T., Bonnefoy, J.Y. and Beck, A., 2001b, A novel bipolar mode of attachment to aluminium-containing adjuvants by BBG2Na, a recombinant RSV subunit vaccine. *Vaccine* **19**:4143-4152.

Doreleijers, J.F., Langedijk, J.P., Hard, K., Boelens, R., Rullmann, J.A., Schaaper, W.M., Van Oirschot, J.T. and Kaptein, R., 1996, Solution structure of the immunodominant region of protein G of bovine respiratory syncytial virus. *Biochemistry* **35**:14684-14688.

Dornmair, K., Kiefer, H. and Jähnig, F., 1990, Refolding of an integral membrane protein: OmpA of *Escherichia coli. J. Biol. Chem.* **265**:18907-18911.

Fernandez, C., Hilty, C., Bonjour, S., Adeishvili, K., Pervushin, K. and Wuthrich, K., 2001, Solution NMR studies of the integral membrane proteins OmpX and OmpA from *Escherichia coli. FEBS Lett.* **504**:173-178.

Food and Drug Administration, 1985, Points to consider in the production and testing of new drugs and biologicals produced by recombinant technology. Center for Drugs and Biologics, Office of Biologics Research and Review.

Fouillard, L., Mouthon, L., Laporte, J.P., Isnard, F., Stachowiak, J., Aoudjhane, M., Lucet, J.C., Wolf, M., Bricourt, F. and Douay, L., 1992, Severe respiratory syncytial virus pneumonia after autologous bone marrow transplantation: a report of three cases and review. *Bone Marrow Transplant.* **9**:97-100.

Goetsch, L., Gonzalez, A., Plotnicky-Gilquin, H., Haeuw, J.F., Aubry, J.P., Beck, A., Bonnefoy, J.Y. and Corvaia, N., 2001, Targeting of nasal mucosa-associated antigen-presenting cells in vivo with an outer membrane protein A derived from *Klebsiella pneumoniae. Infect. Immun.* **69**:6434-6444.

Goetsch, L., Haeuw, J.F., Champion, T., Lacheny, C., Nguyen, T., Beck, A. and Corvaia, N., 2003, Identification of B- and T-Cell Epitopes of BB, a Carrier Protein Derived from the G Protein of Streptococcus Strain G148. *Clin. Diagn. Lab. Immunol.* **10**:125-132.

Ghobrial, I.A., Wong, D.T. and Sharma, B.G., 1997, An immuno-ligand assay for the detction and quantitation of contaminating proteins in recombinant human erythropoeitin (r-HuEPO). *BioTechniques* **10**:42-45.

Haeuw, J.F., Rauly, I., Zanna, L., Libon, C., Andréoni, C., Nguyen, T.N., Baussant, T., Bonnefoy, J.Y. and Beck, A., 1998, The recombinant *Klebsiella pneumoniae* outer membrane protein OmpA has carrier properties for conjugated antigenic peptides. *Eur. J. Biochem.* **255**:446-454.

Haeuw, J.F., Libon, C., Zanna, L., Goetsch, L., Champion, T., Nguyen, T.N., Bonnefoy, J.Y., Corvaia, N. and Beck, A., 2000, Physico-chemical characterization and immunogenicity studies of peptide and polysaccharide conjugate vaccines based on a promising new carrier protein, the recombinant *Klebsiella pneumoniae* OmpA. *Dev. Biol.* **103**:245-250.

Heller, K.B., 1978, Apparent molecular weights of a heat modifiable protein from the outer membrane of *Escherichia coli* in gels with different acrylamide concentrations. *J. Bacteriol.* **134**:1181-1183.

Jeannin, P., Renno, T., Goetsch, L., Miconnet, I., Aubry, J.P., Delneste, Y., Herbault, N., Baussant, T., Magistrelli, G., Soulas, C., Romero, P., Cerottini, J.C. and Bonnefoy, J.Y., 2000, OmpA targets dendritic cells, induces their maturation and delivers antigen into MHC class I presentation pathway. *Nat. Immunol.* **1**:502-509.

Jeannin, P., Magistrelli, G., Goetsch, L., Haeuw, J.F., Thieblemont, N., Bonnefoy, J.Y. and Delneste, Y., 2002, Outer membrane protein A (OmpA): a new pathogen-associated molecular pattern that interacts with antigen presenting cells – impact on vaccine strategies. *Vaccine* **20:**A23-27.

Kawasaki, N., Itoh, S., Ohta, M. and Hayakawa, T., 2003, Microanalysis of N-linked oligosaccharides in a glycoprotein by capillary liquid chromatography/mass spectrometry and liquid chromatography/tandem mass spectrometry. *Anal. Biochem.* **316:**15-22.

Klinguer, C., Beck, A., De-Lys, P., Bussat, M.C., Blaecke, A., Derouet, F., Bonnefoy, J.Y., Nguyen, T., Corvaïa, N. and Velin, D., 2001, Lipophilic quaternary ammonium salt acts as mucosal adjuvant when co-administrated by nasal route with antigens. *Vaccine* **19:**4236-4244.

Klinguer-Hamour, C., Libon, C., Plotnicky-Gilquin, H., Bussat, M.C., Revy, L., Nguyen, T., Bonnefoy, J.Y., Corvaïa, N. and Beck, A., 2002, DDA adjuvant induces a mixed Th1/Th2 immune response when associated with BBG2Na, a respiratory syncytial virus potential vaccine. *Vaccine* **20:**2743-2751.

Libon, C., Corvaïa, N., Haeuw, J.F., Nguyen, T., Bonnefoy, J.Y. and Andréoni, C., 1999, The serum albumin-binding region of Streptococcal protein G (BB) potentiates the immunogenicity of the G130-230 RSV-A protein. *Vaccine* **17:**406-414.

Libon, C., Haeuw, J.F., Crouzet, F., Mugnier, C., Bonnefoy, J.Y., Beck, A. and Corvaia, N., 2002, *Streptococcus pneumoniae* polysaccharide conjugated to the outer membrane protein A from *Klebsiella pneumoniae* elicit protective antibodies. *Vaccine* **20:**2174-2180.

Lokteff, M., Klinguer-Hamour, C., Julien, E., Picot, D., Lannes, L., Nguyen, T.N., Bonnefoy, J.Y. and Beck, A., 2001, Residual DNA quantification in clinical batches of BBG2Na, a recombinant subunit vaccine against human respiratory syncytial virus. *Biologicals* **29:**123-132.

Miconnet, I., Coste, I., Beermann, F., Haeuw, J.F., Cerottini, J.C., Bonnefoy, J.Y., Romero, P. and Renno, T., 2001, Cancer vaccine design: a novel bacterial adjuvant for peptide-specific CTL induction. *J. Immunol.* **166:**4612-4619.

Mlinaric-Galinovic, G., Falsey, A.R. and Walsh, E.E., 1996, Respiratory syncytial virus infection in the elderly. *Eur. J. Clin. Microbiol. Infect. Dis.* **15:**777-781.

Molloy, M.P., Herbert, B.R., Walsh, B.J., Tyler, M.I., Traini, M., Sanchez, J.C., Hochstrasser, D.F., Williams, K.L. and Gooley, A.A., 1998, Extraction of membrane proteins by differential solubilization for separation using two-dimensional gel electrophoresis. *Electrophoresis* **19:**837-844.

Molloy, M.P., Herbert, B.R., Slade, M.B., Rabilloud, T., Nouwens, A.S., Williams, K.L. and Gooley, A.A., 2000, Proteomic analysis of the *Escherichia coli* outer membrane. *Eur. J. Biochem.* **267:**2871-2881.

Nguyen, T.N., Samuelson, P., Sterky, F., Merle-Poitte, C., Robert, A., Baussant, T., Haeuw, J.F., Uhlen, M., Binz, H. and Stahl, S., 1998, Chromosomal sequencing using a PCR-based biotin-capture method allowed isolation of the complete gene for the outer membrane protein A of *Klebsiella pneumoniae*. *Gene* **210:**93-101.

Note for guidance on specifications: test procedures and acceptance criteria for biotechnological/biological products, CPMP/ICH/395/96, ICH Topic Q6B, September 1999.

O'Keefe, D.O., DePhillips, P. and Will, M.L., 1993, Identification of an *Escherichia coli* protein impurity in preparations of a recombinant pharmaceutical. *Pharm. Res.* **10:**975-979.

Pautsch, A. and Schulz, G.E., 1998, Structure of the outer membrane protein A transmembrane domain. *Nat Struct Biol.* **5:**1013-1017.

Pautsch, A. and Schulz, G.E., 2000, High-resolution structure of the OmpA membrane domain. *J Mol Biol.* **298:**273-282.

Piedra, A.P., 2003, Clinical experience with respiratory syncytial virus vaccines. *Pediatr. Infect. Dis. J.* **22**:S94-99.

Plotnicky-Gilquin, H., Huss, T., Aubry, J.P., Haeuw, J.F., Beck, A., Bonnefoy, J.Y., Nguyen, T. and Power, U.F., 1999, Absence of Lung Immunopathology Following RSV Challenge in Mice Immunized with a Recombinant RSV G Protein Fragment. *Virology* **258**:128-140.

Plotnicky-Gilquin, H., Robert, A., Chevalet, L., Haeuw, J.F., Beck, A., Bonnefoy, J.Y., Siegrist, C.A., Nguyen, T. and Power, U.F., 2000, CD4+ T-cell-mediated anti-viral protection of upper respiratory tract in BALB/c mice following parenteral immunisation with a recombinant RSV G protein fragment. *J. Virology* **74**:3455-3463.

Plotnicky-Gilquin, H., Cyblat-Chanal, D., Goetsch, L., Lacheny, C., Libon, C., Champion, T., Beck, A., Pasche, H., Nguyen, T.N., Bonnefoy, J.Y., Bouveret-le-Cam, N. and Corvaia, N., 2002, Passive transfer of serum antibodies induced by BBG2Na, a subunit vaccine, in the elderly protects SCID mouse lungs against respiratory syncytial virus challenge. *Virology* **303**:130-137.

Power, U., Plotnicky-Gilquin, H., Goetsch, L., Champion, T., Beck, A., Haeuw, J.F., Nguyen, T., Bonnefoy, J.Y. and Corvaïa N., 2001, Identification and characterization of multiple linear B cell protectotes in the RSV G protein. *Vaccine* **19**:2345-2351.

Prasadarao, N.V., Wass, C.A., Weiser, J.N., Stins, M.F., Huang, S.H. and Kim, K.S.., 1996, Outer membrane protein A of *Escherichia coli* contributes to invasion of brain microvascular endothelial cells. *Infect. Immun.* **64**:146-153.

Rauly, I., Goetsch, L., Haeuw, J.F., Tardieux, C., Baussant, T., Bonnefoy, J.Y. and Corvaia, N., 1999, Carrier properties of a protein derived from outer membrane protein A of *Klebsiella pneumoniae. Infect. Immun.* **67**:5547-5551.

Raju, T.S., Briggs, J.B., Borge, S.M. and Jones, A.J.S., 2000, Species-specific variation in glycosylation of IgG: evidence for the species-specific sialylation and branch-specific galactosylation and importance for engineering recombinant glycoprotein therapeutics. *Glycobiology* **10**:477-486.

Roberts, G.D., Johnson, W.P., Burman, S., Anumula, K.R. and Carr, S.A., 1995, An integrated strategy for structural characterization of the protein and carbohydrate components of monoclonal antibodies: application to anti-respiratory syncytial virus Mab. *Anal. Chem.* **67**:3613-3625.

Saint, N., De, E., Julien, S., Orange, N. and Molle, G., 1993, Ionophore properties of OmpA of *Escherichia coli. Biochim. Biophys. Acta* **1145**:119-123.

Shay, D.K., Holman, R.C., Newman, R.C., Liu, L.L., Stout, J.W. and Anderson, L.J., 1999, Bronchiolitis associated hospitalisations among children 1980-1996. *J. Am. Med. Assoc.* **282**:1440-1446.

Sonntag, I., Schwartz, H., Hirota, Y. and Henning, U., 1978, Cell envelope and shape of *Escherichia coli*: multiple mutants missing the outer membrane lipoprotein and other major outer membrane proteins. *J. Bacteriol.* **136**:280-285.

Sugawara, E. and Nikaido, H., 1992, Pore forming activity of OmpA protein of *Escherichia coli. J. Biol. Chem.* **267**:2507-2511.

Sugawara, M., Czaplicki, J., Ferrage, J., Haeuw, J.F., Power, U.F., Corvaïa, N., Nguyen, T., Beck, A. and Milon, A., 2002, Structure-antigenicity relationship studies of the central conserved region of human respiratory syncytial virus protein G. *J. Pept. Res.* **60**:271-282.

Vogel, H. and Jähnig, F., 1986, Models of the structure of outer-membrane proteins of *Escherichia coli* derived from Raman spectroscopy and prediction method. *J. Mol. Biol.* **190**:191-199.

Walsh, G., 2003, Pharmaceutical biotechnology products approved within the European Union. *Eur. J. Pharm. Biopharm.* **55**:3-10.

Wan, M., Wang, Y., Rabideau, S., Moreadith, R., Schrimsher, J. and Conn, G., 2002, An enzyme-linked immunosorbent assay for host cell protein contaminants in recombinant PEGylated staphylokinase mutant SY161. *J. Pharm. Biomed. Anal.* **28:**953-963.

Whitmire, M.L and Eaton, L.C., 1997, An immunoligand assay for quantitation of process specific Escherichia coli host cell contaminant proteins in a recombinant bovine somatotropin. *J. Immunoassay* **18:**49-65.

Yuen, C.T., Storring, P.L., Tiplady, R.J., Izquierdo, M., Wait, R., Gee, C.K., Gerson, P., Lloyd, P. and Cremata, J.A., 2003, Relationships between the N-glycan structures and biological activities of recombinant human erythropoietins produced using different culture conditions and purification procedures. *Br. J. Haematol.* **121:**511-526.

Chapter 12

Role of Proteomics in Medical Microbiology

PHILLIP CASH

Department of Medical Microbiology, University of Aberdeen, Foresterhill, Aberdeen AB25 2ZD, Scotland

1.   INTRODUCTION

The proteome is defined as the complete protein complement expressed by a cell or unicellular organism at any one time (Wilkins *et al.*, 1996). The proteome is a highly dynamic entity that can change, qualitatively and quantitatively, as the cell metabolism responds to the intracellular and extracellular environment. Mutations within the genome leading to changes in a protein's primary amino acid sequence can also affect the cellular proteome. Consequently, proteomics plays an important role in functional genomics to investigate gene expression and so complements other technologies that define gene expression at the level of nucleic acid synthesis. In contrast to the latter methods, proteomics can provide information on post-translational events affecting the protein products. At present it is not possible to predict these post-translational modifications reliably from the gene sequence alone and the direct analysis of the protein products themselves is essential. In recent years the technologies that comprise proteomics have been used to look at many different areas of the biological sciences, microbiology in particular has been a fruitful area of study using proteomics (Cash, 1998; VanBogelen *et al.*, 1999a; VanBogelen *et al.*, 1999b). The relative simplicity of microbes and the extensive genomic data, both partial and complete, that are available for many bacterial groups provides the opportunity to obtain a comprehensive picture of the gene expression and metabolism of a free-living organism.

H. Hondermarck (ed.), Proteomics: Biomedical and Pharmaceutical Applications, 279–315.

A significant amount of research in the field of microbiology is concerned with the characterisation of well-defined model systems to investigate specific aspects of microbial gene expression and metabolism. Proteomics has been very successful in providing valuable data for these areas of research (reviewed in (Cash, 1995; Cash, 1998)). The Medical Microbiologist is primarily interested in how pathogenic organisms cause disease and how they interact with their hosts. The long-term goal of these investigations is generally the development of novel approaches for

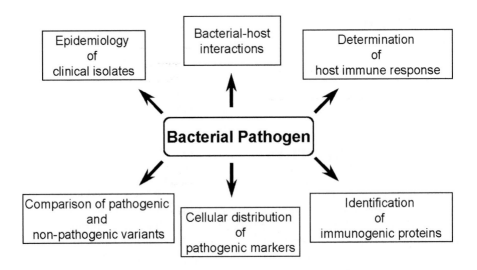

Figure 1. Research areas in Medical Microbiology that utilise proteomic technologies.

treatment and prevention of infectious diseases. A major proportion of these studies will require the analysis of bacterial isolates that have been only poorly characterised at the molecular level by other methods. However, researchers have taken advantage of the large body of data available from more extensively studied bacteria, including in some instances the complete genome sequences, to extrapolate to poorly characterised bacteria that are of medical importance. As described below significant progress has been achieved in defining the proteomes of some major human pathogens. Fig. 1 illustrates the areas in which proteomic technologies have played a role in characterising pathogenic bacteria. This chapter will review the progress that has been made in the characterisation of the proteomes of bacterial pathogens under laboratory conditions as well as the more limited progress achieved in the defining the proteomes of pathogens in association with their hosts. The majority of the examples discussed will be drawn for the investigation of bacterial pathogens but many of the same experimental strategies can be readily applied to other infectious agents.

2. MOLECULAR TYPING OF PATHOGENS

The ability to discriminate between isolates of particular microbial species is an essential requirement for many areas of Medical Microbiology ranging from taxonomic to epidemiological investigations. Studies on microbial pathogenesis that require the comparison of virulent and avirulent isolates to locate molecular determinants of virulence will be considered later. The following section considers primarily the role of proteomics in the molecular epidemiology of infectious agents in which it is essential to detect specific markers that can be used to track the distribution and movement of an infectious agent within the host population. Molecular epidemiology and particularly molecular taxonomic studies of human pathogens rely heavily on the analysis of the organism's genomic nucleic acid due to its relatively lack of response to external stimuli. A variety of techniques can be used to identify suitable markers at the level of the genomic nucleic acid. For example, complete or partial gene sequences can be compared (Amezaga *et al.*, 2002) as well as more rapid methods based on restriction enzyme analysis of the whole genome (Bruce and Jordens, 1991).

The analysis of microbial proteins to locate suitable molecular markers has also been used to discriminate different virus and bacterial isolates. Qualitative changes in electrophoretic mobility can reflect fundamental changes in gene sequence. For example, a characteristic shift in the mobility of the poliovirus structural protein, VP3, correlated not only with an amino acid substitution but also a mutated antigenic epitope (Cash, 1988; Cash, 1995). Similarly, the phosphoprotein of respiratory syncytial (RS) virus has a characteristic electrophoretic mobility as shown by 1-dimensional SDS-PAGE that correlates with the antigenic variation of the virus as shown by a type specific monoclonal antibody (Cash *et al.*, 1977; Gimenez *et al.*, 1986). The shift in the phosphoprotein electrophoretic mobility also correlates with the two major subtypes of RS virus. Although quantitative changes in protein levels may be useful as specific markers, care must be taken since the proteome is readily influenced by intracellular and extracellular stimuli; an issue stressed in a taxonomic study of *Listeria sp.* using 2DGE to separate the bacterial proteins (Gormon and Phan-Thanh, 1995).

The analysis of bacterial proteins by 1-dimensional gel electrophoresis (1DGE) has been applied to a large number of bacterial groups for epidemiological investigations. Typically, the proteins are prepared as a whole cell lysate from bacteria grown *in vitro* under standard conditions. Following electrophoresis, the proteins are detected using either a total protein stain or by *in vivo* radiolabelling of the bacterial proteins and autoradiography. The protein profiles can then be compared either manually or using computer-assisted densitometry to determine the relationships of the bacterial isolates based on co-migrating bands (Bruce and Pennington, 1991). *"Similarity values"* for the bacteria are based on the electrophoretic mobility of the proteins (i.e. the number of protein bands which co-migrate between bacteria) as well as the relative intensity of the co-migrating proteins. A review of 1DGE in bacterial taxonomy has been published by Costas (1992). Refinements to the comparison of global protein synthesis employ a fractionation step to look at particular classes of bacterial proteins. Among the protein fractions analysed for epidemiological and taxonomic studies are the outer membrane proteins (Loeb and Smith, 1980) and ribosomal proteins (Ochiai *et al.*, 1993). Alternatively, the protein detection method can be modified to include only selected protein classes in the analysis. For example, the protein profile can be probed using human immune sera to use only the antigenic bacterial proteins as epidemiological markers (Morgan *et al.*, 1992; Wolfhagen *et al.*, 1993). Similarly, multi-locus enzyme electrophoresis (MLEE), adapted from previous work on Drosophila and humans, has been applied to bacteriological studies. Essentially, protein extracts that retain the necessary enzymatic activity are prepared from the bacteria and separated by electrophoresis on a variety of separation media including starch, polyacrylamide or agarose. The enzymes are then located using appropriate reaction conditions to stain bands expressing the required enzymatic activity (Selander *et al.*, 1986). This approach detects electrophoretic polymorphisms at the loci encoding the enzymes, and by assaying for a range of enzymatic activities bacterial isolates can be grouped into "clones" on the basis of the observed polymorphisms.

One-dimensional electrophoretic separations are limited in their resolving capacity, which can be a particular problem when analysing non-fractionated cellular proteins. The range of proteins open to analysis can be readily extended through the application of 2DGE to analyse either non-fractionated or fractionated cellular proteins. Comparative studies of the non-fractionated cellular proteins resolved by 2DGE to differentiate clinical isolates have been described for various bacterial groups, including *Neisseria* sp (Klimpel and Clark, 1988), *Campylobacter sp* (Dunn *et al.*, 1989), *Haemophilus* sp (Cash *et al.*, 1995) and *Mycoplasma* sp (Watson *et al.*, 1987). These early applications of 2DGE used the technique simply as a sensitive method to differentiate bacterial isolates on the basis of protein charge and molecular weight. It was generally assumed that co-migrating protein spots were functionally equivalent proteins with amino acid homology. In some instances bacterial isolates were shown to exhibit a greater degree of variability by 2DGE than observed by DNA-DNA hybridisation (Rodwell and Rodwell, 1978). Jackson *et al.* (1984) compared the proteins of multiple isolates of *Neisseria gonorrhoea, N. meningitidis* and *Branhamella catarrhalis.* Quantitative and qualitative comparisons of approximately 200 resolved proteins showed that the differences in the 2D protein profiles observed for these bacteria followed their generally recognised taxonomic classification (Jackson *et al.*, 1984). In a study of *Mycoplasma spp.* six strains, representing four different species, showed the equivalent degree of similarity by 2DGE as found when the same isolates were compared by DNA-DNA hybridisation (Andersen *et al.*, 1984).

An extensive study of *Listeria* sp classified twenty-nine isolates, representing 6 different species, on the basis of the cellular proteins resolved by 2DGE (Gormon and Phan-Thanh, 1995). Genetic differences between the bacteria were determined on the basis of the co-migration of proteins resolved by 2DGE. This approach distinguished the *Listeria* species as well as separating the 19 strains of *L. monocytogenes* into two distinct clusters. Up to 600 protein spots were resolved for each of bacterial isolates but relatively few of the protein spots were conserved between the isolates. Fourteen protein spots were common for all 6 *Listeria* species and 75 spots were conserved across all 19 *L. monocytogenes* strains.

Clinical isolates of *H. pylori* show a high degree of variation when compared by 2DGE (Dunn *et al.*, 1989; Jungblut *et al.*, 2000). Enroth *et al.* (2000) used 2DGE to discriminate *H. pylori* isolates on the basis of the non-fractionated cellular proteins resolved by 2DGE. *H. pylori* isolates were collected from patients with different disease symptoms and their proteins compared by 2DGE. In agreement with the extensive heterogeneity observed at the level of the genomic DNA (Taylor *et al.*, 1992) a high-level of protein variation was revealed using 2DGE. Classification of the data demonstrated some clustering of the isolates according to the disease with which the specific bacteria were associated (i.e. duodenal ulcer, gastric cancer and gastritis). The authors suggest that there may be some, as yet unidentified, disease-specific proteins contributing to the clustering of the bacterial isolates. However, the significance of this clustering must remain in doubt since Hazell *et al.* (1996) failed to observe any disease-specific clustering among *H. pylori* isolates which were compared by restriction enzyme digestion analysis of the genomic DNA.

High-resolution mass spectrometry methods have recently been used to discriminate bacterial isolates through the characterisation of specific cellular components. One of these techniques, MALDI-TOF MS, has been used to compare bacterial isolates on the basis of the cell surface proteins. Essentially, intact bacterial cells are mixed with an appropriate UV-absorbing matrix and irradiated with the laser in the MALDI-TOF MS. Under these conditions 500-10000 Da fragments, derived from the cell surface, are ionised and detected (Jarman *et al.*, 2000; Evason *et al.*, 2001; Bernardo *et al.*, 2002). The ionised fragments form characteristic and reproducible profiles for different bacteria. The same analytical method has been used to discriminate between methicillin resistant *Staphylococcus aureus* and methicillin sensitive *S. aureus*. *S. aureus* isolates with these antibiotic resistance profiles differed in peaks with masses in the range of 500-3500 Da (Edwards-Jones *et al.*, 2000; Du *et al.*, 2002). At the present time the identities of the proteins generating the ionised fragments are unknown. However, simply at the level of profile comparisons this approach has the potential for a rapid but very sensitive method to discriminate closely related bacterial isolates. As with all methods based on the analysis of expressed bacterial proteins due care must be paid to standardising the growth conditions of the bacterial isolates (Walker *et al.*, 2002).

3.   COMPREHENSIVE ANALYSIS OF BACTERIAL PROTEOMES

In parallel with the determination of complete bacterial genome sequences there are long-term research programmes aimed at deriving comprehensive proteomic databases for representative bacterial groups. The capacity of the proteome to change in response to external stimuli means that a comprehensive proteome database, detailing both the encoded proteins and their expression profiles, is unlikely to be fully achieved. Even though many of the basic techniques widely used to determine bacterial proteomes were established over 20 years ago, there are still no complete bacterial proteome databases. A basic catalogue of the predicted proteins encoded by an organism can be established *in silico* from the complete gene sequence (see for example (Fleischmann *et al.*, 1995; Bult *et al.*, 1996; Fraser *et al.*, 1997). However the complexity increases as one takes into account the differential expression of the proteome as the bacteria respond to their environment and change their growth phase. Nevertheless proteome databases are being established for specific bacterial groups. As with the determination of a complete genome nucleic acid sequences it is essential to carefully select the most representative isolate from which to derive a proteomic database. In many instances, a laboratory-adapted strain is selected, since it is likely to grow efficiently under laboratory conditions; for similar reasons this may also be the isolate used for the determination of the genome sequence,

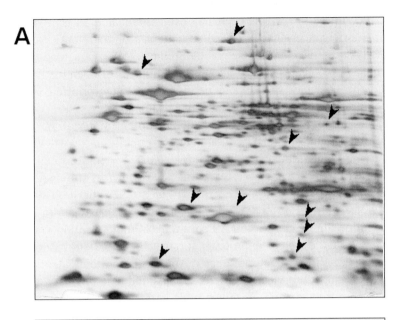

Figure 2. Analysis of cellular proteins from Helicobacter pylori strain 26695 (A) and a recent clinical isolate of H. pylori (B). Bacterial proteins prepared by lysis of the cells with detergent were analysed on small format 2-dimensional gel (Cash and Kroll, 2003) and the resolved proteins stained with silver. Arrows indicate a few of the protein differences detected by manual inspection of the protein profiles.

H. pylori is a major cause of gastrointestinal infections and is of significant clinical interest since it has been implicated as a pre-disposing cause of gastric cancer. The bacterial genome is 1,667,867 base pairs and *in silico* analysis of the genome sequence predicts 1590 open reading frames (Tomb *et al.*, 1997). This size of genome and number of open reading frames is within the scope of 2DGE and peptide mass fingerprinting to achieve almost complete coverage of the proteome (Bumann *et al.*, 2001). A number of research groups have now published 2D protein maps of *H. pylori* on which protein identities have been annotated (Lock *et al.*, 2001; Cho *et al.*, 2002). In a detailed analysis of the *H. pylori* proteome Jungblut et al (Jungblut *et al.*, 2000) compared two sequenced *H. pylori* strains (26695 and J99) as well as a type strain used for animal studies. Up to 1800 protein species were resolved by 2DGE when silver staining was used to detect the proteins following electrophoresis. Peptide mass fingerprinting was used to identify 152 of the detected proteins for the *H. pylori* strain26695, these represented 126 unique bacterial genes. Among the proteins identified were previously documented virulence determinants and antigens. Due to the site of infection in the stomach *H. pylori* must survive extremes of pH and the detailed characterisation of proteins under acid conditions is certain to be a key area of investigation in defining the bacterial proteome in relation to pathogenesis. Initial data reported in this comparative study revealed the pH dependant expression of 5 protein spots. One of the 5 proteins, identified as serine protease HtrA, decreased in abundance when bacteria were grown at pH 5 compared to growth at pH 8. The remaining four proteins were identified as species of the vacuolating cytotoxin, which either decreased in abundance or were completely absent when *H. pylori* was grown at pH 5. Slonczewski *et al* (2000) have also reported on pH-dependant changes in *H. pylori* protein synthesis during growth at either pH 5.7 or pH 7.5. Growth under acidic conditions leads to an increase in the abundance of the urease structural protein UreB, which may represent a survival response for the bacteria under the acidic conditions. The *H. pylori* urease hydrolyses urea to produce ammonia and ultimately carbonic acid. The net result of this reaction is an increase in the pH of surrounding medium. Although not essential for *in vitro* growth, urease is required for *in vivo* colonisation (Eaton *et al.*, 1991). When the growth media was maintained at pH >7 three stress proteins (GroES, GroEL and TsaA) were induced suggesting that these growth conditions lead to a typical stress response for *H. pylori* (Slonczewski *et al.*, 2000). Interestingly, the proteomic studies revealed an uptake of mammalian apolipoprotein A-1 from the bovine serum in the growth medium when the bacteria were grown under acidic conditions. The absence of common acid induced proteins identified by the two groups of workers may be due to a number of factors. Firstly, despite the extensive data available, only a small proportion of the *H. pylori* proteome is actually detected by 2DGE in the two sets of data. Secondly, there are fundamental differences in the experimental conditions used in the two studies. Jungblut *et al* (2000) grew the bacteria on acidified solid media whereas Slonczewski *et al.* (2000) used a broth culture system for growing H. pylori. It is reasonable to expect that different components of the proteome will be expressed under these different growth conditions. Complementary studies to the analyses by 2DGE of the *H. pylori* proteome have provided a protein-protein interaction map for *H. pylori*, which indictes connections between 46.6% of the proteins encoded by the proteome (Rain *et al.*, 2001). Inspection of the protein interactions revealed by this approach is bringing to light specific biological pathways and contributes towards the prediction of protein function.

The proteome of *Mycobacterium tuberculosis* has been extensively characterised to identify potential virulence determinants as well as for specific changes induced during the interaction between bacteria and eukaryotic cells. The genome of *M. tuberculosis* is 4,441,529 base pairs with a predicted coding capacity of 3924 ORFs. Urquhart and colleagues (Urquhart *et al.*, 1997; Urquhart *et al.*, 1998) compared the predicted proteome map of *M. tuberculosis* with that obtained experimentally from a series of six overlapping zoom 2D gels; the latter resolved 493 unique proteins, equivalent to approximately 12% of the expected protein coding. The predicted proteome map displays a bimodal distribution when plotting the predicted pIs and molecular masses for each protein (Urquhart *et al.*, 1998; Mattow *et al.*, 2001). A similar bimodal distribution has also been observed for plots of the theoretical proteomes for *E. coli* (VanBogelen *et al.*, 1997) and *H. influenzae* (Link *et al.*, 1997). Comparisons were made between the predicted and experimentally determined proteome maps for *M. tuberculosis*. When the predicted and observed maps were superimposed outlier protein spots were observed in the observed map that were not present in the predicted proteome. These included 13 proteins with pIs <3.3, i.e. the lowest pI predicted from the genome sequence, as well as a cluster of low molecular mass (≤10kDa) with pIs between 5 and 8. None of these outlier proteins were identified in the reported study, although it was suggested that they might have been derived through fragmentation or post-translational modification of the products of predicted open reading frames. More recently, Jungblut *et al.* (2001) highlighted the ability of proteomic technologies to complement genome sequencing and *in silico* determination of genome coding for *M. tuberculosis* for the identification of low molecular weight proteins. Six genes encoding proteins of <11.5kDa (pI of 4.5-5.9) were identified by 2DGE and peptide mass fingerprinting; these proteins were not predicted from the genome sequence of the H37Rv strain. Partial sequencing by MS/MS of one of the proteins showed that the predicted DNA sequence derived from the peptide was present in the genome but that it had not been assigned in the original determination of predicted open reading frames. Rosenkrands *et al.* (2000b) reported similar observations and identified a 9kDa (pI 4.9) protein that was not predicted from the known genome sequence. In a systematic analysis of low molecular mass proteins 76 proteins with molecular masses between 6 and 15 kDa (pI 4 to 6) were excised and processed for peptide mass fingerprinting (Mattow *et al.*, 2001). Seventy-two of the proteins were identified and were found to comprise approximately 50 structural proteins.

Two detailed studies of the *M. tuberculosis* proteome are underway (Jungblut *et al.*, 1999a; Mollenkopf *et al.*, 1999; Rosenkrands *et al.*, 2000a) both of which can be accessed *via* the Internet. Jungblut and his colleagues (Jungblut *et al.*, 1999a; Mollenkopf *et al.*, 1999) compared two virulent strains of *M. tuberculosis* and two avirulent vaccine *M. bovis* strains using 2DGE and peptide mass mapping. Up to 1800 and 800 proteins were resolved from either cell lysates or culture media respectively. Two hundred and sixty three protein spots were identified by peptide mass fingerprinting. In the second *M. tuberculosis* proteome database currently available Rosenkrands *et al.* (2000a) have looked at the H37Rv strain of *M. tuberculosis* and reported on the identification of 288 and recorded their distribution between culture filtrates and cell lysates.

4.   THE INVESTIGATION OF BACTERIAL PATHOGENESIS AT THE LEVEL OF THE PROTEOME

The long-term aim of many investigations in Medical Microbiology is to understand why and how a particular microbe interacts with its host and causes disease. Then, most importantly, how can one take this information to develop therapeutic and preventive measures against subsequent infection? For any pathogenic microbe the outcome of infection is determined by a number of determinants that act together or in close co-ordination. These pathogenic determinants determine the ability of the organism to colonise the host, to interact and survive within the host as well as the production of proteins or metabolites that may be harmful to the host. Modern molecular technologies now allow researchers to probe these events in fine detail and, as discussed below, proteomics plays an important role in these developments. The investigation of complex phenomena such as pathogenesis on a gene-by-gene basis where the role of each gene is investigated in isolation is unreasonable. The extensive genome sequence data now available for many bacterial pathogens opens up strategies for analysing global gene expression that are more appropriate for investigating polygenic phenomena like pathogenesis. The capacity of proteomics to analyse global protein synthesis covering both gene expression and post-translational modifications makes this approach a powerful tool for identifying and characterising the determinants of bacterial pathogenesis.

4.1 Identification of Pathogenic Determinants for Bacteria grown *in vitro*

A simple approach that has been widely used to investigate pathogenesis is the comparison of the proteomes of pathogenic and non-pathogenic bacterial strains, grown under standard conditions; difference in the proteomes are determined by comparative 2DGE and protein differences that correlate with virulence are subjected to further analysis. However, this approach has some limitations. Firstly, when comparing naturally occurring bacterial variants, protein differences occur that are unrelated to virulence. Secondly, bacteria grown *in vitro* on defined culture media are unlikely to express all of the proteins encoded by the genome at levels characteristic of *in vivo* growth in the bacteria's natural host. This is apparent when comparing the proteins synthesised by facultative intracellular bacterial pathogens grown either in defined culture or in association with eukaryotic cells; many proteins are only synthesised in the latter.

High-resolution 2DGE remains a popular method for identifying pathogenic determinants at the level of the proteome. Early studies that compared virulent and avirulent *Mycoplasma pneumoniae* isolates and identified 3 novel proteins expressed by the virulent isolates that were absent from avirulent strains (Hansen *et al.*, 1979). Comparisons of a virulent parental strain of *M. pneumoniae* with two derived avirulent mutant strains revealed both quantitative and qualitative differences in the protein profiles when analysed by 2DGE (Hansen *et al.*, 1981). At the time that these analyses appeared no data were provided on the identities of these proteins. It might now be a useful time to revisit these data to take advantage of recent progress in defining the *M. pneumoniae* proteome (Regula *et al.*, 2000; Ueberle *et al.*, 2002). Pathogenic determinants in *Brucella abortus* have been investigated using a similar approach to that described for *M. pneumoniae* (Sowa *et al.*, 1992). Up to 935 proteins were resolved by 2DGE for virulent and vaccine strains of *B. abortus*, equivalent to approximately 43% of the expected 2129 proteins predicted for a genome of the size of *B. abortus* (Allardet-Servent *et al.*, 1991). Ninety-two qualitative and quantitative protein differences were found between the 2D protein profiles of the virulent and avirulent strains. It was proposed that the observed differences in protein synthesis in the proteomes of virulent and vaccine isolates may occur due to a variety of factors including point mutations in coding and regulatory sequences as well genome rearrangements or deletions that indirectly affected the expression of bacterial genes required to maintain cellular homeostasis (Sowa *et al.*, 1992).

One of the long term objectives for the development of the Mycobacterial proteome databases discussed above is the elucidation of bacterial pathogenesis. The specific comparisons directed towards identifying these determinants will be considered. Proteomics has been used to identify virulence determinants in *M. tuberculosis* through the comparison of virulent *M. tuberculosis* and vaccine *M. bovis* strains (Mahairas *et al.*, 1996; Sonnenberg and Belisle, 1997; Urquhart *et al.*, 1997; Urquhart *et al.*, 1998; Jungblut *et al.*, 1999a; Mollenkopf *et al.*, 1999). Relatively few differences were detected at the proteome level for these bacteria (Jungblut *et al.*, 1999a; Betts *et al.*, 2000) which was originally thought to be consistent with the limited genetic variability found from the sequence comparison of 26 genes among 842 *M. tuberculosis* complex isolates (Sreevatsan *et al.*, 1997). However, the minimal genetic variability reported at this time may actually have been due to the limited genome sequence available for comparison. More recent studies based on the complete genome sequence have indicated the existence of more extensive genetic variability (Fleischmann *et al.*, 2002). Urquhart *et al.* (1997) used a high-resolution multi-gel system to compare the virulent *M. tuberculosis* H37Rv and *M. bovis* BCG (Pasteur ATCC 35734) strains. Up to 772 protein spots were identified for *M. bovis* BCG and virulent *M. tuberculosis* over a broad range of isoelectric points and apparent molecular weights. No protein identifications were made from these comparisons so their significance in the context of pathogenic determinants is unknown. A more detailed comparison of virulent *M. tuberculosis* (H37Rv and Erdman) and vaccine (BCG Chicago and BCG Copenhagen) strains has been reported that covers the analysis of proteins extracted from both bacterial cell lysates and infected culture media (Jungblut *et al.*, 1999a). The majority of the bacterial proteins detected in either the bacterial cells or released into the culture media were common for all four isolates. Up to 31 variant spots were found in pairwise comparisons. Although the most extensive variation was found in the comparison of virulent and vaccine strains, differences were also observed between the 2 virulent strains as well as between the 2 vaccine strains. Presumably, the latter differences were unrelated to pathogenesis and represented natural variation between isolates. No novel virulence determinants were found for *M. tuberculosis*, but previously identified virulence determinants were assigned to the proteome. From these comparative analyses, 96 spots were specific for either the virulent or avirulent strains (56 and 40 protein spots respectively). Thirty-two of the *M. tuberculosis* specific protein species were identified and corresponded to 27 different proteins. Twelve of the *M. tuberculosis* unique proteins were assigned to genes previously shown to be deleted from avirulent strains. The remaining 20 protein spots were assigned to genes not previously predicted to be deleted from avirulent *M. bovis* strains. The *M. bovis* specific protein species identified in this study were mapped to known *M. tuberculosis* genes. Those proteins specifically synthesised by the virulent *M. tuberculosis* are potential candidates as both pathogenicity determinants and vaccine targets. Betts *et al* (2000) compared the proteomes of two virulent strains of *M. tuberculosis*, the laboratory adapted H37Rv strain and CDC1551, a recent clinical isolate (Valway *et al.*, 1998). The analysis demonstrated that the classic virulent H37Rv strain used as the basis for many proteomic analyses had retained many of the features present in recent virulent *M. tuberculosis* isolates. A total of 1750 intracellular proteins were resolved by 2DGE for each of the strains. Comparable proteomes were observed for both isolates over the 12 day growth curve despite the fact that the CDC 1551 strain entered stationary phase in advance of H37Rv. Comparative studies of the proteomes of the bacteria during their growth cycle revealed just 13 consistent spot differences between the two isolates. Seven protein spots were specific for CDC 1551 whilst 3 were specific for the H37Rv strain. Two other protein spots were increased in abundance for H37Rv compared to CDC 1551. Nine of the proteins exhibiting differences between the isolates were identified by peptide mass mapping. Four of the proteins corresponded to mobility variants of MoxR. The difference in the mobility of the MoxR protein was consistent with a nucleotide change observed in the gene for this protein. One of the *M. tuberculosis* CDC1551 specific proteins identified was a probable alcohol dehydrogenase and one H37Rv specific protein was identified as HisA. The two H37Rv induced spots were electrophoretic variants of alkyl hydroperoxide reductase chain C.

One problem in interpreting the data obtained from comparative studies of potentially genetically distinct bacterial isolates is that some of the genetic and protein variation is unrelated to differences in pathogenesis and represents instead unrelated generic variation between bacterial isolates. Mahairas *et al.* (1996) have used an alternative approach to further characterise *M. tuberculosis*. These researchers located genetic differences between *M. bovis* BCG vaccine and virulent isolates of both *M. bovis* and *M. tuberculosis* through the use of subtractive genomic hybridisation. Three regions were identified that were deleted in BCG vaccine compared to virulent strains. One 9.5kb segment (designated RD1), predicted to encode at least 8 open reading frames, was absent from 6 BCG substrains but present in virulent *M. tuberculosis* and *M bovis* strains as well as in 62 clinical *M.*

tuberculosis isolates analysed. The virulent *M. bovis* and *M. tuberculosis* showed indistinguishable protein profiles when compared by 2DGE. In contrast, the *M. bovis* vaccine strains expressed at least 10 additional proteins and raised expression levels for a number of other proteins. The RD1 genome region was inserted into the BCG genome to generate BCG::RD1. The protein profile, determined by 2DGE, of BCG::RD1 was now indistinguishable from virulent *M. bovis* suggesting that parts of RD1 specifically suppressed protein synthesis in virulent Mycobacteria. Some low molecular weight proteins were identified for BCG::RD1 that were consistent in size with the short ORFs encoded in the RD1 sequence.

A number of studies have looked at sub fractions of bacterial cells to generate bacterial "subproteomes". The majority of these investigations have been concentrated on looking at those proteins located on the cell surface or secreted into the extracellular environment. Proteins present in either of these fractions are likely to play key roles in bacterial pathogenesis, particularly for those pathogens that do not have an intracellular phase in their infection process; these pathogens interact with the host and initiate damage *via* their exposed surface and secreted proteins (McGee and Mobley, 1999). In addition, the proteins located on the surface of the bacterial cells are exposed to the host immune system and many highly immunogenic proteins are found in this position in the bacterial cells (Sabarth *et al.*, 2002). From a practical viewpoint these proteins are also potential targets for vaccine development (Vytvytska *et al.*, 2002). To date, many of these analyses of bacterial subproteomes document the presence of bacterial proteins using the type strain(s) grown *in vitro* but they are sure to be importat areas of research for defining determinants of bacterial pathogenesis,

Depending on their interaction with the host cell *P. aeruginosa* isolates are classed as either invasive or cytotoxic; the latter kill the host cell through the secretion of cytotoxic factors. The membrane and secreted proteomes of these two bacterial types have been examined. Almost 350 proteins were detected, by 2DGE, in purified preparations of membrane from *P. aeruginosa* PAO1, an invasive strain. 189 protein species, equivalent to 104 genes, were identified by peptide mass mapping; 63% of the identified proteins were either Pseudomonas membrane proteins or homologous to membrane proteins from other bacteria. The remaining proteins had not been previously characterised at the functional level and were either unique to *P. aeruginosa* or matched hypothetical proteins in other bacteria (Nouwens *et al.*, 2000). A representative cytotoxic strain (*P. aeruginosa*, 6206) was compared to the PAO1 strain. Membrane proteins prepared from these strains showed minimal variation to each other apart from minor differences likely due to amino acid substitutions. More significant differences were apparent when secreted proteins from the two strains were compared. The profile for the invasive strain PAO1 was dominated by a large number of proteases and there was also evidence for proteolytic degradation of the bacterial surface and secreted proteins possibly by these bacterial proteases. The authors comment on the lack of known virulence factors in their data set. This was likely due to the fact that the bacteria were grown *in vitro* under conditions that do not promote the expression of some virulence factors, which require the bacterial interaction with the host cell (Nouwens *et al.*, 2000).

Extensive data are also available for the membrane and secreted proteomes of *H. pylori*. Membrane proteins of *H. pylori* strain 26695 were identified by biotinylating the external proteins of intact bacteria and then purifying the labelled proteins from a membrane subfraction by affinity chromatography (Sabarth *et al.*, 2002). From these preparations, 82 biotinylated proteins were detected by 2DGE and 18 were identified by peptide mass mapping. Seven of the identified proteins have been previously associated with the *H. pylori* pathogenesis including urease and one protein encoded in the *H. pylori* Cag pathogenicity island. The external cellular environment can influence the expression of the surface proteins. Hynes *et al* (2003) demonstrated that growing *Helicobacter spp* in the presence of bile, an environmental stimulus to which Helicobacter is exposed *in vivo*, leads to different expression patterns of low molecular weight (<8kDa) proteins on the cell surface. Different species of Helicobacter, which colonise different niches in the gut exhibited varying responses to bile that might reflect their specific pattern of pathogenesis, Secreted proteins released by *H. pylori* 26695 were prepared from infected *in vitro* cultures have also been examined by 2DGE (Bumann *et al.*, 2002). Thirty-three reproducible proteins were detected of which 26 were identified. As expected VacA and flagellar proteins were present among the identified proteins as well as 8 proteins of unknown function. No cagA proteins were detected but this was believed due to the absence of the type IV secretion system, responsible for CagA secretion, since it requires the presence of the host eukaryotic cell to induce its expression.

Pseudomonas aeruginosa is a major cause of respiratory infections in persons with the genetic disease cystic fibrosis. During infection in these persons *P. aeruginosa* undergoes a number of structural changes that enable the bacteria to colonise the host (Boucher *et al.*, 1997). Many of these changes can be mimicked *in vitro* by growth of the bacteria at limiting magnesium concentrations. In order to understand the changes in the proteome associated with the ability of the bacteria to adapt to the host *P. aeruginosa* cultures were grown at low magnesium concentration and the protein synthesis analysed using isotope coded affinity tag (ICAT) (Guina *et al.*, 2003). This approach allowed the quantitative and qualitative analysis of 1331 *P. aeruginosa* proteins, estimated to represent approximately 59% of the expressed bacterial proteome. The relative abundance of 546 proteins was determined at the high and low magnesium concentrations and 145 proteins changed in their abundance in response to the magnesium concentration. Many of the proteins showing differences in synthesis were consistent with the previously documented membrane reorganisation reported during *P. aeruginosa* infections of person with CF (Guina *et al.*, 2003). Other proteins showing induced levels of synthesis identified in these experiments were associated with a process known as quorum sensing in *P. aeruginosa*, an essential process for the development of biofilms during colonisation of the lungs in persons with CF. The expression of virulence factors by Gram negative bacteria is partly controlled in a population dependant manner by quorum sensing. Quorum sensing depends on the production of small diffusible N-acyl homoserine lactone (AHL) molecules and their transfer among members of the bacterial population. As the bacterial population density increases the concentration of AHL rises to a critical level at which point there is an activation of the target genes within the bacteria. Proteomics has been used to look at quorum sensing in a second opportunistic bacterial pathogen affecting persons with CF, namely *Burkholderia cepacia*. Protein synthesis was examined in wild-type *B. cepacia*, H111, as well as an AHL-deficient *cepI* mutant, designated H111-I. Pre-fractionation of the cells into cellular, membrane and secreted proteins increased the coverage of the *B. cepacia* proteome (Riedel *et al.*, 2003). Out of the 985 detected proteins, 55 proteins were differentially expressed between H111 and H111-I; these were proposed as genes under the control of the quorum sensing system. Since the complete genome sequence of *B. cepacia* H111 has not been determined, protein identifications were carrying out using N-terminal sequencing rather than the more widely applied peptide mass mapping. Addition of N-octanoylhomoserine lactone, the signal molecule used in the *cep* system, fully restored the H111-I profile to that of the wild type bacteria. The analysis of *B. cepacia* illustrates the successful application of proteomics to characterise bacterial pathogens beyond the well-characterised sequenced model organisms.

4.2   Changes in bacterial proteomes during in vivo infection of eukaryotic cells

As discussed there are limitations to the comparative analyses of pathogenic variants grown *in vitro* on defined culture media. Most importantly is the fact that pathogenic bacteria express different genes when grown on artificial media compared to gene expression when they are in association with their host eukaryotic cells. Specifically, some of the known virulence determinants are only expressed when bacteria are in association with their host (Nouwens *et al.*, 2000; Bumann *et al.*, 2002). Extensive efforts are underway to characterise these differentially expressed gene networks since they may represent highly regulated pathogenic determinants or they may be potential targets for new therapeutic drugs. Recombinant DNA technologies play a key role in identifying those bacterial genes specifically expressed *in vivo*, *i.e.* when the bacteria are in association with the eukaryotic cell. For example, reporter genes linked to bacterial promoter sequences can be used to track the promoter's activation *in vivo* (Hautefort and Hinton, 2001). *In vivo* expression technology (IVET) has played an important role in the high throughput screening of those bacterial promoters specifically activated when facultative bacterial pathogens are grown with their eukaryotic host either in an intact animal or in vitro cell culture (Slauch *et al.*, 1994; Mahan *et al.*, 1995; Wang *et al.*, 1996). Proteomics has the potential to complement these recombinant DNA approaches but the technologies most widely used to characterise bacterial proteomes suffer from a number of limitations for analysing *in vivo* gene expression. Using current methods of 2DGE, proteomics is relatively insensitive and is generally restricted to analysing *in vitro* cell systems. *In vitro* grown cell lines are themselves poor as models for pathogenesis, since they are unable to represent reliably the many interactive networks occurring between differentiated cell types within an intact animal. At present there is no reported means to characterise the proteomes of bacterial pathogens grown *in vivo* with the intact animal. The low recovery of bacterial cells from the host and an absence of a protein amplification method, similar to the polymerase chain reaction for mRNA, present a number of technical difficulties to the investigator.

Some progress has been on analysing *in vivo* expressed bacterial proteomics using selected *in vitro* grown cell lines. One advantage of *in vitro* cell lines is that they can be infected under highly reproducible, controlled conditions and radioactive amino acids can be used to increase the sensitivity of detecting bacterial protein synthesis. Typically, a combination of antibiotics and radioactive amino acids are used to selectively radiolabel the proteins synthesised by intracellular bacteria (Rafie-Kolpin *et al.*, 1996; Yamamoto *et al.*, 1994; Abshire and Neidhardt, 1993; Kovarova *et al.*, 1992). Briefly, eukaryotic cells are infected with bacteria and once the bacteria have entered the cells gentamycin is added to kill extracellular bacteria but not the engulfed intracellular bacteria, which retain their viability. Radioactive amino acid tracers are then added to radiolabel newly synthesised proteins. Cycloheximide is included during this incubation stage to inhibit eukaryotic cell protein synthesis but not bacterial protein synthesis.

The radiolabelled proteins synthesised by the intracellular bacteria can then be compared to radiolabelled proteins prepared from bacteria grown in defined culture media to locate differentially expressed bacterial proteins. Data obtained by this experimental strategy must be considered in the light of two possible factors. Firstly, the antibiotics used during the radiolabelling may interfere with the normal interaction of the intracellular bacteria and eukaryotic cell (Monahan *et al.*, 2001). Secondly, the intracellular and *in vitro* grown bacteria may exhibit different growth kinetics leading to non-specific variation in protein synthesis. For at least one bacterial group it has been possible to overcome these restrictions and eliminate the need for the use of cycloheximide. This has been shown for *M. bovis* infection of macrophage cells (Monahan *et al.*, 2001). In this case radiolabelling was carried out in the absence of cycloheximide and, following radiolabelling, the macrophage cells were lysed with SDS and the bacteria collected by centrifugation and then washed with Tween-80. Appropriate controls were included to show that there was minimal carry over of cellular proteins into the bacterial pellet. How widely applicable this approach is to look at intracellular has yet to be investigated.

The following discussion considers primarily the changes induced in the proteomes of intracellular bacterial. However, it has been shown that exposure of *Campylobacter jejuni* to conditioned culture media lone can lead to altered bacterial protein synthesis (Konkel and Cieplak, 1992). When intracellular bacteria grow within the cell's phagosome they are exposed to a various stress conditions, including extremes of acidity, oxygen and nutrients (Kwaik and Harb, 1999). However, those bacterial pathogens that migrate out of the intracellular vacuoles into the cytoplasm are exposed to a much reduced level of stress. During infection of macrophage cells, *Brucella abortus*, *Salmonella typhimurium*, *Yersinia enterocolitica* and *Legionella pneumophila* remain within the cellular phagosome. A consistent observation for these bacterial pathogens is that the synthesis of specific bacterial proteins can be either induced or repressed in intracellular bacteria compared to bacteria growing in artificial culture media. Moreover, some of the bacterial proteins whose synthesis is induced during intracellular growth also show induced biosynthesis when bacteria are stressed *in vitro*. For example, the synthesis of the bacterial heat shock proteins GroEL and DnaK are induced during *B. abortus* infection of bovine or murine macrophages (Rafie-Kolpin *et al.*, 1996; Lin and Ficht, 1995). A similar pattern of synthesis of these two heat shock proteins is found during *S. typhimurium* infection of macrophages (Buchmeier and Heffron, 1990). Two major bacterial proteins homologous to the stress proteins DnaK and CRPA are also induced during *Y. enterocolitica* infection of J774 cells, a murine macrophage cell line as well as being induced *in vitro* by heat shock and oxidative stress (Yamamoto *et al.*, 1994). Thirty-two out of 67 bacterial proteins induced during *L. pneumophila* infection of the U937 macrophage cell line are also induced by *in vitro* stress conditions (Abu Kwaik *et al.*, 1993). A protein, global stress protein (GspA), is expressed by *L. pneumophila* in response to all stress conditions examined to date as well as during intracellular replication (Abu Kwaik *et al.*, 1997; Kwaik and Harb, 1999). Although the spectrum of proteins exhibiting altered synthesis by intracellular bacteria and bacteria stressed *in vitro* have many similarities, the changes observed for the former are not simply a summation of the *in vitro* stress responses (Abshire and Neidhardt, 1993) suggesting that specific response patterns are induced during the intracellular replication phase.

In contrast to the above data *Listeria monocytogenes* infection of J774 cells shows a different pattern of bacterial protein synthesis. None of the thirty-two proteins observed for intracellular bacteria are induced by typical *in vitro* stress conditions, including heat shock and oxidative stress (Hanawa *et al.*, 1995). This observation would be consistent with the intracellular bacteria rapidly migrating from the phagosome to the cytoplasm during intracellular growth.

During the infection of THP-1 macrophage cells by *M. bovis* BCG strain proteins are synthesised that are not expressed under standard *in vitro* growth conditions (Monahan *et al.*, 2001). These proteins were demonstrated using a combination of radiolabelling and immunoblotting with human *M. tuberculosis* infected sera. During *M. bovis* infection of macrophages at least 20 bacterial proteins were differentially which were either specifically expressed during the infection of macrophage cells or exhibited significantly induced levels of synthesis by intracellular bacteria. Six of the proteins induced at 24 hours post-infection were identified using a combination of peptide mass mapping by MALDI-TOF MS and sequencing by nanoES MS. The identified proteins were the GroEL homologues, GroEL-1/GroEL-2, InhA, 16kDa antigen (α-crystallin Hsp-X), EF-Tu and a 31kDa hypothetical protein. As commented upon by the authors these data were significant since they represent an early successful identification of bacterial proteins recovered directly from intracellular bacteria using MS techniques (Monahan *et al.*, 2001). As might be expected the source of the eukaryotic cell influences the outcome of the bacterial infection. For example, the type and origin of the host cell influences both the initial interaction between bacteria and eukaryotic cell as well as the bacterial growth kinetics for *M. tuberculosis* and *M. smegmatis* (Mehta *et al.*, 1996; Barker *et al.*, 1996). The response of the bacterial proteome to growth in host cells of differing origin has been described for *M. avium* (Sturgill-Koszycki *et al.*, 1997). *M. avium* infection of J774 cells results in the specific induced synthesis of bacterial proteins. The induced bacterial protein synthesis was initiated by 6 hours post-infection (p.i.) and continued until 4 days p.i. None of the proteins induced during intracellular bacterial growth were induced by *in vitro* stress conditions. In contrast, when primary bone macrophages were infected with *M. avium* the bacteria followed different kinetics of replication and protein synthesis. In these cells, intracellular bacteria radiolabelled at 5 and 12 days p.i. showed significant differences in their protein synthesis. The data were consistent with the bacteria initially entering a stasis phase early in infection before commencing a normal replication cycle between 5 and 12 days p.i. At 12 days p.i. the bacteria synthesised a similar range of proteins to those found for *M. avium* growing in J774 cells. The variations in the replication cycle and protein synthesis of the bacteria using macrophage cells of different sources demonstrate the care required in selecting the host cell used to investigate intracellular bacterial replication.

The previous discussion considered the response of bacterial gene expression to the intracellular environment of the host. Similar approaches can, in principal, be used to investigate the host cell response to the bacterial infection, although these have been fairly limited in number. A short communication by Kovarova *et al.* (1992) reported on global changes in macrophage protein synthesis following *Francisella tularensis* infection of mice. Principal component analysis was used to identify significant changes in protein synthesis between infected and uninfected cells. It was possible to distinguish uninfected and infected macrophage cells on the basis of their overall protein patterns obtained using 2DGE. The infected macrophage cells were further subdivided into macrophages collected 3-7 days post-infection and cells collected at 10 days post-infection (Kovarova *et al.*, 1992). No data were provided on the identities of the host proteins altered following infection. The *Nramp1* gene determines cellular resistance to intracellular parasites in mice (Govoni *et al.*, 1996).

The effect of the infecting bacteria on the host cell proteome has also been examined for *M. tuberculosis* specifically at the level of the phagosome. During the intracellular replication of *M. tuberculosis* in macrophage cells the *M. tuberculosis* phagosomal compartments fails to fuse with the lysosomes, the normal fate of intracellular phagosomes. Essentially, the bacteria arrest the development and processing of the phagosome Comparisons of the phagosomal compartments from Mycobacterial infected cells with the same structures from uninfected cells show a number of differences in the cellular proteins present. In a recent review of the *M. tuberculosis* phagosome Fratti *et al.* (2000) proposed that one of the key features of these structures was the exclusion of EEA-1 (Early-endosomal autoantigen), which plays a role in vesicle tethering and also in endosomal fusion. At present the question remains on determining the identity of the protein, or proteins, expressed by *M. tuberculosis*, which leads to the exclusion of EEA-1.

5. ANALYSIS OF THE HOST IMMUNE RESPONSE TO BACTERIAL INFECTION

The basic methods of proteomics used in studies described elsewhere in this chapter can be readily adapted to investigate the immune response of infectious diseases. The data derived from such experiments are clearly of relevance to deriving future vaccines. Two-dimensional electrophoresis combined with immunoblotting can be used to investigate the humoral and cellular immune response against microbial pathogens. The analysis of bacterial proteins by 2DGE improves the resolution of antigenic proteins and increases the number of unique protein species amenable to analysis compared to similar studies using 1-dimensional separation methods. Combining this approach with protein identification by peptide mass mapping of the resolved proteins it is possible to identify the protein(s) reacting with the sera. Thus, one can make rapid progress towards identifying the spectrum of antibody specificities induced during infection.

The global analysis of antigenic proteins using proteomics is a potentially useful approach to identify novel antigenic determinants for inclusion in future vaccines as applied to *H. pylori*. McAttee and colleagues (McAtee *et al.*, 1998a; McAtee *et al.*, 1998b) found over 30 immunogenic protein spots when the bacterial proteins were separated over a broad pH range in the first dimension of 2DGE. The antigenic proteins were identified by N-terminal sequencing and peptide mass mapping. Twenty-nine of the antigenic proteins were identified and 15 were assigned to known ORFs in the *H. pylori* genome sequence; the remainder had homologies with bacterial proteins not annotated in the available *H. pylori* genome sequence. One 30 kDa protein spot contained two protein species both derived by post-translational processing from the *H. pylori* cell binding 2 protein (McAtee *et al.*, 1998c), which had not previously been shown be an immunogenic protein. Comparisons of the proteins expressed by different isolates of *H. pylori* demonstrated that the majority of the antigenic proteins were expressed by all of the isolates, although differences in the expression levels of the flagellin protein and catalase were observed between some isolates (McAtee *et al.*, 1998b). Lock *et al* (2002) used a similar approach to identify 5 major groups of spots, resolved by 2DGE that reacted with IgA and IgG antibodies. These 5 immunoreactive protein clusters were composed of isoforms of 4 *H. pylori* proteins, specifically HspB (2 of the groups), UreB, EF-Tu and FlaA. These same proteins were strongly immunogenic for seven of the *H. pylori* isolates examined with only the flagella proteins, FlaA, exhibiting any variation between the 7 isolates. The latter observation being consistent with that of McAttee *et al.* (1998b). The studies from these two research groups used whole cell protein preparations against which to screen the sera. In contrast, Utt *et al.* (2002) used a membrane-enriched fraction of *H. pylori* cells as the antigen against which to screen sera containing *H. pylori* antibodies. One hundred and eighty six protein species were found to react against the *H. pylori* antibody positive sera. Peptide sequencing identified six highly immunogenic protein spots equivalent to 4 *H. pylori* proteins in the molecular mass region of 25-30kDa. These proteins were cell binding factor 2, urease A, an outer membrane protein (HP1564), and a hypothetical protein (HP0231). One of these proteins, cell binding protein 2, has previously been proposed as a potential vaccine candidate (McAtee *et al.*, 1998c). Haas *et al.* (2002) reported on the association between disease status and the pattern of the immune response to *H. pylori* proteins. Using sera from patients with active *H. pylori* infections compared to controls and patients with unrelated gastric infection, 310 antigenic protein species were recognised by *H. pylori* positive sera. Forty-two protein species corresponding to 32 genes were found to be highly immunogenic in the *H. pylori* positive sera. Differential immune responses, compared to the disease status, were found with 23 protein species (equivalent to 19 genes), which were significantly more immunogenic, based on their detection or signal strength, in patients with duodenal or gastric ulcers compared to patients with gastritis.

The between patient variation of antibody responses has also been reported for *Chlamydia trachomatis* (Sanchez-Campillo *et al.*, 1999). Antibodies were detected against a number of antigenic proteins; including outer the membrane proteins (OMP) and the GroEL-like and DnaK-like heat shock proteins. In addition, at least four novel immunogenic protein spots were detected that reacted with a variable proportion of the sera. The recognition frequencies for these antigens varied among the patients, although all of the sera reacted against OMP2. A limitation of the data was the possible occurrence of cross-reactions with either other Chlamydia species or unrelated bacteria (Sanchez-Campillo *et al.*, 1999). Similar variations in the range of antibody specificities have been observed among patients with different clinical presentations of lyme borrelliosis (Jungblut *et al.*, 1999b).

The identification of Mycobacterial immunogenic proteins that elicit a protective humoral and cellular immune response during infection has received a lot of attention. Attempts to identify immunogenic bacterial proteins using 1DGE in immunoblot assays against patients' sera have largely suffered from the poor protein resolution making it difficult to reliably identify the individual immunogenic proteins (Chakrabarty *et al.*, 1982; Ehrenberg and Gebre, 1987). Improved resolution was achieved through the adoption of 2DGE to separate the proteins prior to immunoblotting. Antibody specificities have been examined for patients infected with *Mycobacterium leprae* with either borderline tuberculoid or lepromatous leprosy. All patients in the two groups have serum antibodies reacting against a major 30kDa antigen (Mahon *et al.*, 1990). However, at least 6 antigenic proteins were preferentially recognised by sera from patients with lepromatous leprosy. Peripheral blood lymphocytes prepared from leprosy patients and stimulated *in vitro* with anti-CD3 monoclonal antibody expressed antibodies recognising a 10kDa *M. leprae* protein. Lymphocytes from healthy controls reacted non-specifically against some *M. leprae* proteins but not the 10kDa protein. No antibodies reacting against the 10kDa protein were detected in the sera collected from the same patients.

The T cell response is a major line of defence in immunity against *M. tuberculosis* and the identification of those bacterial proteins eliciting these responses will be important for future vaccine development. During growth in defined culture media, *M. tuberculosis* releases proteins into the media capable of inducing a protective cellular immune response (Pal and Horwitz, 1992; Roberts *et al.*, 1995; Andersen, 1994) and work is in progress to identify the proteins responsible. Up to 205 protein spots were detected in *M. tuberculosis* culture media by 2DGE (Sonnenberg and Belisle, 1997) and 34 unique proteins were identified in these preparations based on their reactions with specific monoclonal antibodies and amino acid sequencing. One member of the protein cluster at 83-85 kDa, previously recognised as a dominant humoral antigen (Laal *et al.*, 1997), was identified as the *M. tuberculosis* catalase, KatG; at least two other members of this cluster shared an epitope with KatG (Sonnenberg and Belisle, 1997). In these experiments the bacterial proteins were harvested from the culture media during the late logarithmic phase of bacterial growth when some bacterial cytoplasmic proteins might also be present in the media (Sonnenberg and Belisle, 1997; Andersen *et al.*, 1991). To overcome this potential problem, Weldingh *et al.* (1998) analysed the cell-released proteins after only 7 days growth. Six immunogenic proteins from these preparations, recognised by T cells, were assigned to the *M. tuberculosis* genome sequence using N-terminal amino acid sequencing. One of the immunogenic proteins (designated CFP21) mapped onto the RD2 genome segment that is absent from some BCG strains (Mahairas *et al.*, 1996). A 29kDa immunogenic protein found in the media from short-term cultures, as well as in the *M. tuberculosis* membrane, is also present in several mycobacterial species (Rosenkrands *et al.*, 1998). The KatG protein identified by Sonnenberg and Belisle (1997) was not detected in these short-term culture filtrates.

The ability of *M. tuberculosis* proteins present in either infected culture filtrates or bacterial cells to stimulate T cells collected from tuberculosis patients, healthy controls as well as tuberculin positive and negative individuals has been examined (Schoel *et al.*, 1992; Daugelat *et al.*, 1992). The bacterial proteins were resolved by 2DGE and then transferred to the liquid phase by electroblotting (Gulle *et al.*, 1990). The individual protein fractions were then used to stimulate T cells *in vitro*. Although there was a variable response between patients, T cells from tuberculosis patients and tuberculin positive individuals reacted with cell-associated antigens of >30kDa (pI 4.2-6.6) (Schoel *et al.*, 1992). The reaction patterns differed from those found for the tuberculin negative individuals and healthy controls. Similar analyses for antigenic proteins in the culture filtrates revealed major antigenic components of 30-100kDa and pI's of 4-5. T cells prepared from tuberculin negative contacts showed either a weak or no reaction to the bacterial proteins (Daugelat *et al.*, 1992). Although present in the culture media, neither the 38kDa nor the 65kDa major T cell antigens of *M. tuberculosis* migrated in this region (Daugelat *et al.*, 1992). More detailed studies (Covert *et al.*, 2001) have been reported that identify the antigenic proteins themselves. The individual protein fractions prepared as before were assayed for T cell activation based on interferon-γ expression. Thirty-eight fractions induced a significant T-cell response and the proteins present in these fractions were identified peptide mass mapping and partial sequencing. Positive identifications were achieved for 16 proteins from infected culture filtrates and 18 proteins from cell lysates. Although the majority of these proteins correspond to previously described bacterial antigens, 17 proteins were novel T-cell antigens.

6.   THERAPY AND VACCINE DEVELOPMENT THROUGH PROTEOMICS

Proteomics has played a role in the development of novel therapeutic strategies *via* the identification of vaccine and antibiotic targets (Rosamond and Allsop, 2000; Nilsson *et al.*, 2000). Antigenic proteins can be identified by investigating the immune response following either natural or laboratory infection with the bacterium of interest. The immunogenic proteins identified by these means can then be used in the development of vaccine strategies. To optimise the identification of vaccine candidates, specific protein classes can be selectively analysed. It is reasonable to expect that the membrane proteins will be major targets for the host immune response. In order to catalogue potential vaccine candidates among the membrane proteins of *H. pylori* cells were solubilised with n-octylglucoside and the proteins analysed with no further enrichment or fractionation (Nilsson *et al.*, 2000). The bacterial proteins were fractioned by IEF in liquid phase for the 1st dimension and by continuous elution of selected fractions by SDS-PAGE. This separation technology using a liquid phase was specifically aimed to overcome the well-documented limitation of 2DGE to resolve hydrophobic proteins. Fifteen of the forty proteins identified were either membrane or membrane-associated proteins and many had not been previously identified using standard 2DGE. Chakravarti *et al.* (2000) described a bioinformatics approach to the identification of vaccine candidate proteins, which took advantage of the extensive gene sequence data now available. These authors used the *H. influenzae* Rd genome sequence to search for potential vaccine candidate proteins *in silico*. As with the previous "wet bench" approach to locate *H. pylori* vaccine candidates, potential *H. influenzae* candidate proteins were selected on the basis of their possession of characteristics of membrane proteins. A complementary proteomic approach to identify the proteins from soluble outer membrane fractions of *H. pylori* was also described. Although this is a potentially valuable approach the authors did not confirm that the proteins identified by these criteria were in fact immunogenic under natural conditions. Ferroro and Labigne (2001) have reviewed a similar limitation of purely *in silico* analysis of gene sequence data for vaccine candidates for *H. pylori*. De Groot *et al.* (2001) have used *in silico* analysis of the *M. tuberculosis* genome to identify putative protective epitopes likely to be involved in the T cell response to infection. An estimated 3000 peptides were selected on this basis which have to be screened for their immunogenic activity.

The spread of antibiotic and drug resistance among microbial pathogens is a major problem in infection control. An understanding of the mechanism(s) by which drug resistance develops will lead to improvements in extending the efficacy of current anti-microbials. Proteomics can contribute towards determining anti-microbial resistance mechanisms through the capacity to analyse global changes in microbial proteins. Resistance to beta-lactam antibiotics has been investigated for *Pseudomonas aeruginosa*. A reduced expression of a 47kDa (pI 5.2) outer membrane protein has been found for imipenem resistant *P. aeruginosa* (Vurma-Rapp *et al.*, 1990). Michea-Hamzehpour *et al.* (1993) reported equivalent data with the loss of an outer membrane protein with a similar isoelectric point in imipenem resistant *P. aeruginosa*. N-terminal sequencing showed that the protein was homologous to the porin outer membrane protein D (Michea-Hamzehpour *et al.*, 1993). Changes in outer membrane proteins have also been shown among ceftazidime resistant *P. aeruginosa* isolates with the expression of a basic protein homologous to the *ampC* gene product (Michea-Hamzehpour *et al.*, 1993).

Penicillin resistance is on the increase among clinical isolates of *Streptococcus pneumoniae* and resistance to erythromycin, used as an alternative antibiotic to penicillin, has also emerged as a potential problem (Johnson *et al.*, 1996). Two erythromycin resistant phenotypes are recognised for *S. pneumoniae*, specifically the MLS and M phenotypes (Johnson *et al.*, 1996). Erythromycin resistant *S. pneumoniae* possessing the MLS phenotype owe their resistance to the methylation of rRNA (Clewell *et al.*, 1995; Trieu-Cuot *et al.*, 1990). *S. pneumoniae* isolates with the M phenotype are resistant through the expression of *mefE*, which encodes a membrane transporter protein that reduces the intracellular levels of erythromycin (Tait-Kamradt *et al.*, 1997; Sutcliffe *et al.*, 1996). Proteomics has been used to further investigate erythromycin resistance in M phenotype *S. pneumoniae* isolates (Cash *et al.*, 1999). Cellular proteins were prepared from erythromycin resistant (M phenotype) and sensitive isolates of *S. pneumoniae* and analysed using 2DGE. The M phenotype erythromycin resistant isolates showed a characteristic electrophoretic variation in glyceraldehyde phosphate dehydrogenase (GAPDH) (Cash *et al.*, 1999). Three electrophoretic variants of GAPDH that differed in their isoelectric points were resolved for M phenotype resistant isolates two of the three forms were also founding sensitive isolates with the most basic GAPDH form being specific for resistant bacteria. Thus, the abnormal synthesis of GAPDH in the M phenotype isolates may be related to an altered pattern of post-translational modification. The alteration in GAPDH is unlikely to be the primary cause of erythromycin resistance but may play some, as yet, unknown role in the development of resistance.

Antibiotic action can be investigated by locating proteins that show differential expression patterns when bacteria are grown in the presence or absence of antibiotics. The response of the bacterial proteome following their exposure to antibiotics has been investigated for *Staphylococcus aureus* (Singh *et al.*, 2001). *S. aureus* grown in the presence of inhibitory concentrations of oxacilin, a cell-wall active antibiotic, caused the elevated expression of at least 9 proteins. Five of the induced proteins were identified by N-terminal sequencing as methionine sulphoxide reductase, enzyme IIA component of the phosphotransferase system, signal transduction protein (TRAP), GroES and GreA. A similar pattern of induced protein synthesis was found with other antibiotics that act on the bacterial cell wall but not by antibiotics acting on other cellular functions, thus suggesting that the induced protein expression represents a "proteomic signature" for this response (Singh *et al.*, 2001). A similar experimental approach has been used to examine metronidazole induced gene expression in *H. pylori* (McAtee *et al.*, 2001). Metronidazole resistance depends on the mutation of the NADH reductase; higher levels of resistance are due to the loss of function in additional reductase genes. When functional, the reductase converts Metronidazole from a harmless drug to mutagenic and bactericidal products with the generation of reactive oxygen metabolites. In the presence of sublethal concentrations of metronidazole 19 protein spots exhibited differential expression in *H. pylori*. Among the proteins showing an increased expression level in the presence of the antibiotic were identified as alkylhydroperoxide reductase (AHP) and aconitase B. AHP is known to protect against oxygen toxicity and it was proposed that the increased expression of AHP in metronidazole resistant *H. pylori* is important in the generation of the resistance phenotype.

7. CONCLUDING REMARKS

Proteomics has made an important impact in many areas of the biosciences providing a complementary approach to nucleic acid based technologies in the area of functional genomics. These benefits are also apparent in those areas of research that are of relevance to the analysis of medically important bacteria. As discussed in this chapter, the practical applications of proteomic technologies are diverse both complementing and expanding studies at the level of the microbial genome nucleic acid. Technical limitations of the original proteomic technology of 2DGE concerning sensitivity and protein classes amenable to analysis can be largely overcome by modifying the 2DGE method or by adopting new analytical approaches. Initial problems relating to the post-electrophoretic identification of proteins by peptide mass mapping can be overcome due to the burgeoning gene sequence data available and the development of algorithms for cross-species identification of proteins (Lester and Hubbard, 2002). Thus, poorly characterised bacteria can be investigated by these methods based on the data derived from other bacterial groups. We can now rapidly identify immunogenic proteins, many of which are problematic hydrophobic membrane proteins, which may serve as candidates for future vaccines. Proteomics may either serve as the primary screening method for antigen detection or as a complementary method to confirm bioinformatic predictions of immunogenic at the "wet bench".

It is in the area of microbial pathogenesis that the application of proteomics faces its greatest challenges and potential successes. The identification and characterisation of *in vitro* expressed virulence determinants is relatively straightforward providing care is taken in interpreting the significance of the data. However, difficulties arise as one attempts to look at specific *in vivo* expressed bacterial genes and their protein products. Progress can be made using of model systems of bacterial infection (i.e. bacterial infections of *in vitro* grown cell lines) but these can never reliably mimic the intact animal. The widely used current methods for defining bacterial proteomes, namely 2DGE and peptide mass mapping, just do not have the sensitivity for the direct analysis of bacterial proteomes during the *in vivo* infection of whole animals or humans. Strategies must be developed to either enrich for the bacteria present in the infected tissue, to allow the analysis of the proteome by 2DGE, or new analytical methods with improved detection sensitivities will need to be adopted for the direct analysis of the total bacterial proteome with no prior enrichment. The most likely technologies currently available for the latter are those based on modern techniques of high-throughput mass spectrometry. This approach has already been demonstrated by the use of ICAT for the analysis of *Pseudomonas aeruginosa*. (Guina *et al.*, 2003). Once this and similar technology become widely available to the medical microbiology research community then there is certain to be an increase in the progress for analysing the proteomes of pathogens *in vivo*.

REFERENCES

Abshire, K.Z. and Neidhardt, F.C. 1993, Analysis of proteins synthesized by Salmonella typhimurium during growth within a host macrophage. *J Bacteriol* **175**:3734-3743.

Abu Kwaik, Y., Eisenstein, B.I., and Engleberg, N.C. 1993, Phenotypic modulation by Legionella pneumophila upon infection of macrophages. *Infection & Immunity* **61**:1320-1329.

Abu Kwaik, Y., Gao, L.Y., Harb, O.S., and Stone, B.J. 1997, Transcriptional regulation of the macrophage-induced gene (gspA) of Legionella pneumophila and phenotypic characterization of a null mutant. *Mol Microbiol* **24**:629-642.

Allardet-Servent, A., Carles-Nurit, M.J., Bourg, G., Michaux, S., and Ramuz, M. 1991, Physical map of the Brucella melitensis 16 M chromosome. *J Bacteriol* **173**:2219-2224.

Amezaga, M.R., Carter, P.E., Cash, P., and McKenzie, H. 2002, Molecular Epidemiology of Erythromycin Resistance in Streptococcus pneumoniae Isolates from Blood and Noninvasive Sites. *J Clin Microbiol* **40**:3313-3318.

Andersen, H., Christiansen, G., and Christiansen, C. 1984, Electrophoretic analysis of proteins from Mycoplasma capricolum and related serotypes using extracts from intact cells and from minicells containing cloned mycoplasma DNA. *J Gen Microbiol* **130**:1409-1418.

Andersen, P., Askgaard, D., Ljungqvist, L., Bennedsen, J., and Heron, I. 1991, Proteins released from Mycobacterium tuberculosis during growth. *Infection & Immunity* **59**:1905-1910.

Andersen, P. 1994, Effective vaccination of mice against Mycobacterium tuberculosis infection with a soluble mixture of secreted mycobacterial proteins. *Infection & Immunity* **62**:2536-2544.

Barker, K., Fan, H., Carroll, C., Kaplan, G., Barker, J., Hellmann, W., and Cohn, Z.A. 1996, Nonadherent cultures of human monocytes kill Mycobacterium smegmatis, but adherent cultures do not. *Infection & Immunity* **64**:428-433.

Bernardo, K., Pakulat, N., Macht, M., Krut, O., Seifert, H., Fleer, S., Hunger, F., and Kronke, M. 2002, Identification and discrimination of Staphylococcus aureus strains using matrix-assisted laser desorption/ionization-time of flight mass spectrometry. *Proteomics* **2**:747-753.

Betts, J.C., Dodson, P., Quan, S., Lewis, A.P., Thomas, P.J., Duncan, K., and McAdam, R.A. 2000, Comparison of the proteome of mycobacterium tuberculosis strain H37Rv with clinical isolate CDC 1551. *Microbiology* **146**:3205-3216.

Boucher, J.C., Yu, H., Mudd, M.H., and Deretic, V. 1997, Mucoid Pseudomonas aeruginosa in cystic fibrosis: characterization of muc mutations in clinical isolates and analysis of clearance in a mouse model of respiratory infection. *Infection & Immunity* **65**:3838-3846.

Bruce, K.D. and Jordens, J.Z. 1991, Characterization of noncapsulate Haemophilus influenzae by whole-cell polypeptide profiles, restriction endonuclease analysis, and rRNA gene restriction patterns. *J Clin Microbiol* **29**:291-296.

Bruce, K.D. and Pennington, T.H. 1991, Clonal analysis of non-typable Haemophilus influenzae by sodium dodecyl sulphate-polyacrylamide gel electrophoresis of whole cell polypeptides. *J Med Microbiol* **34**:277-283.

Buchmeier, N.A. and Heffron, F. 1990, Induction of Salmonella stress proteins upon infection of macrophages. *Science* **248**:730-732.

Bult, C.J., White, O., Olsen, G.J., Zhou, L., Fleischmann, R.D., Sutton, G.G., Blake, J.A., FitzGerald, L.M., Clayton, R.A., Gocayne, J.D., Kerlavage, A.R., Dougherty, B.A., Tomb, J.F., Adams, M.D., Reich, C.I., Overbeek, R., Kirkness, E.F., Weinstock, K.G., Merrick, J.M., Glodek, A., Scott, J.L., Geoghagen, N.S.M., and Venter, J.C. 1996, Complete genome sequence of the methanogenic archaeon, *Methanococcus jannaschii*. *Science* **273**:1058-1073.

Bumann, D., Meyer, T.F., and Jungblut, P.R. 2001, Proteome analysis of the common human pathogen Helicobacter pylori. *Proteomics* **1**:473-479.

Bumann, D., Aksu, S., Wendland, M., Janek, K., Zimny-Arndt, U., Sabarth, N., Meyer, T.F., and Jungblut, P.R. 2002, Proteome analysis of secreted proteins of the gastric pathogen Helicobacter pylori. *Infection & Immunity* **70**:3396-3403.

Cash, P., Wunner, W.H., and Pringle, C.R. 1977, A comparison of the polypeptides of human and bovine respiratory syncytial viruses and murine pneumonia virus. *Virology* **82**:369-379.

Cash, P. 1988, Structural and antigenic variation of the structural protein VP3 in serotype 1 poliovirus isolated from vaccinees. *Canadian Journal of Microbiology* **34**:802-806.

Cash, P. 1995, Protein mutations revealed by two-dimensional electrophoresis. *J Chromatogr A* **698**:203-224.

Cash, P., Argo, E., and Bruce, K.D. 1995, Characterisation of Haemophilus influenzae proteins by 2-dimensional gel electrophoresis. *Electrophoresis* **16**:135-148.

Cash, P., Argo, E., Langford, P.R., and Kroll, J.S. 1997, Development of an *Haemophilus* two-dimensional protein database. *Electrophoresis* **18**:1472-1482.

Cash, P. 1998, Characterisation of bacterial proteomes by two-dimensional electrophoresis. *Analytica Chimica Acta* **37**:121-146.

Cash, P., Argo, E., Ford, L., Lawrie, L., and McKenzie, H. 1999, A proteomic analysis of erythromycin resistance in Streptococcus pneumoniae. *Electrophoresis* **20**: :2259-2268.

Cash, P. and Kroll, J.S. 2003, Protein characterization by two-dimensional gel electrophoresis. *Methods in Molecular Medicine* **71**:101-118.

Chakrabarty, A.K., Maire, M.A., and Lambert, P.H. 1982, SDS-PAGE analysis of M. leprae protein antigens reacting with antibodies from sera from lepromatous patients and infected armadillos. *Clin Exp Immunol* **49**: :523-531.

Chakravarti, D.N., Fiske, M.J., Fletcher, L.D., and Zagursky, R.J. 2000, Application of genomics and proteomics for identification of bacterial gene products as potential vaccine candidates. *Vaccine* 19: :601-612.

Cho, M.J., Jeon, B.S., Park, J.W., Jung, T.S., Song, J.Y., Lee, W.K., Choi, Y.J., Choi, S.H., Park, S.G., Park, J.U., Choe, M.Y., Jung, S.A., Byun, E.Y., Baik, S.C., Youn, H.S., Ko, G.H., Lim, D., and Rhee, K.H. 2002, Identifying the major proteome components of Helicobacter pylori strain 26695. *Electrophoresis* 23: :1161-1173.

Clewell, D.B., Flannagan, S.E., and Jaworski, D.D. 1995, Unconstrained bacterial promiscuity: the Tn916-Tn1545 family of conjugative transposons. *Trends Microbiol* 3: :229-236.

Costas, M. 1992, Classification, identification and typing of bacteria by the analysis of their one-dimensional polyacrylamide gel electrophoretic protein patterns. *Advances in Electrophoresis* 5: :351-408.

Covert, B.A., Spencer, J.S., Orme, I.M., and Belisle, J.T. 2001, The application of proteomics in defining the T cell antigens of Mycobacterium tuberculosis. *Proteomics* 1: :574-586.

Daugelat, S., Gulle, H., Schoel, B., and Kaufmann, S.H. 1992, Secreted antigens of Mycobacterium tuberculosis: characterization with T lymphocytes from patients and contacts after two-dimensional separation. *J Infect Dis* 166: :186-190.

De Groot, A.S., Bosma, A., Chinai, N., Frost, J., Jesdale, B.M., Gonzalez, M.A., Martin, W., and Saint-Aubin, C. 2001, From genome to vaccine: in silico predictions, ex vivo verification. *Vaccine* 19: :4385-4395.

Du, Z., Yang, R., Guo, Z., Song, Y., and Wang, J. 2002, Identification of Staphylococcus aureus and determination of its methicillin resistance by matrix-assisted laser desorption/ionization time-of-flight mass spectrometry. *Anal Chem* 74: :5487-5491.

Dunn, B.E., Perez-Perez, G.I., and Blaser, M.J. 1989, Two-dimensional gel electrophoresis and immunoblotting of Campylobacter pylori proteins. *Infection & Immunity* 57: :1825-1833.

Eaton, K.A., Brooks, C.L., Morgan, D.R., and Krakowka, S. 1991, Essential role of urease in pathogenesis of gastritis induced by Helicobacter pylori in gnotobiotic piglets. *Infection & Immunity* 59: :2470-2475.

Edwards-Jones, V., Claydon, M.A., Evason, D.J., Walker, J., Fox, A.J., and Gordon, D.B. 2000, Rapid discrimination between methicillin-sensitive and methicillin-resistant Staphylococcus aureus by intact cell mass spectrometry. *J Med Microbiol* 49: :295-300.

Ehrenberg, J.P. and Gebre, N. 1987, Analysis of the antigenic profile of Mycobacterium leprae: cross-reactive and unique specificities of human and rabbit antibodies. *Scand J Immunol* 26: :673-681.

Enroth, H., Akerlund, T., Sille, A., and Engstrand, L. 2000, Clustering of Clinical Strains of Helicobacter pylori Analyzed by Two-Dimensional Gel Electrophoresis. *Clin Diag Lab Immunol* 7: :301-306.

Evason, D.J., Claydon, M.A., and Gordon, D.B. 2001, Exploring the limits of bacterial identification by intact cell-mass spectrometry. *Journal of the American Society for Mass Spectrometry* 12: :49-54.

Ferrero, R.L. and Labigne, A. 2001, Helicobacter pylori vaccine development in the post-genomic era: can in silico translate to in vivo. [Review] [56 refs]. *Scand J Immunol* 53: :443-448.

Fleischmann, R.D., Adams, M.D., White, O., Clayton, R.A., Kirkness, E.F., Kerlavage, A.R., Bult, C.J., Tomb, J.F., Dougherty, B.A., Merrick, J.M., McKenney, K., Sutton, G., FitzHugh, W., Fields, C., Gocayne, J.D., Scott, J., Shirley, R., Liu, L.I., Glodek, A., Kelley, J.M., Weidman, J.F., Phillips, C.A., Spriggs, T., Hedblom, E., Cotton, M.D., Utterback, T.R., Hanna, M.C., Nguyen, D.T., Saudek, D.M., Brandon, R.C., Fine, L.D.,

Fritchman, J.L., Fuhrmann, J.L., Geoghagen, N.S.M., Gnehm, C.L., McDonald, L.A., Small, K.V., Fraser, C.M., Smith, H.O., and Venter, J.C. 1995, Whole-genome random sequencing and assembly of *Haemophilus influenzae* Rd. *Science* 269: :496-511.

Fleischmann, R.D., Alland, D., Eisen, J.A., Carpenter, L., White, O., Peterson, J., DeBoy, R., Dodson, R., Gwinn, M., Haft, D., Hickey, E., Kolonay, J.F., Nelson, W.C., Umayam, L.A., Ermolaeva, M., Salzberg, S.L., Delcher, A., Utterback, T., Weidman, J., Khouri, H., Gill, J., Mikula, A., Bishai, W., Jacobs Jr, W.R., Jr., Venter, J.C., and Fraser, C.M. 2002, Whole-genome comparison of Mycobacterium tuberculosis clinical and laboratory strains. *J Bacteriol* 184: :5479-5490.

Fraser, C.M., Casjens, S., Huang, W.M., Sutton, G.G., Clayton, R., Lathigra, R., White, O., Ketchum, K.A., Dodson, R., Hickey, E.K., Gwinn, M., Dougherty, B., Tomb, J.F., Fleischmann, R.D., Richardson, D., Peterson, J., Kerlavage, A.R., Quackenbush, J., Salzberg, S., Hanson, M., van Vugt, R., Palmer, N., Adams, M.D., Gocayne, J., and Venter, J.C. 1997, Genomic sequence of a Lyme disease spirochaete, Borrelia burgdorferi [see comments]. *Nature* 390: :580-586.

Fratti, R.A., Vergne, I., Chua, J., Skidmore, J., and Deretic, V. 2000, Regulators of membrane trafficking and Mycobacterium tuberculosis phagosome maturation block. *Electrophoresis* 21: :3378-3385.

Gimenez, H.B., Hardman, N., Keir, H.M., and Cash, P. 1986, Antigenic variation between human respiratory syncytial virus isolates. *J Gen Virol* 67: :863-870.

Gormon, T. and Phan-Thanh, L. 1995, Identification and classification of Listeria by two-dimensional protein mapping. *Res Microbiol* 146: :143-154.

Govoni, G., Vidal, S., Gauthier, S., Skamene, E., Malo, D., Gros, P., and P. 1996, The Bcg/Ity/Lsh locus: genetic transfer of resistance to infections in C57BL/6J mice transgenic for the Nramp1 Gly169 allele. *Infection & Immunity* 64: :2923-2929.

Guina, T., Purvine, S.O., Yi, E.C., Eng, J., Goodlett, D.R., Aebersold, R., and Miller, S.I. 2003, Quantitative proteomic analysis indictes increased synthesis of a quinolone by Pseudomonas aeruginosa isolates cystic fibrosis airways. *Proc Natl Acad Sci U S A* 100: :2771-2776.

Gulle, H., Schoel, B., and Kaufmann, S.H. 1990, Direct blotting with viable cells of protein mixtures separated by two-dimensional gel electrophoresis. *J Immunol Methods* 133: :253-261.

Haas, G., Karaali, G., Ebermayer, K., Metzger, W.G., Lamer, S., Zimny-Arndt, U., Diescher, S., Goebel, U.B., Vogt, K., Roznowski, A.B., Wiedenmann, B.J., Meyer, T.F., Aebischer, T., and Jungblut, P.R. 2002, Immunoproteomics of Helicobacter pylori infection and relation to gastric disease. *Proteomics* 2: :313-324.

Hanawa, T., Yammamoto, T., and Kamiya, S. 1995, *Listeria monocytogenes* can grow in macrophages without the aid of proteins induced by environmental stresses. *Infection & Immunity* 63: :4595-4599.

Hansen, E.J., Wilson, R.M., and Baseman, J.B. 1979, Two-dimensional gel electrophoretic comparison of proteins from virulent and avirulent strains of *Mycoplasma pneumoniae*. *Infection & Immunity* 24: :468-475.

Hansen, E.J., Wilson, R.M., Clyde, W.A., Jr., and Baseman, J.B. 1981, Characterization of hemadsorption-negative mutants of Mycoplasma pneumoniae. *Infection & Immunity* 32: :127-136.

Hautefort, I. and Hinton, J.C. 2001, Measurement of bacterial gene expression in vivo. *Philosophical Transactions of the Royal Society of London - Series B: Biological Sciences* 355: :601-611.

Hazell, S.L., Andrews, R.H., Mitchell, H.M., Daskalopoulous, G., Jiang, Q., Hiratsuka, K., and Taylor, D.E. 1996, Variability of gene order in different Helicobacter pylori strains contributes to genome diversity. *FEMS Microbiol Lett* 20: :833-842.

Hecker, M. and Engelmann, S. 2000, Proteomics, DNA arrays and the analysis of still unknown regulons and unknown proteins of Bacillus subtilis and pathogenic gram-positive bacteria. *Int J Med Microbiol* 290: :123-134.

Hynes, S.O., McGuire, J., Falt, T., and Wadstrom, T. 2003, The rapid detection of low molecular mass proteins differentially expressed under biological stress for four Helicobacter spp. using ProteinChip technology. *Proteomics* 3: :273-278.

Ilver, D., Arnqvist, A., Ogren, J., Frick, I.M., Kersulyte, D., Incecik, E.T., Berg, D.E., Covacci, A., Engstrand, L., and Boren, T. 1998, Helicobacter pylori adhesin binding fucosylated histo-blood group antigens revealed by retagging. *Science* 279: :373-377.

Jackson, P., Thornley, M.J., and Thompson, R.J. 1984, A study by high resolution two-dimensional polyacrylamide gel electrophoresis of relationships between Neisseria gonorrhoeae and other bacteria. *J Gen Microbiol* 130: :3189-3201.

Jarman, K.H., Cebula, S.T., Saenz, A.J., Petersen, C.E., Valentine, N.B., Kingsley, M.T., and Wahl, K.L. 2000, An algorithm for automated bacterial identification using matrix-assisted laser desorption/ionization mass spectrometry. *Anal Chem* 72: :1217-1223.

Johnson, A.P., Speller, D.C., George, R.C., Warner, M., Domingue, G., and Efstratiou, A. 1996, Prevalence of antibiotic resistance and serotypes in pneumococci in England and Wales: results of observational surveys in 1990 and 1995. *BMJ* 312: :1454-1456.

Jungblut, P.R., Schaible, U.E., Mollenkopf, H.J., Zimny-Arndt, U., Raupach, B., Mattow, J., Halada, P., Lamer, S., Hagens, K., and Kaufmann, S.H. 1999a, Comparative proteome analysis of Mycobacterium tuberculosis and Mycobacterium bovis BCG strains: towards functional genomics of microbial pathogens. *Mol Microbiol* 33: :1103-1117.

Jungblut, P.R., Zimny-Arndt, U., Zeindl-Eberhart, E., Stulik, J., Koupilova, K., Pleissner, K.P., Otto, A., Muller, E.C., Sokolowska-Kohler, W., Grabher, G., and Stoffler, G. 1999b, Proteomics in human disease: cancer, heart and infectious diseases. *Electrophoresis* 20: :2100-2110.

Jungblut, P.R., Bumann, D., Haas, G., Zimny-Arndt, U., Holland, P., Lamer, S., Siejak, F., Aebischer, A., and Meyer, T.F. 2000, Comparative proteome analysis of Helicobacter pylori. *Mol Microbiol* 36: :710-725.

Jungblut, P.R., Muller, E.C., Mattow, J., and Kaufmann, S.H. 2001, Proteomics reveals open reading frames in Mycobacterium tuberculosis H37Rv not predicted by genomics. *Infection & Immunity* 69: :5905-5907.

Klimpel, K.W. and Clark, V.L. 1988, Multiple protein differences exist between Neisseria gonorrhoeae type 1 and type 4. *Infection & Immunity* 56: :808-814.

Konkel, M.E. and Cieplak, W. 1992, Altered synthetic response of Campylobacter jejuni to cocultivation with human epithelial cells is associated with enhanced internalization. *Infection & Immunity* 60: :4945-4949.

Kovarova, H., Stulik, J., Macela, A., Lefkovits, I., and Skrabkova, Z. 1992, Using two-dimensional gel electrophoresis to study immune response against intracellular bacterial infection. *Electrophoresis* 13: :741-742.

Kwaik, Y.A. and Harb, O.S. 1999, Phenotypic modulation by intracellular bacterial pathogens. *Electrophoresis* 20: :2248-2258.

Laal, S., Samanich, K.M., Sonnenberg, M.G., Zolla-Pazner, S., Phadtare, J.M., and Belisle, J.T. 1997, Human humoral responses to antigens of Mycobacterium tuberculosis: immunodominance of high-molecular-mass antigens. *Clinical & Diagnostic Laboratory Immunology* 4: :49-56.

Langen, H., Gray, C., Roder, D., Juranville, J.F., Takacs, B., and Fountoulakis, M. 1997, From genome to proteome: Protein map of *Haemophilus influenzae. Electrophoresis* 18: :1184-1192.

Langen, H., Takacs, B., Evers, S., Berndt, P., Lahm, H.W., Wipf, B., Gray, C., and Fountoulakis, M. 2000, Two-dimensional map of the proteome of Haemophilus influenzae. *Electrophoresis* 21: :411-429.

Lester, P.J. and Hubbard, S.J. 2002, Comparative bioinformatic analysis of complete proteomes and protein parameters for cross-species identification in proteomics. *Proteomics* 2: :1392-1405.

Lin, J. and Ficht, T.A. 1995, Protein synthesis in Brucella abortus induced during macrophage infection. *Infection & Immunity* 63: :1409-1414.

Link, A.J., Hays, L.G., Carmack, E.B., and Yates, J.R. 1997, Identifying the major proteome components of Haemophilus influenzae type-strain NCTC 8143. *Electrophoresis* 18: :1314-1334.

Lock, R.A., Cordwell, S.J., Coombs, G.W., Walsh, B.J., and Forbes, G.M. 2001, Proteome analysis of Helicobacter pylori: major proteins of type strain NCTC 11637. *Pathology* 33: :365-374.

Lock, R.A., Coombs, G.W., McWilliams, T.M., Pearman, J.W., Grubb, W.B., Melrose, G.J., and Forbes, G.M. 2002, Proteome analysis of highly immunoreactive proteins of Helicobacter pylori. *Helicobacter* 7: :175-182.

Loeb, M.R. and Smith, D.H. 1980, Outer membrane protein composition in disease isolates of Haemophilus influenzae: pathogenic and epidemiological implications. *Infection & Immunity* 30: :709-717.

Mahairas, G.G., Sabo, P.J., Hickey, M.J., Singh, D.C., and Stover, C.K. 1996, Molecular analysis of genetic differences between Mycobacterium bovis BCG and virulent M. bovis. *J Bacteriol* 178: :1274-1282.

Mahan, M.J., Tobias, J.W., Slauch, J.M., Hanna, P.C., Collier, R.J., and Mekalanos, J.J. 1995, Antibiotic-based selection for bacterial genes that are specifically induced during infection of a host. *Proc Natl Acad Sci U S A* 92: :669-673.

Mahon, A.C., Gebre, N., and Nurlign, A. 1990, The response of human B cells to Mycobacterium leprae. Identification of target antigens following polyclonal activation in vitro. *International Immunology* 2: :803-812.

Mattow, J., Jungblut, P.R., Muller, E.C., and Kaufmann, S.H. 2001, Identification of acidic, low molecular mass proteins of Mycobacterium tuberculosis strain H37Rv by matrix-assisted laser desorption/ionization and electrospray ionization mass spectrometry. *Proteomics* 1: :494-507.

McAtee, C.P., Fry, K.E., and Berg, D.E. 1998a, Identification of potential diagnostic and vaccine candidates of Helicobacter pylori by "proteome" technologies. *Helicobacter* 3: :163-169.

McAtee, C.P., Lim, M.Y., Fung, K., Velligan, M., Fry, K., Chow, T., and Berg, D.E. 1998b, Identification of potential diagnostic and vaccine candidates of Helicobacter pylori by two-dimensional gel electrophoresis, sequence analysis, and serum profiling. *Clinical & Diagnostic Laboratory Immunology* 5: :537-542.

McAtee, C.P., Lim, M.Y., Fung, K., Velligan, M., Fry, K., Chow, T.P., and Berg, D.E. 1998c, Characterization of a Helicobacter pylori vaccine candidate by proteome techniques. *J Chromatogr B Biomed Sci Appl* 714: :325-333.

McAtee, C.P., Hoffman, P.S., and Berg, D.E. 2001, Identification of differentially regulated proteins in metronidozole resistant Helicobacter pylori by proteome techniques. *Proteomics* 1: :516-521.

McGee, D.J. and Mobley, H.L. 1999, Mechanisms of Helicobacter pylori infection: bacterial factors. *Curr Top Microbiol Immunol* 241: :155-180.

Mehta, P.K., King, C.H., White, E.H., Murtagh, J.J., and Quinn, F.D. 1996, Comparison of in vitro models for the study of Mycobacterium tuberculosis invasion and intracellular replication. *Infection & Immunity* 64: :2673-2679.

Michea-Hamzehpour, M., Sanchez, J.C., Epp, S.F., Paquet, N., Hughes, G.J., Hochstrasser, D., and Pechere, J.C. 1993, Two-dimensional polyacrylamide gel electrophoresis isolation and microsequencing of Pseudomonas aeruginosa proteins. *Enzyme Protein* 47: :1-8.

Mollenkopf, H.J., Jungblut, P.R., Raupach, B., Mattow, J., Lamer, S., Zimny-Arndt, U., Schaible, U.E., and Kaufmann, S.H. 1999, A dynamic two-dimensional polyacrylamide gel electrophoresis database: the mycobacterial proteome via Internet. *Electrophoresis* 20: :2172-2180.

Monahan, I.M., Betts, J., Banerjee, D.K., and Butcher, P.D. 2001, Differential expression of mycobacterial proteins following phagocytosis by macrophages. *Microbiology* 147: :459-471.

Morgan, M.G., McKenzie, H., Enright, M.C., Bain, M., and Emmanuel, F.X. 1992, Use of molecular methods to characterize Moraxella catarrhalis strains in a suspected outbreak of nosocomial infection. *Eur J Clin Microbiol Infect Dis* 11: :305-312.

Nilsson, C.L., Larsson, T., Gustafsson, E., Karlsson, K.A., and Davidsson, P. 2000, Identification of protein vaccine candidates from Helicobacter pylori using a preparative two-dimensional electrophoretic procedure and mass spectrometry. *Anal Chem* 72: :2148-2153.

Nouwens, A.S., Cordwell, S.J., Larsen, M.R., Molloy, M.P., Gillings, M., Willcox, M.D., and Walsh, B.J. 2000, Complementing genomics with proteomics: the membrane subproteome of Pseudomonas aeruginosa PAO1. *Electrophoresis* 21: :3797-3809.

Ochiai, K., Uchida, K., and Kawamoto, I. 1993, Similarity of ribosomal proteins studied by two-dimensional coelectrophoresis for identification of gram-positivie bacteria. *International Journal of Systematic Bacteriology* 43: :69-76.

Ohlmeier, S., Scharf, C., and Hecker, M. 2000, Alkaline proteins of Bacillus subtilis: first steps towards a two-dimensional alkaline master gel. *Electrophoresis* 21: :3701-3709.

Pal, P.G. and Horwitz, M.A. 1992, Immunization with extracellular proteins of Mycobacterium tuberculosis induces cell-mediated immune responses and substantial protective immunity in a guinea pig model of pulmonary tuberculosis. *Infection & Immunity* 60: :4781-4792.

Rafie-Kolpin, M., Essenberg, R.C., and Wyckoff, J.H., 3rd. 1996, Identification and comparison of macrophage-induced proteins and proteins induced under various stress conditions in Brucella abortus. *Infection & Immunity* 64: :5274-5283.

Rain, J.C., Selig, L., De Reuse, H., Battaglia, V., Reverdy, C., Simon, S., Lenzen, G., Petel, F., Wojcik, J., Schachter, V., Chemama, Y., Labigne, A., and Legrain, P. 2001, The protein-protein interaction map of Helicobacter pylori. *Nature* 409: :211-215.

Regula, J.T., Ueberle, B., Boguth, G., Gorg, A., Schnolzer, M., Herrmann, R., and Frank, R. 2000, Towards a two-dimensional proteome map of Mycoplasma pneumoniae. *Electrophoresis* 21: :3765-3780.

Riedel, K., Arevalo-Ferro, C., Reil, G., Gorg, A., Lottspeich, F., and Eberl, L. 2003, Analysis of the quorum-sensing regulon of the opportunistic pathogen Burkholderia cepacia H111 by proteomics. *Electrophoresis* 24: :740-750.

Roberts, A.D., Sonnenberg, M.G., Ordway, D.J., Furney, S.K., Brennan, PJ, Belisle, J.T., and Orme, I.M. 1995, Characteristics of protective immunity engendered by vaccination of mice with purified culture filtrate protein antigens of Mycobacterium tuberculosis. *Immunology* 85: :502-508.

Rodwell, A.W. and Rodwell, E.S. 1978, Relationships between strains of Mycoplasma mycoides subspp mycoides and capri studied by two-dimensional gel electrophoresis of cell proteins. *J Gen Microbiol* 109: :259-263.

Rosamond, J. and Allsop, A. 2000, Harnessing the power of the genome in the search for new antibiotics [see comments]. [Review] [35 refs]. *Science* 287: :1973-1976.

Rosenkrands, I., Rasmussen, P.B., Carnio, M., Jacobsen, S., Theisen, M., and Andersen, P. 1998, Identification and characterization of a 29-kilodalton protein from mycobacterium tuberculosis culture filtrate recognized by mouse memory effector cells. *Infect Immun* 66: :2728-2735.

Rosenkrands, I., King, A., Weldingh, K., Moniatte, M., Moertz, E., and Andersen, P. 2000a, Towards the proteome of Mycobacterium tuberculosis. *Electrophoresis* 21: :3740-3756.

Rosenkrands, I., Weldingh, K., Jacobsen, S., Hansen, C.V., Florio, W., Gianetri, I., and Andersen, P. 2000b, Mapping and identification of Mycobacterium tuberculosis proteins by two-dimensional gel electrophoresis, microsequencing and immunodetection. *Electrophoresis* 21: :935-948.

Sabarth, N., Lamer, S., Zimmy-Arndt, U., Jungblut, P.R., Meyer, T.F., and Bumann, D. 2002, Identification of surface proteins of Helicobacter pylroi by selective biotinylation, affinity purification and two-dimensional gel electrophoresis. *J Biol Chem* 31: :27896-27902.

Sanchez-Campillo, M., Bini, L., Comanducci, M., Raggiaschi, R., Marzocchi, B., Pallini, V., and Ratti, G. 1999, Identification of immunoreactive proteins of Chlamydia trachomatis by Western blot analysis of a two-dimensional electrophoresis map with patient sera. *Electrophoresis* 20: :2269-2279.

Sazuka, T., Yamaguchi, M., and Ohara, O. 1999, Cyano2Dbase updated: linkage of 234 protein spots to corresponding genes through N-terminal microsequencing. *Electrophoresis* 20: :2160-2171.

Sazuka, T. and Ohara, O. 1997, Towards a proteome project of cyanobacterium Synechocytis sp. strain PCC6803: linking 130 protein spots with their respective genes. *Electrophoresis* 18: :1252-1258.

Schoel, B., Gulle, H., and Kaufmann, S.H. 1992, Heterogeneity of the repertoire of T cells of tuberculosis patients and healthy contacts to Mycobacterium tuberculosis antigens separated by high-resolution techniques. *Infection & Immunity* 60: :1717-1720.

Selander, R.K., Caugant, D.A., Ochman, H., Musser, J.M., Gilmour, M.N., and Whittam, T.S. 1986, Methods of multilocus enzyme electrophoresis for bacterial population genetics and systematics. *Applied & Environmental Microbiology* 51: :873-884.

Singh, V.K., Jayaswal, R.K., and Wilkinson, B.J. 2001, Cell wall-active antibiotic induced proteins of Staphylococcus aureus identified using a proteomic approach. *FEMS Microbiol Lett* 199: :79-84.

Slauch, J.M., Mahan, M.J., and Mekalanos, J.J. 1994, In vivo expression technology for selection of bacterial genes specifically induced in host tissues. *Methods Enzymol* 235: :481-492.

Slonczewski, J.L., McGee, D.J., Phillips, J., Kirkpatrick, C., and Mobley, H.L. 2000, pH-dependent protein profiles of Helicobacter pylori analyzed by two-dimensional gels. *Helicobacter* 5: :240-247.

Sonnenberg, M.G. and Belisle, J.T. 1997, Definition of Mycobacterium tuberculosis culture filtrate proteins by two-dimensional polyacrylamide gel electrophoresis, N-terminal amino acid sequencing, and electrospray mass spectrometry. *Infection & Immunity* 65: :4515-4524.

Sowa, B.A., Kelly, K.A., Ficht, T.A., and Adams, L.G. 1992, Virulence associated proteins of Brucella abortus identified by paired two-dimensional gel electrophoretic comparisons of virulent, vaccine and LPS deficient strains. *Appl Theor Electrophor* 3: :33-40.

Sreevatsan, S., Pan, X., Stockbauer, K.E., Connell, N.D., Kreiswirth, B.N., Whittam, T.S., and Musser, J.M. 1997, Restricted structural gene polymorphism in the Mycobacterium tuberculosis complex indicates evolutionarily recent global dissemination. *Proc Natl Acad Sci U S A* **94**: :9869-9874.

Sturgill-Koszycki, S., Haddix, P.L., and Russell, D.G. 1997, The interaction between Mycobacterium and the macrophage analyzed by two-dimensional polyacrylamide gel electrophoresis. *Electrophoresis* 18: :2558-2565.

Sutcliffe, J., Tait-Kamradt, A., and Wondrack, L. 1996, *Streptococcus pneumoniae* and *Streptococcus pyogenes* resistant to macrolides but sensitive to clindamycin: a common resistance pattern mediated by an efflux system. *Antimicrob Agents Chemother* 40: :1817-1824.

Tait-Kamradt, A., Clancy, J., Cronan, M., Dib-Hajj, F., Wondrack, L., Yuan, W., and Sutcliffe, J. 1997, mefE is necessary for the erythromycin-resistant M phenotype in Streptococcus pneumoniae. *Antimicrob Agents Chemother* 41: :2251-2255.

Taylor, D.E., Eaton, M., Chang, N., and Salama, S.M. 1992, Construction of a Helicobacter pylori genome map and demonstration of diversity at the genome level. *J Bacteriol* 174: :6800-6806.

Tomb, J.F., White, O., Kerlavage, A.R., Clayton, R.A., Sutton, G.G., Fleischmann, R.D., Ketchum, K.A., Klenk, H.P., Gill, S., Dougherty, B.A., Nelson, K., Quackenbush, J., Zhou, L., Kirkness, E.F., Peterson, S., Loftus, B., Richardson, D., Dodson, R., Khalak, H.G., Glodek, A., McKenney, K., Fitzegerald, L.M., Lee, N., Adams, M.D., Hickey, E.K., Berg, D.E., Gocayne, J.D., Utterback, T.R., Peterson, J.D., Kelley, J.M., Cotton, M.D., Weidman, J.M., Fujii, C., Bowman, C., Watthey, L., Wallin, E., Hayes, W.S., Borodovsky, M., Karp, P.D., Smith, H.O., Fraser, C.M., and Venter, J.C. 1997, The complete genome sequence of the gastric pathogen *Helicobacter pylori*. *Nature* 388: :539-547.

Tonella, L., Walsh, B.J., Sanchez, J.C., Ou, K., Wilkins, M.R., Tyler, M., Frutiger, S., Gooley, A.A., Pescaru, I., Appel, R.D., Yan, J.X., Bairoch, A., Hoogland, C., Morch, FS, Hughes, G.J., Williams, K.L., and Hochstrasser, D.F. 1998, '98 Escherichia coli SWISS-2DPAGE database update. *Electrophoresis* 19: :1960-1971.

Trieu-Cuot, P., Poyart-Salmeron, C., Carlier, C., and Courvalin, P. 1990, Nucleotide sequence of the erythromycin resistance gene of the conjugative transposon Tn1545. *Nucleic Acids Res* 18: :3660

Ueberle, B., Frank, R., and Herrmann, R. 2002, The proteome of the bacterium Mycoplasma pneumoniae: comparing predicted open reading frames to identified gene products. *Proteomics* 2: :754-764.

Urquhart, B.L., Atsalos, T.E., Roach, D., Basseal, D.J., Bjellqvist, B., Britton, W.L., and Humphery-Smith, I. 1997, 'Proteomic contigs' of Mycobacterium tuberculosis and Mycobacterium bovis (BCG) using novel immobilised pH gradients. *Electrophoresis* 18: :1384-1392.

Urquhart, B.L., Cordwell, S.J., and Humphery-Smith, I. 1998, Comparison of predicted and observed properties of proteins encoded in the genome of Mycobacterium tuberculosis H37Rv. *Biochem Biophys Res Commun* 253: :70-79.

Utt, M., Nilsson, I., Ljungh, A., and Wadstrom, T. 2002, Identification of novel immunogenic proteins of Helicobacter pylori by proteome technology. *J Immunol Methods* 259: :1-10.

Valway, S.E., Sanchez, M.P., Shinnick, T.F., Orme, I., Agerton, T., Hoy, D., Jones, J.S., Westmoreland, H., and Onorato, I.M. 1998, An outbreak involving extensive transmission of a virulent strain of Mycobacterium tuberculosis. *N Engl J Med* 338: :633-639.

VanBogelen, R.A., Abshire, K.Z., Pertsemlidis, A., Clark, R.L., and Neidhardt, F.C. 1996, Gene-Protein database of *Escherichia coli* K-12, Edition 6. In *Escherichia coli* and

Salmonella: Cellular and Molecular Biology.(F.C. Neidhardt, R. Curtiss, J.L. Ingraham, E.C.C. Lin, K.B. Low, B. Magasanik, W.S. Reznikoff, M. Riley, M. Schaechter and H.E. Umbarger eds) ASM Press, Washington, D.C. pp. 2067-2117.

VanBogelen, R.A., Abshire, K.Z., Moldover, B., Olson, E.R., and Neidhardt, F.C. 1997, Escherichia coli proteome analysis using the gene-protein database. *Electrophoresis* 18: :1243-1251.

VanBogelen, R.A., Greis, K.D., Blumenthal, R.M., Tani, T.H., and Matthews, R.G. 1999a, Mapping regulatory networks in microbial cells. *Trends Microbiol* 7: :320-328.

VanBogelen, R.A., Schiller, E.E., Thomas, J.D., and Neidhardt, F.C. 1999b, Diagnosis of cellular states of microbial organisms using proteomics. *Electrophoresis* 20: :2149-2159.

Vurma-Rapp, U., Kayser, F.H., Hadorn, K., and Wiederkehr, F. 1990, Mechanism of imipenem resistance acquired by three Pseudomonas aeruginosa strains during imipenem therapy. *Eur J Clin Microbiol Infect Dis* 9: :580-587.

Vytvytska, O., Nagy, E., Bluggel, M., Meyer, H.E., Kurzbauer, R., Huber, L.A., and Klade, C.S. 2002, Identification of vaccine candidate antigens of Staphylococcus aureus by serological proteome analysis. *Proteomics* 2: :580-590.

Walker, J., Fox, A.J., Edwards-Jones, V., and Gordon, D.B. 2002, Intact cell mass spectrometry (ICMS) used to type methicillin-resistant Staphylococcus aureus: media effects and inter-laboratory reproducibility. *Journal of Microbiological Methods* 48: :117-126.

Wang, J., Mushegian, A., Lory, S., and Jin, S. 1996, Large-scale isolation of candidate virulence genes of Pseudomonas aeruginosa by in vivo selection. *Proc Natl Acad Sci U S A* 93: :10434-10439.

Watson, H.L., Davidson, M.K., Cox, N.R., Davis, J.K., Dybvig, K., and Cassell, G.H. 1987, Protein variability among strains of Mycoplasma pulmonis. *Infect Immun* 55: :2838-2840.

Weldingh, K., Rosenkrands, I., Jacobsen, S., Rasmussen, P.B., Elhay, M.J., and Andersen, P. 1998, Two-dimensional electrophoresis for analysis of Mycobacterium tuberculosis culture filtrate and purification and characterization of six novel proteins. *Infection & Immunity* 66: :3492-3500.

Wilkins, M.R., Pasquali, C., Appel, R.D., Ou, K., Golaz, O., Sanchez, J.C., Yan, J.X., Gooley, A.A., Hughes, G., Humphery-Smith, I., Williams, K.L., and Hochstrasser, D.F. 1996, From proteins to proteomes: Large scale protein identification by two-dimensional electrophoresis and amino acid analysis. *Biotechnology* 14: :61-65.

Wolfhagen, M.J., Fluit, A.C., Torensma, R., Jansze, M., Kuypers, A.F., Verhage, E.A., and Verhoef, J. 1993, Comparison of typing methods for Clostridium difficile isolates. *J Clin Microbiol* 31: :2208-2211.

Yamamoto, T., Hanawa, T., and Ogata, S. 1994, Induction of Yersinia enterocolitica stress proteins by phagocytosis with macrophage. *Microbiol Immun* 38: :295-300.

Chapter 13

The Immunoproteome of *H. pylori*

TONI AEBISCHER[*], ALEXANDER KRAH[*#], DIRK BUMANN[*],
PETER R. JUNGBLUT[#] , and THOMAS F. MEYER[*,]
*Department of Molecular Biology, [#] Proteomics Unit, Max-Planck-Institute for
Infection Biology, Schumannstrasse 21/22, D-10117 Berlin, Germany*

1.   INTRODUCTION

The description of spiral-shaped, gram negative bacilli colonizing the stomach mucosa of patients suffering from severe gastroduodenal diseases two decades (Marshall and Warren, 1984) ago signified a turning point in the way we view the aetiology of these illnesses. They are now recognized to be infectious and caused by *Helicobacter pylori* (Ernst and Gold, 2000). It was quickly realized that half of the globe's population is infected which makes *H. pylori* the second most common bacterial infection in man (Feldman *et al.,* 1998). Infection invariably causes gastritis but only about 10% of infected people develop clinical symptoms ranging from gastric or duodenal ulcer, to atrophic gastritis and even gastric adenocarinoma or mucosa-associated lymphoid tissue (MALT) lymphoma (Ernst and Gold, 2000).

The reasons for this remarkable heterogeneity in outcome are still unexplained but host and bacterial factors have been implicated (Blaser and Berg, 2001;Peek, Jr. and Blaser, 2002). On the host side, genetic polymorphisms in the IL-1β and IL-1 receptor loci may impact on gastric acid secretion after *H. pylori* infection which is thought to affect the extent of gastritis and the progression towards gastric carcinoma (El-Omar *et al.,* 2001;Figueiredo *et al.,* 2002;Machado *et al.,* 2001). The bacteria, on the other side, were found to be extremely heterogeneous genetically and it was proposed that this variability corresponds with clinical presentations (Alm and Trust, 1999;Joyce *et*

H. Hondermarck (ed.), Proteomics: Biomedical and Pharmaceutical Applications, 317–338.
© 2004 *Kluwer Academic Publishers. Printed in the Netherlands.*

al., 2002). Numerous virulence factors have been identified so far, e.g. a functional type IV secretion apparatus encoded mostly in a pathogenicity island called *cag*-PAI (Censini *et al.,* 1996), a vacuolating toxin VacA (Schmitt and Haas, 1994;Telford *et al.,* 1994), proteins involved in motility like flagellins (Andrutis *et al.,* 1997;Eaton *et al.,* 1996) , adhesins like BabA and SabA (Gerhard *et al.,* 1999;Ilver *et al.,* 1998;Mahdavi *et al.,* 2002), outer membrane proteins like the *oiPA* gene product (Yamaoka *et al.,* 2000) and enzymes such as Urease involved in neutralizing gastric acid (Hu *et al.,* 1992). However, expression of many of these factors varies widely from isolate to isolate and their relative contribution to the pathology is still unclear (Graham and Yamaoka, 2000).

The fact that a bacterium caused gastroduodenal disease lead to the rapid implementation of antibiotic therapy (Unge, 1999). The success of this treatment has further strengthened the link between *H. pylori* infection and severe disease like duodenal ulcer but has also cast doubt on the absolute benefit of eradicating the bacteria because of a concomitant increase in diseases of the upper gastro-intestinal tract such as Barrett's oesophagus caused by reflux (Loffeld and van der Hulst, 2002;Peek, Jr. and Blaser, 2002) that suggests a protective role for *H. pylori* in these instances. In addition, the treatment has only a success rate of 80-90% linked to problems with patient's compliance and increasing incidence of antibiotic resistance. Therefore, improvements to disease management would be highly desirable.

Better disease management will require precise diagnosis with the goal to identify people at risk of developing severe pathology and the development of new therapies or strategies to prevent infection. Vaccination as a means to prevent or treat *H. pylori* infection (Del Giudice *et al.,* 2001) has shown considerable promise in animal models but there is still a need to identify candidate antigens that are specific for this bacterium and the clinical trials conducted are still inconclusive (DiPetrillo *et al.,* 1999;Kotloff *et al.,* 2001;Kreiss *et al.,* 1996;Michetti *et al.,* 1999).

Immunoproteomics, i.e. the comprehensive identification of immunogenic proteins from an organism such as *H. pylori*, is an attractive approach to determine i) antigens that could be used to increase the sensitivity and specificity of non-invasive diagnostic tools based on serology and ii) vaccine candidates. Here we discuss the current information available on the *H. pylori* immunoproteome and its potential for application in vaccination or diagnosis.

2. THE *H. PYLORI* PROTEOME

2.1 Genotypic diversity causes variability

H. pylori is thought to be an ancient companion of humans and the species can be classified into four modern groups, based on sequence variation in a set of seven genes, with distinct geographic distribution over the globe reflecting the migratory behaviour of our species (Falush *et al.,* 2003). The extreme genetic variability of *H. pylori* is most instructively reflected in the genomes of two strains, 26695 and J99, that had been sequenced (Alm *et al.,* 1999;Tomb *et al.,* 1997). Roughly 1600 ORFs were predicted and 6-7% of these genes were found in one but not the other strain. The genome structure of the species has since been studied using DNA-microarrays representing the complement of predicted ORFs of both strains (Salama *et al.,* 2000). The results indicate that 1200-1300 ORFs were present in all strains analysed so far and 352 (~22% of the total predicted ORFs) were genotype specific genes, the majority of which appear to cluster in so called plasticity zones. As for many other bacteria these studies have lead to the concept of a species genome being composed of a core and a flexible gene pool (Joyce *et al.,* 2002;Lan and Reeves, 2000). Additional variability arises from non-synonymous mutations that reduce the average identity of orthologous ORF products in 26695 and J99 to 93% at the amino acid level (Alm and Trust, 1999). The expression of several genes, e.g. fucosyltransferases (Appelmelk *et al.,* 1999), phospholipase A (Tannaes *et al.,* 2001), is likely to be modulated by a "slipped-strand repair mechanism" due to the presence of homopolymeric tracts or dinucleotide repeat sequences. Proteins of a particular strain mapped according to their pI and Mr will reproduce this genetic diversity and will display changes in their coordinates (Jungblut *et al.,* 2000a).

2.2 Phenotypic variation

Further variability between strains will arise from differences in gene expression. Gene regulation appears to be comparably simple in *H. pylori* for which only three σ-factors, six two component regulators and four transcriptional regulators are described (Alm and Trust, 1999;Beier and Frank, 2000;Bereswill *et al.,* 2000;Scarlato *et al.,* 2001). These regulators may be complemented however by slipped-strand repair mechanisms leading to phase variation (see above) and differential methylation (Donahue *et al.,* 2002). Studies investigating the transcriptional response of *H. pylori* to changes in the culture conditions indicate that gene expression is extensively modulated (Thompson *et al.,* 2002) and the proteome will be a dynamic mirror of this regulation. Indeed changes in protein composition were described for *H. pylori* strains after change in culture conditions such as pH shifts and long term culture or after exposure to chemical stress (see for a recent review (Bumann *et al.,* 2001). Additional complexity is created by post-translational modification and processing of gene products which may account for the often multiple protein species on 2-DE gel analyses that correspond to one particular ORF.

2.3   Proteome maps of H. pylori

The combination of high spatial resolution of proteins based on pI and Mr with mass spectrometry of the separated polypeptides, has become a common method to analyse the protein species present in whole organisms (Jungblut *et al.,* 1996). In theory the method is able to resolve all the gene products of a bacterium like *H. pylori* on a single gel and if 100ng of an individual protein species can be recovered identification is currently possible. In practice individual protein species of this organism will be present in a concentration range from one to several million copies per cell and many may therefore escape detection. In addition, the aforementioned dynamic nature of the protein composition due to gene regulation precludes the presence of all proteins in one biological sample. The method was quickly adopted to characterize protein patterns of *H. pylori* isolates (cf. (Jungblut *et al.,* 2000a), and (Bumann *et al.,* 2001) and (Nilsson and Utt, 2002) for review) and, not surprisingly, the patterns were found to be highly variable. Although we cannot make precise statements about the in vivo proteome, the recent progress in the description of in vivo expressed genes using selective capture of in situ transcribed sequences (Graham *et al.,* 2002) suggests that most of the genes are transcribed in vivo and in vitro hence the proteome most likely will be highly similar but, in addition, a number of genes (approximately 3-7%) were specifically turned on *in vivo* and their respective products need to be considered. Indeed none of the products of the latter genegroup were so far identified in 2-DE analyses.

2.4   The H. pylori proteome database

The availability of the sequence of two genomes greatly facilitated the identification of proteins by peptide mass fingerprinting. Therefore we started to establish a public database of the sequenced *H. pylori* strain 26695 where our 2-DE gel patterns and results of protein species identification are made available on the internet (http://mpiib-berlin.mpg.de/2D-PAGE). The database currently holds information on more than 1800 protein species of strain 26695 resolved by 2-DE (Fig. 1) of which 455 ((Jungblut *et al.,* 2000a;Schmidt *et al.,* 2003)) have been identified. During this project, it became evident that much of the variability of the *H. pylori* protein maps stems from differences in the ORF sequence and the relative coordinates on 2-DE gels can be predicted using relative distances to reference marker proteins (Jungblut *et al.,* 2000b). We (Bumann *et al.,* 2002a;Sabarth *et al.,* 2002b) and others (Kim *et al.,* 2002;Utt *et al.,* 2002) have established protocols to enrich for proteins with a particular localization in order to increase the sensitivity of the 2-DE analysis. The database holds the information on these subproteomes of *H. pylori,* such as proteins regulated by decreasing the pH of in vitro cultures, or proteins with particular topology, i.e. released (Secretome) or localized to the outer membrane of the bacterial cell. Thereby 33 protein species were detected in the supernatant of *H. pylori* grown to mid log phase in brain heart infusion broth with 1% β-cyclodextrin (Bumann *et al.,* 2002a). Twenty-six of these were identified by peptide mass fingerprinting with a sequence coverage between 17% and 67%. For some of these proteins secretion was confirmed or could be predicted based on their primary sequence, e.g. presence of a signal peptide, or homologies to other secreted proteins but for others such as the newly described hypothetical proteins HP0906 and HP0367 the secretion pathways are unknown. The release of many of the above proteins into culture supernatant was confirmed in a recent study using a different *H. pylori* strain. In this work additional released proteins, HP0129, HP0305, HP0721, HP0827, HP0902, HP0912, HP0913 were identified (Kim *et al.,* 2002). The only known protein secreted via the *cag* PAI encoded type VI secretion system, CagA, was not found in the supernatants of our cultures. These were harvested during the mid-log phase and the upregulation of CagA messenger RNA only in late log phase/ stationary cultures (Thompson *et al.,* 2002) could in part explain this unanticipated result. CagA is an abundant protein species in total lysates though and therefore secretion rather than expression might be regulated. Independent of the explanation the results indicate that the secretome is also subject to regulation.

Although surface localization of *H. pylori* can be predicted based on particular sequence motifs (Alm *et al.,* 2000), predictions may not be comprehensive and do not warrant expression. We used surface biotinylation of the sequenced strain 26695 and affinity purification to localize 82 label-accessible protein species of which we identified 18 (Sabarth *et al.,* 2002b). This group contained at least 9 proteins that were also found in culture supernatant (Bumann *et al.,* 2002a;Kim *et al.,* 2002). The surface localization for 9 of these had been previously found by other methods. However, only two corresponded to predicted surface proteins and we did not find members of the hypothetical outer membrane proteins in this analysis of HP26695. This may be related to the HP26695 strain that does not express BabA (Ilver *et al.,* 1998) and several other members of this outer membrane protein family.

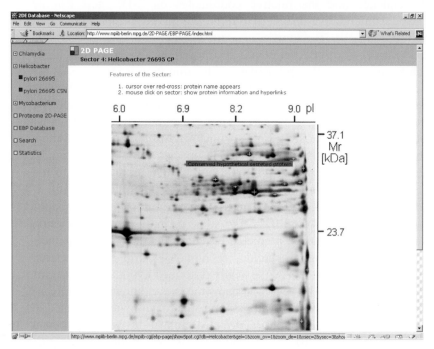

Figure 1. European bacterial proteome 2D-PAGE database (Screenshot): Proteins detected in one sector of the Helicobacter pylori 26695 cell protein 2-DE pattern. Identified antigenic species are marked by yellow crosses. The database is available in the WWW under http://www.mpiib-berlin.mpg.de/2D-PAGE/.

3.   IMMUNOPROTEOME OF *H. PYLORI*:

3.1   Immunoproteomics:

It was found early on that infection with *H. pylori* triggered an immune response and lead to the formation of specific antibodies (for review see (Zevering *et al.,* 1999). Individual or pooled sera from infected patients have since been used to define and characterize the respective immunogenic proteins either in complex bacterial lysates (Andersen *et al.,* 1995;Bazillou *et al.,* 1994;Crabtree *et al.,* 1991;Guruge *et al.,*

1990;Mitchell *et al.,* 1996) or via candidate approaches using purified (Donati *et al.,* 1997;O'Toole *et al.,* 1991) or recombinant test antigens (Xiang *et al.,* 1993). Early studies used one dimensional gels to separate bacterial lysates for detection of immunoreactive proteins but 2-DE gel analyses were introduced not much later (McAtee *et al.,* 1998a;McAtee *et al.,* 1998b) and were extended by many groups since (Enroth *et al.,* 2000;Haas *et al.,* 2002;Jungblut *et al.,* 2000b;Kimmel *et al.,* 2000;Nilsson *et al.,* ;Utt *et al.,* 2002). Immunogenic proteins found in 2-DE analyses can be identified either by peptide mass fingerprinting or sequencing the protein isolated from preparative gels or if the respective protein had been identified before by simple assignment (Fig. 2). For practical reasons the antigens often were prepared from a single (Haas *et al.,* 2002;Jungblut *et al.,* 2000b;Kimmel *et al.,* 2000;McAtee *et al.,* 1998a;Nilsson *et al.,*) and in few instances from a small number of test strains (Enroth *et al.,* 2000;McAtee *et al.,* 1998c;McAtee *et al.,* 1998b).

Several potential limitations of the current approach have been raised in the past which may or may not be relevant:

- A potential shortcoming of the current technology is our ignorance of many of the products of the flexible gene pool which makes the choice of the test strain(s) aleatoric. Ideally the test strain should contain and express the species genome, i.e. the core and flexible gene pool but this is unfortunately not the real situation. This may be a concern for non-immunodominant proteins because immunodominant proteins appear to be expressed in all strains tested so far (McAtee *et al.,* 1998b).

The origin of the serum seems also to be an important parameter in comparative studies. The seroreactivity of patients infected with strains likely to belong to one of the four major geographical groups (Falush *et al.,* 2003) was analysed on strains from the same or different geographical origin and the major immunogenic proteins seen by a particular group of sera were expressed by most strains while the recognition patterns were apparently linked to the different geographical/ethnic origin of the serum donors (Hook-Nikanne *et al.,* 1997). This may be indicative of the influence of host genetics; of crossreactivity due to exposure to different microorganisms that may show distinctive geographic distribution; or of different immunogenicity of *H. pylori* subgroups.

- Patient sera are easy to obtain and non-invasive diagnostic tests based on serology would of course be the preferred format of a diagnostic tool. Such an assay will have to discriminate infection with *H. pylori* from other common infectious agents and hence should include *H. pylori*-specific antigens. To define these specific antigens cross-reactivities against other common pathogens will have to be explored.

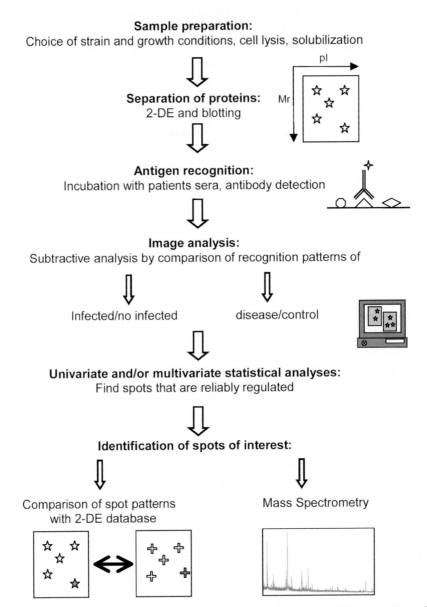

Figure 2. Flowchart of the procedures combining 2-DE separation of proteins, immunoplot and subsequent image analysis to identify antigenic proteins of interest.

- *H. pylori* colonizes the stomach mucosa and defining immunogenic proteins with serum antibodies may not reveal relevant immunogenic proteins. IgA constitutes the major class of mucosal antibodies and could see a different set of antigens. However, apparently this repertoire of antigens is smaller than the set of antigens recognized by other isotypes (Blanchard *et al.,* 1999). A recent study has further addressed this issue and found no evidence for a significant experimental bias when using serum vs gastric secretions for immunodetection using a set of seven bacterial test strains (Lock *et al.,* 2002). Serum antibodies should therefore allow to detect most if not all relevant antigens.

- Conformational epitopes will be lost in many instances but it is unlikely that a relevant antigen will not be detected for this reason in the current immunoblot analyses, since most immunogenic proteins induce polyclonal responses that likely recognize conformational and sequential epitopes.

3.2 Immunoproteome of *H. pylori*

Several groups defined immunogenic proteins from *H. pylori* using a proteomics approach. Pooled sera from 14 patients with undescribed disease states were originally used to define 30 antigenic protein species in *H. pylori ATCC 43504* strain that were identified by N-terminal sequencing and corresponded to at least 23 ORFs (McAtee *et al.,* 1998b). Most of the antigenic proteins were also present in eight other strains tested in the same study. In another report 29 proteins from *H. pylori G27* were identified and a first comparative analysis of their recognition by sera from patients presenting with different clinical symptoms was performed (Kimmel *et al.,* 2000). The 29 protein species could be assigned to a corresponding number of ORFs of which nine had been found previously (McAtee *et al.,* 1998b). The identified antigens corresponded in many cases to abundant proteins.

In order to increase the sensitivity of the approach, it was complemented with biochemical enrichment of primarily surface exposed proteins by acid extraction and heparin-affinity chromatography (Nilsson *et al.,* 2000;Utt *et al.,* 2002). The resulting test antigen was resolved on 2-DE gels and analysed with sera from *H. pylori* positive and negative patients. A total of 186 antigenic species was detected (as compared to 141 by silver staining). Six protein species with basic pI were selected for identification via MS/MS and corresponded to products of four ORFs, HP0073 (*UreA* subunit gene), HP0175 (cell binding factor *gene*) and HP0231 (a so far hypothetical protein *gene*) and the gene encoding a predicted outer membrane protein HP1564.

We followed a strategy without prior enrichment and after establishing the technique using the sequenced strain HP26695 (Jungblut *et al.,* 2000a) analysed first a total of 42 sera (Haas *et al.,* 2002) and more recently complemented this study with the analysis of a further 60 sera (Krah et al. , manuscript submitted). Overall 611 antigenic species (Fig. 3) were detected. The number of antigens recognized by individual sera ranged from 7 to 153 (median of 52) (Haas *et al.,* 2002) and from 24-391(median 141) (Krah et al., manuscript submitted) in the two studies. Forty-two proteins were recognized with frequently by patient sera and were determined to be encoded by 32 ORFs (Haas *et al.,* 2002). More than two thirds of these proteins had been also found in other studies (Kimmel *et al.,* 2000;McAtee *et al.,* 1998b;Utt *et al.,* 2002). The latter indicates that a core response to at least 19 antigens of *H. pylori* can be defined. These included *H. pylori*-specific antigens such as Cag 3, 16 and 26 antigens, products of HP0175, HP0231 and HP0410 ORFs which are surface exposed, or hypothetical proteins such as gene products of HP0305 and proteins that share homologies to proteins of other bacteria but are localized to the bacterial surface or secreted such as products of HP0010 (*groEL*), HP0072 and HP0073 (*ureB/A*), HP0109 (*dnaK*) HP0243 (*napA*), HP0601 (*flaA*) HP0875 (Catalase *gene*), HP1019 (*htrA*), HP1186 (carbonic anhydrase *gene*), HP1125 (*omp18*), HP1285 and HP1564 (outer membrane protein *genes*) (cf. (Sabarth *et al.,* 2002b).

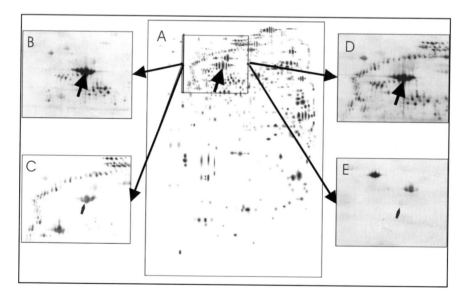

Figure 3. Recognition patterns of individual sera fall into distinct groups. (A) Master blot showing all 611antigenic spots recognized by 60 sera of H. pylori infected patients. (B-E) Characteristics of individual recognition groups were most obvious in a subregion of the 2-DE map (blue arrows refer to the main antigenic species of GroEL for orientation).

In summary, the immunoproteome of *H. pylori*, as it is known to date, consists of at least 611 antigenic species that can be separated on 2-DE gels. This number is close to 30% of the protein species that were detected by silver staining (Jungblut *et al.*, 2000b) although not all antigens corresponded to a protein species detected by silver staining (based on our current analysis, we estimate that the 1800 protein species detected by silver staining correspond to the products of some 1100 genes; AK and PJ unpublished). It can be predicted that many of these antigens are shared between many organisms and it is difficult to ascertain their immunogenic capacity due to *H. pylori* infection alone. However, a core group of antigens was repeatedly identified and although it included antigens with homologues in other bacteria some were specific for *H. pylori*. The latter group may be attractive candidates to increase the specificity of serological tests to assess *H. pylori* infection.

3.3 Surrogate markers for disease

Some of the studies discussed were intended to find surrogate markers for clinical outcomes by describing immunogenic proteins. The genetic variability of *H. pylori* is believed by many groups to be linked to the diverse clinical presentation and results from animal models support this concept (Watanabe *et al.*, 1998) {Bjorkholm, Guruge, et al. 2002 4486 /id}. Others, however, favour the idea that host and environmental factors are more important and this view is mostly supported by epidemiological and population genetic studies (El-Omar *et al.*, ;Ernst and Gold, 2000). The immunoproteome of *H. pylori* will most likely not provide evidence for or against one of these hypotheses because of sensitivity limitations or because the respective bacterial factor is possibly not an immunogenic protein because to date no candidate has been found. Nevertheless, since both a more rigorous host response as well as a bacterial strain with particular pathogenic potential will be interpreted by a patient's immune system, the approach offers the opportunity to identify bacterial antigens with diagnostic value and reveal prospective markers for disease.

A first study aimed to correlate seroreactivity with disease status, reported recognition patterns of sera from patients with chronic gastritis (6), peptic ulcer (5), gastric cancer (4) or MALT lymphoma (1) collected from two geographically distinct regions. These patterns varied considerably and from 20 proteins only flagellins were commonly detected and no conclusion with respect to serological markers for disease could be drawn (Kimmel *et al.*, 2000).

We concentrated on analysing the immunoreactivity of *H. pylori* infected patients from one geographic region since the geographic or ethnic origin of the patients appeared to be an important variable (Hook-Nikanne *et al.,* 1997). Sera from 24 *H. pylori* infected patients of which 9 patients were diagnosed with ulcers, 6 with gastric cancer including one from a MALT lymphoma patient and 12 control samples from patients with unrelated gastric disorders were compared (Haas *et al.,* 2002). Overall 310 antigenic species were detected and a subgroup of 116 antigens were only recognized by sera from *H. pylori*-positive patients and the majority of these antigens still await identification. Sera from patients with more severe clinical disease, i.e. ulcer and cancer patients, in general recognized more antigens and reacted stronger with most of the recognized protein species. The results further suggested a trend towards more intense recognition of some antigenic proteins by sera from cancer patients.

To address the relevance of this finding more thoroughly, we screened the recognition patterns of sera from *H. pylori* infected patients with either duodenal ulcer or gastric adenocarcinoma from one geographic region and uniform ethnic background. We reasoned that these collectives would be best suited to reveal disease-associated markers since the two clinical presentations appear to be mutually exclusive (Ernst and Gold, 2000). The recognition patterns of sera from 30 patients of each clinical condition were analysed on high-resolution immunoblots. Sera from GC patients recognized on average a 49% higher number of antigenic species and the mean total reactivity of the sera was increased by 75% confirming the trend noted in the first study (Haas *et al.,* 2002). Hierarchical agglomerative clustering (Pun *et al.,* 1988) revealed that recognition patterns fell into distinct groups. These groups did not correlate with clinical symptoms and therefore the respective antigens were unlikely to contain surrogate markers of disease. However, the diverse recognition patterns could be interesting from an epidemiological point of view because they may reflect infection with particular groups of *H. pylori* strains since similar data were reported for protein patterns of 12 *H. pylori* isolates from swedish patients suffering from either gastric cancer, duodenal ulcer or gastritis that could be grouped by clustering algorithms (Enroth *et al.,* 2000). The different patterns might correspond to coinherited groups of genes for which evidence has been found in genotype analyses using gene-chip technology (Salama *et al.,* 2000).

The relatively high number of 30 sera per clinical condition (Krah et al., submitted) allowed a more thorough statistical analysis of the recognition of individual antigenic proteins. Thereby 14 protein species were found to be differentially recognized. However not all of these 14 antigens were always detected. We probed whether the recognition of particular subsets of these 14 antigens would accurately predict disease and cancer patients were detected with a sensitivity of 77% and a specificity of 83%. The 14 antigens comprised mostly proteins of ORFs that are highly conserved not only in *H. pylori*. One of the proteins was GroEL which was found previously to be more intensely recognized by sera from cancer patients (Haas *et al.,* 2002;Iaquinto *et al.,* 2000). This result may be indicative of a higher exposure to *H. pylori* (probably a combination of colonization density and duration of infection) in cancer patients. Cancer patients were older on average than patients in the duodenal ulcer group and this fact may confound the data. However, age did not explain all the differences since age matched subgroups revealed a subgroup of the 14 antigens to be correlated with disease but not age.

The fact that no association of disease outcome with recognition of known virulence factors was found, agrees with the current view, that the presence of the genes encoding known virulence factors such as VacA, CagA, IceA does not correlate with clinical status (Graham and Yamaoka, 2000) (probably with the exception of *oipA* (Yamaoka *et al.,* 2002). The remarkable increase in total reactivity though is in line with the hypothesis that proinflammatory alleles of the IL-1β locus were most consistently associated with increased gastric cancer-risk (Gonzalez *et al.,* 2002). Increased inflammation is likely to boost the antibody production which provides a rational for the immunoproteome results.

3.4   Identification of vaccine candidates

Vaccination would be a welcome option to manage *H. pylori* infection and could overcome drawbacks of the current antibiotic therapy and prevent reinfection. The feasibility of vaccination has originally been demonstrated in mice immunized with a whole cell derived vaccine which were protected from a challenge with *H. felis* (Chen *et al.,* 1992). Since then several groups have shown that vaccination with purified and/or recombinant proteins could also induce protective immunity against *H. felis* or *H. pylori* in the mouse model and the current status of vaccine development has been the subject of several excellent recent reviews (Del Giudice *et al.,* 2001;Ferrero and Labigne, 2001;Kleanthous *et al.,* 1998). Both whole bacterial cell and subunit (*H. pylori* Urease) vaccines have also been tested in small clinical trials but so far without a clear indication of, in these cases, therapeutic efficacy. Urease as well as most other vaccine antigens were selected based on abundance or on their role in virulence but additional criteria are required to select more specific candidate antigens. These criteria will be different from the ones relevant for disease markers since a future vaccine should protect against infection with any *H. pylori* strain. Therefore, candidate antigens should be immunogenic, *H. pylori*-specific but ideally conserved between strains.

The knowledge of the genome made a novel approach to vaccine development possible, called reverse vaccinology, which combines in silico selection of potential candidates with high throughput methods for production and testing of the selected antigens. This has been successfully applied to define new vaccine candidates against meningitis caused by serogroup B meningococci (Pizza *et al.,*). The approach has since been extended to other pathogens including *H. pylori* always with similar hit frequencies of 2-3% (Ferrero and Labigne, 2001). Immunity to *H. pylori* in the mouse model is cell-mediated, requires CD4$^+$ T cells and can be induced in the absence of antibodies. The selection criteria were set to identify surface localized or secreted proteins in silico. This was a reasonable strategy but the algorithms developed cannot cope with 'misbehaving' proteins, a drawback of the method that has been acknowledged (Ferrero and Labigne, 2001). Nevertheless, 400 ORFs were selected, in vitro expressed and used to immunize mice. The approach identified 10 protective antigens from 400 selected. Urease, catalase or groEL which are known protective antigens and can be found at the bacterial surface were not predicted to be there and hence missed in the screen. Immunoproteomics using complex antisera offers an alternative or complementary approach to identify vaccine antigens. The approach using antibody reactivity is justified because all protective antigens known to date can be detected without exception by antibodies from infected patients or animal hosts. However, the reverse is not always correct and indicates the need for additional criteria.

Most studies on vaccines against *H. pylori* were conducted in mice but the relevance of this model is debated. Mice are not a natural host for *H. pylori* and severe pathology does not normally develop although gastritis, lymphoid follicle formation and mucosal damage can be observed (Kamradt *et al.,* 2000;Lee *et al.,* 1997;Lee, 1999). However, the immune system faces the same challenge in mice and man to detect the presence of *H. pylori* in the gastric lumen. We aimed to test this hypothesis and performed a comparative analysis of the immunogenic *H. pylori* proteins in mice infected with the mouse adapted strain SS-1 and published lists of antigens recognized by patient sera (Bumann *et al.,* 2002b). For 31 protein species out of 587 antigens recognized specifically by sera from infected C57BL/6 mice the corresponding complement of 21 ORFs was identified. Similarly, 10 proteins were identified that were also recognized by sera from non-infected mice. More than half of these defined proteins were identified before with the help of patient sera (Haas *et al.,* 2002;Kimmel *et al.,* 2000;McAtee *et al.,* 1998b) indicating that they were immunogenic in natural human infections. Importantly the concordance between the human and mouse studies was greater than between the individual human studies. This strongly suggests that the mouse model is indeed a valid model to identify or test immunogenic proteins for vaccination purposes.

Immunogenicity is not a sufficient criterion though for the identification of a candidate vaccine molecule. Analysis of the current immunoproteome data revealed that abundant proteins and proteins secreted or surface exposed were overrepresented as compared to their frequency in the genome (Bumann *et al.,* 2002b;Jungblut *et al.,* 2000b). The immune system may more often sample abundant *H. pylori* proteins or those exhibiting the aforementioned localizations because of budding bacterial membrane vesicles that penetrate the mucosa (Mai *et al.,* 1992). Thus, overall abundance and surface exposure or secretion are valid additional criteria for candidate selection.

As mentioned before, a future vaccine should prevent infection or pathology caused by many if not all strains present in the population to be vaccinated. Candidate antigens should therefore be conserved but specific for *H. pylori*. We recently combined several parameters to select such candidates. Fifteen molecules of which seven had been previously shown to be protective could be identified by combining information from the immunoproteome, proteome (abundance), surface localisation or release, sequence conservation and species restriction (genome) (Sabarth *et al.,* 2002a). These were GroEL, GroES, Urease A and B, catalase, NapA and CagA of which only the last is specific to *H. pylori* while all other are highly conserved among various bacteria. A further two of the 15, HP0231 and HP0410 were conserved in *H. pylori* but had only very low homologies to proteins from other organisms. These predicted candidates were expressed as recombinant proteins and shown to be protective (Sabarth *et al.,* 2002a). Another immunogenic protein detected by patient sera is Lpp20 (Opazo *et al.,* 1999), a lipoprotein that was also shown to be protective in the mouse model (Keenan *et al.,* 2000). Therefore, from 16 candidates, that passed these criteria, at least 10 are protective indicating that multiparameter selection starting from the immunoproteome is very promising. The collective of input antigens in this analysis was not exhaustive since only 72 immunogenic proteins are identified to date. The immunoproteome holds more than 600 immunogenic species and many unidentified candidates may fullfill the criteria.

A recent comparative analysis of selectively captured transcribed sequences from *H. pylori* (Graham *et al.,* 2002) from human biopsies or gerbil stomachs suggests that 3-7% of genes may be expressed in vivo that cannot be detected in logarithmically growing bacteria in vitro. A set of 14 conserved genes specific for *H. pylori* was described in this study that was uniquely expressed in vivo. None of these had been identified in proteomic screens to date for obvious reasons. These genes may encode immunogenic proteins and thereby additional vaccine candidates.

4.   CONCLUSIONS

The analysis of the immunoproteome of *H. pylori* has already lead to the identification of candidate molecules for vaccine development and of putative diagnostic value. Only 10% of the 600 immunogenic molecules are currently identified. The near future will see this number growing rapidly and is likely to lead to the identification of more useful candidates. From a basic science point of view, the cumulative data will yield insight into general properties of the immune system with regard to its detection threshold when confronted with a microbial pathogen of similar complexity. This information will be valuable for improving current algorithms e.g. for vaccine candidate selection.
Technical advances will extend the analysis to the study of conformational epitopes and protein chip technologies may be used to compile the *H. pylori* species proteome. We anticipate that our understanding of the immunoproteome will help to reduce the complexity of such protein arrays while keeping their likely power for the development of new diagnostic tools.

ACKNOWLEDGEMENTS

We wish to thank our colleagues who have helped in the production of this chapter and the current members of the department involved in the H. pylori proteome research.

REFERENCES

Alm, R. A., Bina, J., Andrews, B. M., Doig, P., Hancock, R. E. and Trust, T. J. 2000, Comparative genomics of Helicobacter pylori: analysis of the outer membrane protein families. *Infect.Immun.*, **68:** 4155-4168
Alm, R. A., Ling, L. S., Moir, D. T., King, B. L., Brown, E. D., Doig, P. C., Smith, D. R., Noonan, B., Guild, B. C., deJonge, B. L., Carmel, G., Tummino, P. J., Caruso,

A., Uria-Nickelsen, M., Mills, D. M., Ives, C., Gibson, R., Merberg, D., Mills, S. D., Jiang, Q., Taylor, D. E., Vovis, G. F. and Trust, T. J. 1999, Genomic-sequence comparison of two unrelated isolates of the human gastric pathogen Helicobacter pylori. *Nature*, **397**: 176-180

Alm, R. A. and Trust, T. J. 1999, Analysis of the genetic diversity of Helicobacter pylori: the tale of two genomes. *J Mol Med*, **77**: 834-846

Andersen, L. P., Espersen, F., Souckova, A., Sedlackova, M. and Soucek, A. 1995, Isolation and preliminary evaluation of a low-molecular-mass antigen preparation for improved detection of Helicobacter pylori immunoglobulin G antibodies. *Clin.Diagn.Lab Immunol.*, **2**: 156-159

Andrutis, K. A., Fox, J. G., Schauer, D. B., Marini, R. P., Li, X., Yan, L., Josenhans, C. and Suerbaum, S. 1997, Infection of the ferret stomach by isogenic flagellar mutant strains of Helicobacter mustelae. *Infect.Immun.*, **65**: 1962-1966

Appelmelk, B. J., Martin, S. L., Monteiro, M. A., Clayton, C. A., McColm, A. A., Zheng, P., Verboom, T., Maaskant, J. J., van den Eijnden, D. H., Hokke, C. H., Perry, M. B., Vandenbroucke-Grauls, C. M. and Kusters, J. G. 1999, Phase Variation in Helicobacter pylori Lipopolysaccharide due to Changes in the Lengths of Poly(C) Tracts in alpha3-Fucosyltransferase Genes. *Infect.Immun.*, **67**: 5361-5366

Bazillou, M., Fendri, C., Castel, O., Ingrand, P. and Fauchere, J. L. 1994, Serum antibody response to the superficial and released components of Helicobacter pylori. *Clin.Diagn.Lab Immunol.*, **1**: 310-317

Beier, D. and Frank, R. 2000, Molecular characterization of two-component systems of Helicobacter pylori. *J.Bacteriol.*, **182**: 2068-2076

Bereswill, S., Greiner, S., van Vliet, A. H., Waidner, B., Fassbinder, F., Schiltz, E., Kusters, J. G. and Kist, M. 2000, Regulation of ferritin-mediated cytoplasmic iron storage by the ferric uptake regulator homolog (Fur) of Helicobacter pylori. *J.Bacteriol.*, **182**: 5948-5953

Bjorkholm, B. M., Guruge, J. L., Oh, J. D., Syder, A. J., Salama, N., Guillemin, K., Falkow, S., Nilsson, C., Falk, P. G., Engstrand, L. and Gordon, J. I. 2002, Colonization of germ-free transgenic mice with genotyped Helicobacter pylori strains from a case-control study of gastric cancer reveals a correlation between host responses and HsdS components of type I restriction-modification systems. *J. Biol. Chem.*,**277**: 34191-34197

Blanchard, T. G., Nedrud, J. G., Reardon, E. S. and Czinn, S. J. 1999, Qualitative and quantitative analysis of the local and systemic antibody response in mice and humans with Helicobacter immunity and infection. *J.Infect.Dis.*, **179**: 725-728

Blaser, M. J. and Berg, D. E. 2001, Helicobacter pylori genetic diversity and risk of human disease. *J.Clin.Invest*, **107**: 767-773

Bumann, D., Aksu, S., Wendland, M., Janek, K., Zimny-Arndt, U., Sabarth, N., Meyer, T. F. and Jungblut, P. R. 2002a, Proteome analysis of secreted proteins of the gastric pathogen Helicobacter pylori. *Infect.Immun.*, **70**: 3396-3403

Bumann, D., Holland, P., Siejak, F., Koesling, J., Sabarth, N., Lamer, S., Zimny-Arndt, U., Jungblut, P. R. and Meyer, T. F. 2002b, A comparison of murine and human immunoproteomes of Helicobacter pylori validates the preclinical murine infection model for antigen screening. *Infect.Immun.*, **70**: 6494-6498

Bumann, D., Meyer, T. F. and Jungblut, P. R. 2001, Proteome analysis of the common human pathogen Helicobacter pylori. *Proteomics.*, **1**: 473-479

Censini, S., Lange, C., Xiang, Z., Crabtree, J. E., Ghiara, P., Borodovsky, M., Rappuoli, R. and Covacci, A. 1996, cag, a pathogenicity island of Helicobacter pylori, encodes type I-specific and disease-associated virulence factors. *Proc.Natl.Acad.Sci.U.S.A*, **93:** 14648-14653

Chen, M., Lee, A. and Hazell, S. 1992, Immunisation against gastric helicobacter infection in a mouse/Helicobacter felis model. *Lancet*, **339:** 1120-1121

Crabtree, J. E., Taylor, J. D., Wyatt, J. I., Heatley, R. V., Shallcross, T. M., Tompkins, D. S. and Rathbone, B. J. 1991, Mucosal IgA recognition of Helicobacter pylori 120 kDa protein, peptic ulceration, and gastric pathology. *Lancet*, **338:** 332-335

Del Giudice, G., Covacci, A., Telford, J. L., Montecucco, C. and Rappuoli, R. 2001, The design of vaccines against Helicobacter pylori and their development. *Annual Review of Immunology*, **19:**523-563

DiPetrillo, M. D., Tibbetts, T., Kleanthous, H., Killeen, K. P. and Hohmann, E. L. 1999, Safety and immunogenicity of phoP/phoQ-deleted Salmonella typhi expressing Helicobacter pylori urease in adult volunteers. *Vaccine*, **18:** 449-459

Donahue, J. P., Israel, D. A., Torres, V. J., Necheva, A. S. and Miller, G. G. 2002, Inactivation of a Helicobacter pylori DNA methyltransferase alters dnaK operon expression following host-cell adherence. *FEMS Microbiol.Lett.*, **208:** 295-301

Donati, M., Moreno, S., Storni, E., Tucci, A., Poli, L., Mazzoni, C., Varoli, O., Sambri, V., Farencena, A. and Cevenini, R. 1997, Detection of serum antibodies to CagA and VacA and of serum neutralizing activity for vacuolating cytotoxin in patients with Helicobacter pylori-induced gastritis. *Clin.Diagn.Lab Immunol.*, **4:** 478-482

Eaton, K. A., Suerbaum, S., Josenhans, C. and Krakowka, S. 1996, Colonization of gnotobiotic piglets by Helicobacter pylori deficient in two flagellin genes. *Infect.Immun.*, **64:** 2445-2448

El-Omar, E. M., Carrington, M., Chow, W. H., McColl, K. E., Bream, J. H., Young, H. A., Herrera, J., Lissowska, J., Yuan, C. C., Rothman, N., Lanyon, G., Martin, M., Fraumeni, J. F. J. and Rabkin, C. S., 2001, Interleukin-1 polymorphisms associated with increased risk of gastric cancer. *Nature*, **404:** 398-402

Enroth, H., Akerlund, T., Sillen, A. and Engstrand, L. 2000, Clustering of clinical strains of Helicobacter pylori analyzed by two-dimensional gel electrophoresis. *Clin.Diagn.Lab Immunol.*, **7:** 301-306

Ernst, P. B. and Gold, B. D. 2000, The disease spectrum of Helicobacter pylori: the immunopathogenesis of gastroduodenal ulcer and gastric cancer. *Annu.Rev.Microbiol.*, **54:** 615-640

Falush, D., Wirth, T., Linz, B., Pritchard, J. K., Stephens, M., Kidd, M., Blaser, M. J., Graham, D. Y., Vacher, S., Perez-Perez, G. I., Yamaoka, Y., Megraud, F., Otto, K., Reichard, U., Katzowitsch, E., Wang, X., Achtman, M. and Suerbaum, S. 2003, Traces of human migrations in Helicobacter pylori populations. *Science*, **299:** 1582-1585

Feldman, R. A., Eccersley, A. J. and Hardie, J. M. 1998, Epidemiology of Helicobacter pylori: acquisition, transmission, population prevalence and disease-to-infection ratio. *Br.Med.Bull.*, **54:** 39-53

Ferrero, R. L. and Labigne, A. 2001, Helicobacter pylori vaccine development in the post-genomic era: can in silico translate to in vivo. *Scand.J.Immunol.*, **53:** 443-448

Figueiredo, C., Machado, J. C., Pharoah, P., Seruca, R., Sousa, S., Carvalho, R., Capelinha, A. F., Quint, W., Caldas, C., van Doorn, L. J., Carneiro, F. and Sobrinho-Simoes, M. 2002, Helicobacter pylori and interleukin 1 genotyping: an opportunity to identify high-risk individuals for gastric carcinoma. *J.Natl.Cancer Inst.*, **94:** 1680-1687

Gerhard, M., Lehn, N., Neumayer, N., Boren, T., Rad, R., Schepp, W., Miehlke, S., Classen, M. and Prinz, C. 1999, Clinical relevance of the Helicobacter pylori gene for blood-group antigen-binding adhesin. *Proc.Natl.Acad.Sci.U.S.A*, **96:** 12778-12783

Gonzalez, C. A., Sala, N. and Capella, G. 2002, Genetic susceptibility and gastric cancer risk. *Int.J.Cancer*, **100:** 249-260

Graham, D. Y. and Yamaoka, Y. 2000, Disease-specific Helicobacter pylori virulence factors: the unfulfilled promise. *Helicobacter.*, **5 Suppl:** S3-S9

Graham, J. E., Peek, R. M., Jr., Krishna, U. and Cover, T. L. 2002, Global analysis of Helicobacter pylori gene expression in human gastric mucosa. *Gastroenterology*, **123:** 1637-1648

Guruge, J. L., Schalen, C., Nilsson, I., Ljungh, A., Tyszkiewicz, T., Wikander, M. and Wadstrom, T. 1990, Detection of antibodies to Helicobacter pylori cell surface antigens. *Scand.J.Infect.Dis.*, **22:** 457-465

Haas, G., Karaali, G., Ebermayer, K., Metzger, W. G., Lamer, S., Zimny-Arndt, U., Diescher, S., Goebel, U. B., Vogt, K., Roznowski, A. B., Wiedenmann, B. J., Meyer, T. F., Aebischer, T. and Jungblut, P. R. 2002, Immunoproteomics of Helicobacter pylori infection and relation to gastric disease. *Proteomics.*, **2:** 313-324

Hook-Nikanne, J., Perez-Perez, G. I. and Blaser, M. J. 1997, Antigenic characterization of Helicobacter pylori strains from different parts of the world. *Clin.Diagn.Lab Immunol.*, **4:** 592-597

Hu, L. T., Foxall, P. A., Russell, R. and Mobley, H. L. 1992, Purification of recombinant Helicobacter pylori urease apoenzyme encoded by ureA and ureB. *Infect.Immun.*, **60:** 2657-2666

Iaquinto, G., Todisco, A., Giardullo, N., D'Onofrio, V., Pasquale, L., De Luca, A., Andriulli, A., Perri, F., Rega, C., De Chiara, G., Landi, M., Taccone, W., Leandro, G. and Figura, N. 2000, Antibody response to Helicobacter pylori CagA and heat-shock proteins in determining the risk of gastric cancer development. *Dig.Liver Dis.*, **32:** 378-383

Ilver, D., Arnqvist, A., Ogren, J., Frick, I. M., Kersulyte, D., Incecik, E. T., Berg, D. E., Covacci, A., Engstrand, L. and Boren, T. 1998, Helicobacter pylori adhesin binding fucosylated histo-blood group antigens revealed by retagging. *Science*, **279:** 373-377

Joyce, E. A., Chan, K., Salama, N. R. and Falkow, S. 2002, Redefining bacterial populations: a post-genomic reformation. *Nat.Rev.Genet.*, **3:** 462-473

Jungblut, P., Thiede, B., Zimny-Arndt, U., Muller, E. C., Scheler, C., Wittmann-Liebold, B. and Otto, A. 1996, Resolution power of two-dimensional electrophoresis and identification of proteins from gels. *Electrophoresis*, **17:** 839-847

Jungblut, P. R., Bumann, D., Haas, G., Zimny-Arndt, U., Holland, P., Lamer, S., Siejak, F., Aebischer, A. and Meyer, T. F. 2000a, Comparative proteome analysis of Helicobacter pylori. *Mol.Microbiol.*, **36:** 710-725

Jungblut, P. R., Bumann, D., Haas, G., Zimny-Arndt, U., Holland, P., Lamer, S., Siejak, F., Aebischer, A. and Meyer, T. F. 2000b, Comparative proteome analysis of helicobacter pylori. *Mol.Microbiol.*, **36:** 710-725

Kamradt, A. E., Greiner, M., Ghiara, P. and Kaufmann, S. H. 2000, Helicobacter pylori infection in wild-type and cytokine-deficient C57BL/6 and BALB/c mouse mutants. *Microbes.Infect.*, **2:** 593-597

Keenan, J., Oliaro, J., Domigan, N., Potter, H., Aitken, G., Allardyce, R. and Roake, J. 2000, Immune response to an 18-kilodalton outer membrane antigen identifies lipoprotein 20 as a Helicobacter pylori vaccine candidate. *Infect.Immun.*, **68:** 3337-3343

Kim, N., Weeks, D. L., Shin, J. M., Scott, D. R., Young, M. K. and Sachs, G. 2002, Proteins released by Helicobacter pylori in vitro. *J.Bacteriol.*, **184:** 6155-6162

Kimmel, B., Bosserhoff, A., Frank, R., Gross, R., Goebel, W. and Beier, D. 2000, Identification of immunodominant antigens from Helicobacter pylori and evaluation of their reactivities with sera from patients with different gastroduodenal pathologies. *Infect.Immun.*, **68:** 915-920

Kleanthous, H., Lee, C. K. and Monath, T. P. 1998, Vaccine development against infection with Helicobacter pylori. *Br.Med Bull.*, **54:** 229-241

Kotloff, K. L., Sztein, M. B., Wasserman, S. S., Losonsky, G. A., DiLorenzo, S. C. and Walker, R. I. 2001, Safety and immunogenicity of oral inactivated whole-cell Helicobacter pylori vaccine with adjuvant among volunteers with or without subclinical infection. *Infect.Immun.*, **69:** 3581-3590

Kreiss, C., Buclin, T., Cosma, M., Corthesy-Theulaz, I. and Michetti, P. 1996, Safety of oral immunisation with recombinant urease in patients with Helicobacter pylori infection . *Lancet*, **347:** 1630-1631

Lan, R. and Reeves, P. R. 2000, Intraspecies variation in bacterial genomes: the need for a species genome concept. *Trends Microbiol.*, **8:** 396-401

Lee, A. 1999, Animal models of Helicobacter infection. *Mol Med Today*, **5:** 500-501

Lee, A., O'Rourke, J., De Ungria, M. C., Robertson, B., Daskalopoulos, G. and Dixon, M. F. 1997, A standardized mouse model of Helicobacter pylori infection: introducing the Sydney strain. *Gastroenterology*, **112:** 1386-1397

Lock, R. A., Coombs, G. W., McWilliams, T. M., Pearman, J. W., Grubb, W. B., Melrose, G. J. and Forbes, G. M. 2002, Proteome analysis of highly immunoreactive proteins of Helicobacter pylori. *Helicobacter*, **7:** 175-182

Loffeld, R. J. and van der Hulst, R. W. 2002, Helicobacter pylori and gastro-oesophageal reflux disease: association and clinical implications. To treat or not to treat with anti-H. pylori therapy? *Scand.J.Gastroenterol.Suppl*, 15-18

Machado, J. C., Pharoah, P., Sousa, S., Carvalho, R., Oliveira, C., Figueiredo, C., Amorim, A., Seruca, R., Caldas, C., Carneiro, F. and Sobrinho-Simoes, M. 2001, Interleukin 1B and interleukin 1RN polymorphisms are associated with increased risk of gastric carcinoma. *Gastroenterology*, **121:** 823-829

Mahdavi, J., Sonden, B., Hurtig, M., Olfat, F. O., Forsberg, L., Roche, N., Angstrom, J., Larsson, T., Teneberg, S., Karlsson, K. A., Altraja, S., Wadstrom, T., Kersulyte, D., Berg, D. E., Dubois, A., Petersson, C., Magnusson, K. E., Norberg, T., Lindh, F., Lundskog, B. B., Arnqvist, A., Hammarstrom, L. and Boren, T. 2002, Helicobacter pylori SabA adhesin in persistent infection and chronic inflammation. *Science*, **297:** 573-578

Mai, U. E., Perez-Perez, G. I., Allen, J. B., Wahl, S. M., Blaser, M. J. and Smith, P. D. 1992, Surface proteins from Helicobacter pylori exhibit chemotactic activity for human leukocytes and are present in gastric mucosa. *J.Exp.Med.*, **175:** 517-525

Marshall, B. J. and Warren, J. R. 1984, Unidentified curved bacilli in the stomach of patients with gastritis and peptic ulceration. *Lancet*, **1:** 1311-1315

McAtee, C. P., Fry, K. E. and Berg, D. E. 1998a, Identification of potential diagnostic and vaccine candidates of Helicobacter pylori by "proteome" technologies. *Helicobacter.*, **3:** 163-169

McAtee, C. P., Lim, M. Y., Fung, K., Velligan, M., Fry, K., Chow, T. and Berg, D. E. 1998b, Identification of potential diagnostic and vaccine candidates of Helicobacter pylori by two-dimensional gel electrophoresis, sequence analysis, and serum profiling. *Clin.Diagn.Lab Immunol.*, **5:** 537-542

McAtee, C. P., Lim, M. Y., Fung, K., Velligan, M., Fry, K., Chow, T. P. and Berg, D. E. 1998c, Characterization of a Helicobacter pylori vaccine candidate by proteome techniques. *J.Chromatogr.B Biomed.Sci.Appl.*, **714:** 325-333

Michetti, P., Kreiss, C., Kotloff, K. L., Porta, N., Blanco, J. L., Bachmann, D., Herranz, M., Saldinger, P. F., Corthesy-Theulaz, I., Losonsky, G., Nichols, R., Simon, J., Stolte, M., Ackerman, S., Monath, T. P. and Blum, A. L. 1999, Oral immunization with urease and Escherichia coli heat-labile enterotoxin is safe and immunogenic in Helicobacter pylori-infected adults. *Gastroenterology*, **116**: 804-812

Mitchell, H. M., Hazell, S. L., Li, Y. Y. and Hu, P. J. 1996, Serological response to specific Helicobacter pylori antigens: antibody against CagA antigen is not predictive of gastric cancer in a developing country. *Am.J.Gastroenterol.*, **91**: 1785-1788

Nilsson, C. L., Larsson, T., Gustafsson, E., Karlsson, K. A. and Davidsson, P. Identification of protein vaccine candidates from Helicobacter pylori using a preparative two-dimensional electrophoretic procedure and mass spectrometry. *Anal.Chem.2000.May.1.;72.(9.):2148.-53.*, **72**: 2148-2153

Nilsson, I. and Utt, M. 2002, Separation and surveys of proteins of Helicobacter pylori. *J.Chromatogr.B Analyt.Technol.Biomed.Life Sci.*, **771**: 251-260

Nilsson, I., Utt, M., Nilsson, H. O., Ljungh, A. and Wadstrom, T. 2000, Two-dimensional electrophoretic and immunoblot analysis of cell surface proteins of spiral-shaped and coccoid forms of Helicobacter pylori. *Electrophoresis*, **21**: 2670-2677

O'Toole, P. W., Logan, S. M., Kostrzynska, M., Wadstrom, T. and Trust, T. J. 1991, Isolation and biochemical and molecular analyses of a species-specific protein antigen from the gastric pathogen Helicobacter pylori. *J.Bacteriol.*, **173**: 505-513

Opazo, P., Muller, I., Rollan, A., Valenzuela, P., Yudelevich, A., Garcia-de, l. G., Urra, S. and Venegas, A. 1999, Serological response to Helicobacter pylori recombinant antigens in Chilean infected patients with duodenal ulcer, non-ulcer dyspepsia and gastric cancer. *APMIS*, **107**: 1069-1078

Peek, R. M., Jr. and Blaser, M. J. 2002, Helicobacter pylori and gastrointestinal tract adenocarcinomas. *Nat.Rev.Cancer*, **2**: 28-37

Pizza, M., Scarlato, V., Masignani, V., Giuliani, M. M., Arico, B., Comanducci, M., Jennings, G. T., Baldi, L., Bartolini, E., Capecchi, B., Galeotti, C. L., Luzzi, E., Manetti, R., Marchetti, E., Mora, M., Nuti, S., Ratti, G., Santini, L., Savino, S., Scarselli, M., Storni, E., Zuo, P., Broeker, M., Hundt, E., Knapp, B., Blair, E., Mason, T., Tettelin, H., Hood, D. W., Jeffries, A. C., Saunders, N. J., Granoff, D. M., Venter, J. C., Moxon, E. R., Grandi, G. and Rappuoli, R. Identification of Vaccine Candidates Against Serogroup B Meningococcus by Whole-Genome Sequencing. *Science*, **287**: 1816-1820

Pun, T., Hochstrasser, D. F., Appel, R. D., Funk, M., Villars-Augsburger, V. and Pellegrini, C. 1988, Computerized classification of two-dimensional gel electrophoretograms by correspondence analysis and ascendant hierarchical clustering. *Appl.Theor.Electrophor.*, **1**: 3-9

Sabarth, N., Hurwitz, R., Meyer, T. F. and Bumann, D. 2002a, Multiparameter selection of Helicobacter pylori antigens identifies two novel antigens with high protective efficacy. *Infect.Immun.*, **70**: 6499-6503

Sabarth, N., Lamer, S., Zimny-Arndt, U., Jungblut, P. R., Meyer, T. F. and Bumann, D. 2002b, Identification of surface proteins of Helicobacter pylori by selective biotinylation, affinity purification, and two-dimensional gel electrophoresis. *J. Biol. Chem.*, **277**: 27896-27902

Salama, N., Guillemin, K., McDaniel, T. K., Sherlock, G., Tompkins, L. and Falkow, S. 2000, A whole-genome microarray reveals genetic diversity among Helicobacter pylori strains. *Proc.Natl.Acad.Sci.U.S.A*, **97**: 14668-14673

Scarlato, V., Delany, I., Spohn, G. and Beier, D. 2001, Regulation of transcription in Helicobacter pylori: simple systems or complex circuits? *Int.J.Med.Microbiol.*, **291:** 107-117

Schmidt, F., Schmid, M., Mattow, J., Pleissner, K.-P. and Jungblut, P. R. 2003, Iterations are the key for exhaustive analysis of peptide mass fingerprints from proteins separated by two dimensional elecrophoresis. *JASMS*, **2003:** in press

Schmitt, W. and Haas, R. 1994, Genetic analysis of the Helicobacter pylori vacuolating cytotoxin: structural similarities with the IgA protease type of exported protein. *Mol.Microbiol.*, **12:** 307-319

Tannaes, T., Dekker, N., Bukholm, G., Bijlsma, J. J. and Appelmelk, B. J. 2001, Phase variation in the Helicobacter pylori phospholipase A gene and its role in acid adaptation. *Infect.Immun.*, **69:** 7334-7340

Telford, J. L., Ghiara, P., Dell'Orco, M., Comanducci, M., Burroni, D., Bugnoli, M., Tecce, M. F., Censini, S., Covacci, A., Xiang, Z. and . 1994, Gene structure of the Helicobacter pylori cytotoxin and evidence of its key role in gastric disease. *J.Exp.Med.*, **179:** 1653-1658

Thompson, L. J., Guillemin, K., Falkow, S. and Lee, A. 2002, DNA microarray analysis of the global gene expression of Helicobacter pylori during the growth cycle. *4th Western pacific Helicobacter Congress*, 8.2.4

Tomb, J. F., White, O., Kerlavage, A. R., Clayton, R. A., Sutton, G. G., Fleischmann, R. D., Ketchum, K. A., Klenk, H. P., Gill, S., Dougherty, B. A., Nelson, K., Quackenbush, J., Zhou, L., Kirkness, E. F., Peterson, S., Loftus, B., Richardson, D., Dodson, R., Khalak, H. G., Glodek, A., McKenney, K., Fitzegerald, L. M., Lee, N., Adams, M. D. and Venter, J. C. 1997, The complete genome sequence of the gastric pathogen Helicobacter pylori. *Nature* , **388:** 539-547

Unge, P. 1999, Antibiotic treatment of Helicobacter pylori infection. *Curr.Top.Microbiol.Immunol.*, **241:** 261-300

Utt, M., Nilsson, I., Ljungh, A. and Wadstrom, T. 2002, Identification of novel immunogenic proteins of Helicobacter pylori by proteome technology. *J.Immunol.Methods*, **259:** 1-10

Watanabe, T., Tada, M., Nagai, H., Sasaki, S. and Nakao, M. 1998, Helicobacter pylori infection induces gastric cancer in mongolian gerbils. *Gastroenterology*, **115:** 642-648

Xiang, Z., Bugnoli, M., Ponzetto, A., Morgando, A., Figura, N., Covacci, A., Petracca, R., Pennatini, C., Censini, S., Armellini, D. and . 1993, Detection in an enzyme immunoassay of an immune response to a recombinant fragment of the 128 kilodalton protein (CagA) of Helicobacter pylori. *Eur.J.Clin.Microbiol.Infect.Dis.*, **12:** 739-745

Yamaoka, Y., Kikuchi, S., el Zimaity, H. M., Gutierrez, O., Osato, M. S. and Graham, D. Y. 2002, Importance of Helicobacter pylori oipA in clinical presentation, gastric inflammation, and mucosal interleukin 8 production. *Gastroenterology*, **123:** 414-424

Yamaoka, Y., Kwon, D. H. and Graham, D. Y. 2000, A M(r) 34,000 proinflammatory outer membrane protein (oipA) of Helicobacter pylori. *Proc.Natl.Acad.Sci.U.S.A*, **97:** 7533-7538

Zevering, Y., Jacob, L. and Meyer, T. F. 1999, Naturally acquired human immune responses against *Helicobacter pylori* and implications for vaccine development. *Gut*, **45:** 465-474

Chapter 14

Proteomics of HIV-1 Virion

SHOGO MISUMI, NOBUTOKI TAKAMUNE and SHOZO SHOJI*
* Department of Biochemistry, Faculty of Pharmaceutical Sciences, Kumamoto University,
5-1 Oe-Honmachi, Kumamoto 862-0973, Japan

1.   INTRODUCTION

Acquired immunodeficiency syndrome (AIDS) was first recognized in 1981 when a common pattern of symptoms was observed among a small number of homosexual men in the USA (Brennan and Durack, 1981, Gottlieb et al.1981). AIDS cases were soon reported in other groups, including intravenous drug users and haemophiliacs (CDC, 1982, Davis et al., 1983,Masur et al., 1981). In 1983, Dr. Luc Montagnier and colleagues at the Institute Pasteur in Paris discovered the virus that causes AIDS (Barre-Sinoussi et al., 1983). This virus is named lymphadenopathy-associated virus (LAV). Dr. Robert Gallo's group also identified a retrovirus in 1984 called HTLV-III (Popovic et al., 1984). Furthermore, an independent group in San Francisco identified a retrovirus as AIDS-related virus (ARV) (Levy et al.1984). All these viruses are essentially the same and were named human immunodeficiency virus type 1 (HIV-1) in 1986.

In the two decades since the first reports of AIDS, this disease has become pandemic. Worldwide, more than 42 million adults and children are now living with HIV-1, in spite of implementation of the highly active antiretroviral therapy (HAART) (Hammer et al., 1997, Gulick et al.,1997), which has been shown to reduce viral load and increase CD4 lymphocytes in persons infected with HIV-1, delay onset of AIDS and prolong survival with AIDS. According to the United Nations Program on HIV/AIDS (UNAIDS) and the World Health Organization (WHO), 5.0 million people were

H. Hondermarck (ed.), Proteomics: Biomedical and Pharmaceutical Applications, 339–365.
© 2004 Kluwer Academic Publishers. Printed in the Netherlands.

newly infected with HIV-1 in 2002 alone and AIDS-related deaths reached a record 3.1 million. Among those who died, 610,000 were children.

HIV-1 Life Cycle

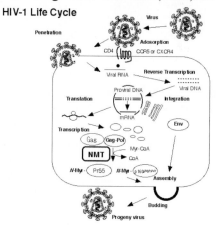

HIV-1 Gene Products

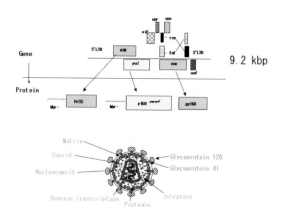

Scheme – HIV-1 Life cycle and gene products.
(See also Colour Plate Section, page 389)

The continued spread of the HIV-1 epidemic in both industrialized and developing countries highlights the need for novel and cost-effective AIDS therapies. Although the introduction of HAART has resulted in significant increases in survival for HIV-1-infected individuals (Gulick *et al.*, 1997), the impact will be largely confined to the industrialized countries, which at present constitute less than 10% of the worldwide HIV-1-infected population. Globally, most people who are carrying the AIDS virus live in countries with a very limited budget for health care. This means that in practice, there is little or no money for HIV-1 testing, condoms, and STI (sexually transmitted infection) treatment and prevention. In settings like this, a vaccine would be very cost-effective. Furthermore, historically, vaccines have been the most efficient and cost-effective means for the control of infectious diseases, as shown by the eradication of smallpox and the control of polio, yellow fever, and measles. Therefore, an effective and widely available preventive vaccine for HIV-1 may be the best option to control the global pandemic.

HIV-1 must essentially introduce its genetic material into the cytoplasm of a CD4(+) T-cell to productively infect it (Maddon *et al.*, 1986) (scheme). The process of viral entry involves fusion of the viral membrane with the host cell membrane and requires the specific interaction of envelope proteins with specific cell surface receptors. The two viral envelope proteins, gp120 and gp41, are conformationally associated to form a trimeric functional unit consisting of three molecules of gp120 exposed on the virion surface and associated with three molecules of gp41 inserted into the viral lipid membrane. Trimeric gp120 on the surface of a virion binds to CD4 on the surface of the target cell (Wyatt *et al.*, 1998), inducing a conformational change in the envelope proteins that in turn allows binding of the virion to the cell surface chemokine receptors (CCR5 or CXCR4), which contain seven membrane-spanning domains and normally transduce signals through G proteins (Kwong *et al.*, 1998, D'Souza and Harden, 1996). CCR5 binds to macrophage-tropic, non-syncytium-inducing (R5) viruses, which are associated with mucosal and intravenous transmission of HIV-1 infection. The importance of CCR5 for HIV-1 infection in vivo was shown by the discovery that approximately 1% of Caucasians lack CCR5, and that these individuals are highly resistant but not entirely immune to virus infection (Dean *et al.*, 1996, Liu *et al.*, 1996, Michael *et al.*, 1997, Samson *et al.*, 1996). On the other hand, CXCR4 binds to T-cell-tropic, syncytium-inducing (X4) viruses, which are frequently found during the later stages of the disease (Scarlatti *et al.*, 1997). These observations suggest that AIDS vaccines that prevent HIV-1 interaction with CCR5 or CXCR4, which are being developed in our laboratory (Misumi *et al.*, 2001a, Misumi *et al.*, 2001b), might form a promising new class of anti-AIDS therapy that is effective against multiple clades of HIV-1, especially those predominant in developing countries. Following fusion of the virus with the host cell, HIV-1 enters the cell. The viral RNA is released and undergoes reverse transcription. The viral reverse transcriptase is necessary for catalyzing this conversion of viral RNA into DNA. Once the viral RNA has been reverse-transcribed to DNA, this viral DNA enters the host cell nucleus where it can be integrated into the genetic material of the cell by the viral integrase. The genes of HIV-1 encode at least nine proteins that are divided into three classes: i) the major structural proteins (Gag, Pol, and Env), ii) the regulatory proteins (Tat and Rev): and iii) the accessory proteins (Vpu, Vpr, Vif, and Nef) (Gallo *et al.*, 1988) (Scheme). These viral proteins function in the replication of HIV-1 that includes the following steps: viral entry, reverse transcription, integration, gene expression, assembly, budding, and maturation.

However, only such actions of viral proteins could not completely explain how HIV-1 can efficiently replicate in a susceptible host. To date, several studies of purified HIV-1 virions have shown that in addition to proteins encoded by HIV-1, cellular proteins from the host are found in these virions (Ott, 1997). Cellular proteins, such as cyclophilin A (CyPA), are included in HIV-1 virions as a result of their interaction with Gag proteins during assembly and release (Ott *et al.*, 1995, Franke *et al.*, 1994, Thali *et al.*, 1994). CyPA is an abundant cytosolic protein ubiquitously expressed in eukaryotic cells (Ryffel *et al.*, 1991) and binds to the interaction sites that are located in the *N*-terminal domain of $p24^{gag}$ around Gly89-Pro90 for packaging and in the C-terminal domain of $p24^{gag}$ around Gly156-Pro157 and Gly223-Pro224 for destabilization of the capsid cone (Endrich *et al.*, 1999, Gamble *et al.*, 1996). Moreover, additional evidence of the role of CyPA in the HIV-1 life cycle has recently emerged. CyPA mediates HIV-1 attachment to target cells via heparans (Saphire *et al.*, 1999). The actions of the cellular protein CyPA support HIV-1 replication.

Some viral proteins with co/post-translational modifications, such as myristoylation and phosphorylation, also support HIV-1 replication. *N*-myristoylation of $p17^{gag}$ of HIV-1 is essential for structural assembly and replication (Bryant and Ratner, 1990, Shiraishi *et al.*, 2001, Tashiro *et al.*, 1989). Phosphorylation of $p17^{gag}$ is related to its dissociation from the membrane during the early post-entry step of HIV-1 (Bukrinskaya *et al.*, 1996, Gallay *et al.*, 1995). These results suggest that the identification of new cellular proteins and co/post-translational modifications of viral and cellular proteins that are indispensable to viral replication is needed to accomplish future AIDS-therapeutic breakthroughs.

In this study, a purified HIV-1_{LAV-1} preparation was directly analyzed by proteomics using two-dimensional (2D) gel electrophoresis and MALDI TOF-MS without using molecular biological approaches in order not to disclose the artificial biological phenomena of HIV-1.

2.   A PROTEOMIC APPROACH FOR HUMAN IMMUNODEFICIENCY VIRUS TYPE 1

2.1   Proteome Analysis Strategy for HIV-1 virion

The proteom analysis of HIV-1 virion is a five-step sequential strategy as shown in Fig. 1. The five steps of this strategy are:

(i) Preparation of HIV-1 virion. An important step in the characterization of HIV-1 is its purification. Purification can generally be achieved either by column chromatography or by sucrose density gradient centrifugation. The use of column chromatography avoids the use of a needle to aspirate a virus-containing material. We prepared the purified HIV-1 virion according to

Section 2.2.

(ii) Separation of proteins by 2D polyacrylamide gel electrophoresis. Proteome analysis is performed using a combination of 2D polyacrylamide gel electrophoresis and matrix-assisted laser desorption/ionization time of flight mass spectrometry (MALDI TOF-MS). Separation of proteins by 2D polyacrylamide gel electrophoresis is an accepted and common technique and is still one of the most effective methods for the separation of a complex mixture of proteins. Normally, over 3000-5000 proteins in a biological mixture can be separated and visualized by a staining technique as spots on a single analytical run.

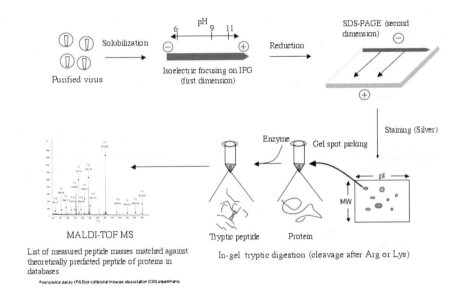

Figure 1. Strategy for the identification of proteins and its co/post-translational modifications that are indispensable for HIV-1 replication.

(iii) Determination of masses of peptides derived from proteins of interest. Each spot on the gel is excised and placed in a tube for digestion. The digested peptides are then analyzed by MALDI TOF-MS, providing sets of mass values of peptides (peptide-mass fingerprinting). Theoretically, peptide mass finger printing can identify any protein if its DNA sequence is present in a reference database. Using the tryptic hydrolysates, MALDI TOF-MS data are obtained.

(iv) Determination of partial sequences using post source decay (PSD) experiments. For proteins that cannot be identified by the pattern matching techniques described above, PSD experiments are conducted. This technique often allows us to obtain sequence tags.

(v) Analysis of co/post-translational modifications. For peptides that cannot be identified by the peptide-mass fingerprinting, PSD or collisional induced dissociation experiments are conducted. Those experiments allow the identification of the site and the type of co/post-translational modification.

2.2 Virus purification and Subtilisin treatment

The study of proteins inside HIV-1 was complicated by the potential contamination of the HIV-1 preparation with nonviral particles called microvesicles (Ott *et al.*, 1995, Bess *et al.*, 1997, Gluschankof *et al.*, 1997) (Fig. 2A). As shown in Fig. 2B, both microvesicle-contaminated HIV-1 preparation and microvesicle alone were eluted at the same positions as those of fractions 7, 8, and 9 on a Sepharose CL-4B^{TM} gel using PBS (-). In this study, the HIV-1$_{LAV-1}$ preparation was digested with subtilisin to remove nonviral particles using the protocol of Ott et al. (Ott *et al.*, 1995) with modifications. Typical results of the subtilisin-treated HIV-1$_{LAV-1}$ preparation and the microvesicles are shown in Figs. 2C and 2D, respectively. As expected, more than 95% of microvesicle-associated proteins were removed by the subtilisin treatment (Fig. 2C, lane 4). In contrast, proteins inside the virion, such as p17^{gag} and p24^{gag}, were not digested, although gp120 was completely digested (Fig. 2D, lanes 2 and 4). In addition, the HIV-1 reverse transcriptase activity was not decreased by the subtilisin treatment (Fig. 2E). The proteins inside the virion were not digested by the subtilisin treatment since they were protected by the virion membrane. Therefore, the subtilisin digestion procedure was used to identify unknown cellular and viral proteins inside HIV-1$_{LAV-1}$ indispensable to its replication. The subtilisin-treated HIV-1$_{LAV-1}$ preparation was finally purified by column chromatography prior to proteome analysis.

2.2.1   Cell culture

A chronically HIV-1$_{LAV-1}$ infected T-cell line (CEM/LAV-1) was maintained at 37 °C in RPMI-1640 medium supplemented with 10% fetal calf serum (FCS) containing 100 IU/ml penicillin and 100 µg/ml streptomycin in 5% CO2.

2.2.2   Purification of HIV-1$_{LAV-1}$ and microvesicles

A microvesicle-contaminated HIV-1$_{LAV-1}$ preparation was prepared using the protocol of McGrath *et al.* (1978) with modifications. The supernatant from the culture medium of CEM/LAV-1 was filtered through a 0.22-μm disposable filter (Asahi Techno Glass Corp., Tokyo, Japan) and then centrifuged at 43,000 x *g* for 3 hr at 4 °C. The pellet was resuspended in PBS(-) and then centrifuged at 100,000 x *g* for 1 hr at 4 °C. The resulting pellet was resuspended in 10 mM Tris-HCl buffer (pH 8.0). Virion digestion was carried out as previously described (Ott *et al.*, 1995, Ott *et al.*, 2000), using 1 mg/ml subtilisin (ICN Biomedicals Inc., Costa Mesa, CA, USA) in 10 mM Tris-HCl (pH 8.0) and 1 mM EDTA·2Na. Virions from the digested preparation were then purified by column chromatography using a 10-ml Sepharose CL-4B™ column at room temperature in a biosafety hood with PBS(-) as the column buffer. Fractions containing the digested virus were monitored by protein measurement using bicinchoninic acid, pooled and centrifuged at 100,000 x *g* for 1 hr at 4 °C. The resulting pellet was boiled for 1 min and lysed in 200 μl of lysis buffer (8 M Urea, and 4%(w/v) CHAPS in 2% IPG buffer (pH 6-11)). Microvesicles were produced in the culture medium of uninfected CEM by a similar procedure.

2.2.3   HIV-1 reverse transcriptase activity assay

Subtilisin-treated or nontreated HIV-1$_{LAV-1}$ preparation was lysed with lysis buffer (50 mM Tris-HCl (pH 7.8) 80 mM KCl, 2.5 mM dithiothreitol, 0.75 mM ethylenediamine-*N, N, N', N'*-tetraacetic acid disodium salt, and 0.5% Triton X-100), and the lysate (40 μl) was subjected as an enzyme solution to an HIV-1 RT activity assay using a Reverse Transcriptase Assay nonradioactive kit (Roche Diagnostics Corp., Tokyo, Japan) in accordance with the manufacturer's instructions.

2.2.4   Sodium dodecyl sulfate-polyacrylamide gel electrophoresis (SDS-PAGE) and Western immunoblot analysis

The sample was separated by SDS-PAGE (Laemmli, 1970) (Multigel 4/20, Daiichi Pure Chemicals Co., Ltd., Tokyo, Japan) and the separated proteins were subsequently electroblotted onto a polyvinylidene difluoride membrane (Immobilon, Millipore Corporation, Bedford, MA, USA). Antigens were probed with HIV-1-positive plasma and 0.5beta (a kind gift from Dr. Shuzo Matsushita of Kumamoto University, AIDS Research Institute, Kumamoto, Japan). The bands were visualized by means of chemiluminescence detection (NEN Life Science Products, Inc., Boston, MA, USA).

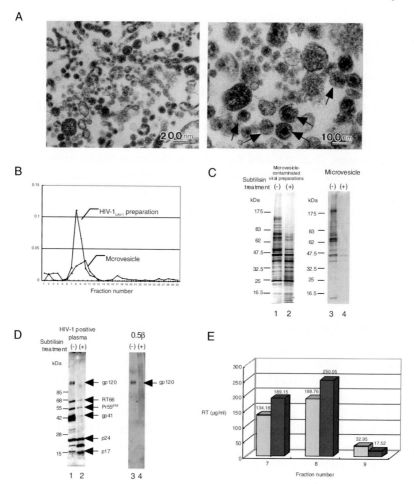

Figure 2. Elution profiles of both microvesicle-contaminated HIV-1$_{LAV-1}$ preparation and microvesicle, protein profiles of subtilisin-treated HIV-1$_{LAV-1}$ preparation and microvesicles, and HIV-1 reverse transcriptase activity assay.

(A) HIV-1 particles are indicated by arrows. (B) Elution profiles of microvesicle-contaminated HIV-1$_{LAV-1}$ preparation (solid circles), and microvesicle (solid squares) using a 10-ml column of Sepharose CL-4B™. (C) SDS-PAGE of microvesicle-contaminated HIV-1$_{LAV-1}$ preparation, and microvesicles derived from CEM cells. Lanes containing subtilisin-untreated and subtilisin-treated materials are denoted by (-) and (+), respectively. (D) Western immunoblot analysis of subtilisin-treated HIV-1$_{LAV-1}$ preparation. Antibodies used for analyses are displayed above each of the blots. The bands were visualized by means of chemiluminescence detection (NEN Life Science Products, Inc., Boston, MA, USA). Lanes containing subtilisin-untreated and subtilisin-treated HIV-1$_{LAV-1}$ materials are denoted by (-) and (+), respectively. SDS-PAGE and western immunoblot analysis of the digests using a sample containing 1 μg of starting total protein. (E) Reverse transcriptase activity of each fraction containing subtilisin-untreated (open column) or subtilisin-treated HIV-1$_{LAV-1}$ materials (closed column) was determined as described in section 2.2.3.

2.3 Two-dimensional gel separation of HIV-1$_{LAV-1}$ lysate and characterization of the proteins inside the virion

The nontreated HIV-1$_{LAV-1}$ preparation, the subtilisin-treated HIV-1$_{LAV-1}$ preparation, and the microvesicles were individually analyzed by 2D polyacrylamide gel electrophoresis under the same conditions to create a 2D image of HIV-1 virion proteins. Spots that overlapped in the protein profile of the subtilisin-untreated HIV-1$_{LAV-1}$ preparation and that of the subtilisin–treated HIV-1$_{LAV-1}$ preparation are emphasized in red except for the spots newly produced by the subtilisin treatment. This result suggests that 24 spots are identified as proteins inside the virion (Fig. 3). In addition, spots that overlapped in the protein profile of the subtilisin-untreated HIV-1$_{LAV-1}$ preparation and that of the microvesicles are also emphasized in light blue (Fig. 3). Sixty-one spots were identified as proteins derived from microvesicles. The other spots were identified as proteins outside the virion. The spots corresponding to the proteins inside the virion were preferentially excised and subjected to tryptic peptide mass fingerprinting using MALDI TOF-MS. The masses of peptides from each spot were used to search for their match from the on-line SWISS-PROT and TrEMBL databases, and protein matches were assigned by referring to theoretical fingerprints derived from published data. For example, spot 6 was excised and subjected to tryptic peptide-mass fingerprinting using MALDI TOF-MS (Fig. 4A). The peptide masses from spot 6 were used to search for their match from on-line SWISS-PROT and TrEMBL databases (Table I), and protein matches were assigned by referring to theoretical fingerprints derived from published data. Consequently, spot 6 was assigned to viral HIV-1 p24^{gag} and also confirmed by mass spectrometric peptide sequence determination of the peak (T8) with m/z 1295.94 in Fig. 4A (Fig. 4B). Furthermore, proteins corresponding to spots 7 and 8 (with pIs of 6.40 and 6.53, respectively) were assigned to two isoforms of cellular CyPA inside the virion, which we, therefore, designated as CyPA$_{6.40}$ and CyPA$_{6.53}$, respectively (Table II). At present, 9 of the 24 spots are identified as the major viral structural proteins (spots 1-6 and 9-11), and 2 of the 24 spots are identified as cellular proteins (spots 7 and 8) (Table III). Only gp120 is assigned based on the 2D image by western immunoblot analysis of the subtilisin-untreated HIV-1$_{LAV-1}$ preparation using 0.5beta because proteins outside the virion are completely digested by the subtilisin treatment (Fig. 3). Characterization of other proteins inside the virion (spots 12-24 in Fig. 3) is ongoing.

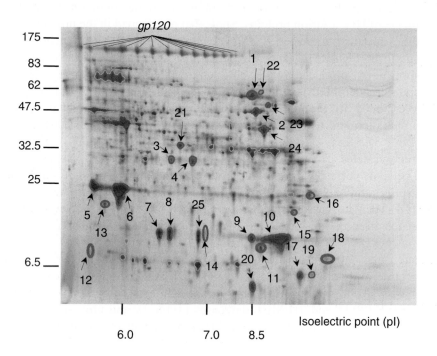

Figure 3. 2D gel image of HIV-1$_{LAV-1}$. Horizontal axis: protein separation by isoelectric focusing. Vertical axis: protein separation by molecular weight. The gel was subjected to silver staining. 2D SDS-PAGE standards (Nippon Bio-Rad Laboratories, Yokohama, Japan) were used for reproducibility of 2D experiments. The most striking advantage of these standards is that they allow valid comparison of 2D electrophoresis patterns of different samples. The 2D gel pattern shows the results obtained after superimposing the patterns of the subtilisin-treated HIV-1$_{LAV-1}$ preparation and the microvesicles on that of the subtilisin-untreated HIV-1$_{LAV-1}$ preparation. Spots of proteins inside the virion are emphasized in red. Spots derived from microvesicles are emphasized in light blue.

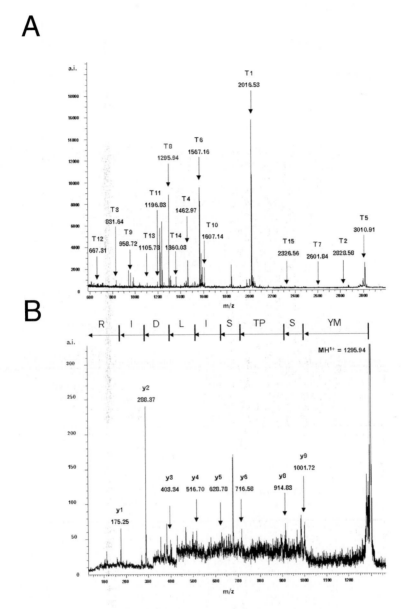

Figure 4. Tryptic peptide digest spectrum of spot 6 and post source decay spectrum of a tryptic peptide (T8), MYSPTSILDIR, with mass at m/z 1295.94 from the protein at spot 6. (A) MALDI reflector mass spectrum of tryptic peptide mixture derived from spot 6. (B) Mass spectrometric determination of partial peptide sequence of a tryptic peptide (T8). Y" series ions were observed.

Table I. Comparison of Theoretical Masses and Observed Masses Derived from Tryptic Digests of Spot 6

Spot	Tryptic peptide number	Theoretical mass (m/z)	Observed mass (m/z)	Corresponding sequence
spot 6	T1 (1-18)	2016.08	2016.53	PIVQNIQGQMVHQAISPR
	T2 (1-25)	2828.54	2828.58	PIVQNIQGQMVHQAISPRTLNAWVK
	T3 (19-25)	831.47	831.64	TLNAWVK
	T4 (71-82)	1462.65	1462.97	ETINEEAAEWDR
	T5 (71-97)	3010.46	3010.91	ETINEEAAEWDRVHPVHAGPIAPGQMR
	T6 (83-97)	1566.83	1567.16	VHPVHAGPIAPGQMR
	T7 (133-154)	2601.50	2601.84	WIILGLNKIVRMYSPTSILDIR
	T8 (144-154)	1295.67	1295.94	MYSPTSILDIR
	T9 (155-162)	958.51	958.72	QGPKEPFR
	T10 (155-16)	1606.80	1607.14	QGPKEPFRDYVDR
	T11 (159-16)	1196.57	1196.83	EPFRDYVDR
	T12 (163-16)	667.31	667.31	DYVDR
	T13 (163-17)	1105.53	1105.78	DYVDRFYK
	T14 (171-18)	1359.73	1360.03	TLRAEQASQEVK
	T15 (204-22)	2326.10	2326.56	ALGPAATLEEMMTACQGVGGPGHK

Tryptic peptide numbers correspond to that of the MALDI spectrum in Fig. 4A.

Table II. Comparison of Theoretical Masses and Observed Masses Derived from Tryptic Digests of Spots 7, 8, and 25

Spot	Tryptic peptide number	Theoretical mass (m/z)	Observed mass (m/z)	Corresponding sequence	Number of missed cleavages
spot 7	T1 (1-18)	1946.00	1946.30	VNPTVFFDIAVDGEPLGR	0
	T2 (19-30)	1379.76	1380.00	VSFELFADKVPK	1
	T3 (31-36)	737.36	737.35	TAENFR	0
	T4 (76-90)	1831.91	1832.20	SIYGEKFEDENFILK	1
	T5 (82-90)	1154.57	1153.58	FEDENFILK	0
	T6 (118-124)	848.41	848.35	TEWLDGK	0
	T7 (118-130)	1515.80	1515.57	TEWLDGKHVVFGK	1
	T8 (118-132)	1742.96	1742.91	TEWLDGKHVVFGKVK	2
spot 8	T1 (1-18)	1988.01	1988.26	AcVNPTVFFDIAVDGEPLGR*	0
	T2 (31-36)	737.36	737.37	TAENFR	0
	T3 (37-43)	705.38	705.42	ALSTGEK	0
	T4 (55-68)	1541.72	1542.00	IIPGFMCQGGDFTR	0
	T5 (76-90)	1831.91	1832.20	SIYGEKFEDENFILK	1
	T6 (82-90)	1154.57	1153.58	FEDENFILK	0
	T7 (118-124)	848.41	848.35	TEWLDGK	0
	T8 (118-130)	1515.80	1515.57	TEWLDGKHVVFGK	1
	T9 (131-143)	1505.75	1506.00	VKEGMNIVEAMER	1
	T10 (133-143)	1278.58	1278.80	EGMNIVEAMER	0
spot 25	T1 (1-18)	1946.00	1945.70	VNPTVFFDIAVDGEPLGR	0
	T2 (31-36)	737.36	737.46	TAENFR	0
	T3 (55-68)	1541.72	1541.50	IIPGFMCQGGDFTR	0
	T4 (76-90)	1831.91	1832.29	SIYGEKFEDENFILK	1
	T5 (82-90)	1154.57	1153.75	FEDENFILK	0
	T6 (118-124)	848.41	848.36	TEWLDGK	0
	T7 (118-130)	1515.80	1515.58	TEWLDGKHVVFGK	1
	T8 (131-143)	1505.75	1505.50	VKEGMNIVEAMER	1
	T9 (133-143)	1278.58	1278.40	EGMNIVEAMER	0

*Ac: Acetyl

Table III. Data for 25 Protein Spots Excised from 2D Gel

No.[1]	Protein name	AC[2]	%seq[3]	Gel feature[4] pI	Mr(kDa)	Predicted[5] pI	Mw (kDa)
Proteins inside the virion.							
01	Reverse transcriptase/RNase H (Fragment)	O40174	10.7	8.40	58	8.63	65
02	POL (Fragment)	Q9IDF2	27.2	8.59	49	8.64	60
03	Intergase (Fragment)	O92844	21.9	6.54	30	8.16	32
04	Intergase (Fragment)	O92844	28.1	6.80	30	7.75	32
05	CHAIN 2: CORE PROTEIN P24 (CORE ANTIGEN)	P03348	11.3	5.69	24	6.26	26
06	CHAIN 2: CORE PROTEIN P24 (CORE ANTIGEN)	P03348	54.5	5.94	24	6.26	26
07	Peptidyl-prolyl cis-trans isomerase A (Cyclophilin A)	P05092	40.2	6.40	18	7.82	18
08	Peptidyl-prolyl cis-trans isomerase A (Cyclophilin A)	P05092	41.5	6.53	18	7.82	18
09	CHAIN 1: CORE PROTEIN P17 (MATRIX PROTEIN)	P03348	25.2	8.41	17	9.28	15
10	CHAIN 1: CORE PROTEIN P17 (MATRIX PROTEIN)	P03348	35.9	9.72	17	9.28	15
11	CHAIN 1: CORE PROTEIN P17 (MATRIX PROTEIN)	P03348	58.8	8.85	12	9.28	15
Protein outside the virion.							
25	Peptidyl-prolyl cis-trans isomerase A (Cyclophilin A)	P05092	48.2	6.88	18	7.82	17.9

1) Number key to spot labels in Fig. 3.
2) SWISS PROT/TrEMBL database.
3) Ratio of the length of peptide sequences actually detected to the total sequence length of theoretical gene product, expressed as a percentage.
4) pI and Mr of feature calculated directly from 2D gel.
5) pI and Mw predicted for theoretical gene product.

2.3.1   2D polyacrylamide gel electrophoresis

2D polyacrylamide gel electrophoresis was carried out using the protocol of Görg *et al.* (1997) with modifications. The sample was loaded on the gel by anodic cup loading using an 18-cm Immobiline Drystrip (pH 6-11) (Amersham Biosciences Corp., Buckinghamshire, UK). The gel was run in the gradient mode: the voltage was raised from 500 V to 3500 V in 8 hr, and then maintained at 3500 V for 25 hr. After the 2D electrophoresis (12-14% ExcelGel XL SDS, Amersham Biosciences Corp., Buckinghamshire, UK), the gel was stained by silver staining.

2.3.2   Protein visualization and analysis

Proteins separated in the slab gel were visualized by silver staining using the protocol of Shevchenko *et al.* (1996) and Wilm *et al.* (1996). The 2D gel was analyzed using Bio-Rad Melanie II software (Nippon Bio-Rad Laboratories, Yokohama, Japan).

2.3.3   Peptide-mass fingerprinting

Silver-stained gel pieces were excised from the gel manually and destained prior to enzymatic digestion using the protocol of Gharahdaghi *et al.* (1999). These pieces were immersed in 100 µl of acetonitrile, dried under vacuum centrifugation for 60 min, rehydrated in 50 µl of trypsin solution (25 ng/µl trypsin in 100 mM ammonium bicarbonate), and then incubated for 45 min on ice. Unabsorbed trypsin solution was removed and the gel pieces were immersed in 20 µl of 100 mM ammonium bicarbonate (pH 8.0), and further incubated for 12 hr at 37 °C. Tryptic peptides were purified using ZipTip C18 according to the manufacturer's instructions (literature number, TN 224, Nihon Millipore Ltd., Tokyo, Japan), and then analyzed by MALDI TOF-MS with the FAST accessory for PSD (Burker Daltonik GmbH, Bremen, Germany). A nitrogen laser (337 nm) was used to irradiate the sample with alpha-cyano-4-hydroxy-cinnamic acid as the matrix. Peaks were manually assigned using the XMASS software (Burker Daltonik GmbH, Bremen, Germany).

The peptide masses obtained were subjected to search to determine their match from the SWISS-PORT and TrEMBL databases using the program PeptIdent (http://expasy.proteome.org.au/tools/peptident.html). Confidence in a given match was based on: (i) the percentage of nucleotide sequence covered versus the size (in kDa) of the matched protein, combined with (ii) the number of matching peptides versus the number of nonmatching peptides from a particular database entry. The remaining nonmatching peptide masses were subjected to search using the FindMod program (http://www.expasy.ch/tools/findmod) in order to identify probable co/post-translational modifications (Wilkins *et al.*, 1999).

2.3.4   Mass spectrometric determination of partial peptide sequence

The decay products were focused on by means of a set of reflectron lenses whose voltages were reduced in 14 steps under the control of the FAST software (Burker Daltonik GmbH, Bremen, Germany). y"-Dominant fragment ions were defined according to the nomenclature of Biemann (Biemann *et al.*, 1990). Mass calibration was performed using fragment ions from adrenocorticotropic hormone (ACTH) residues 18-39.

2.4   Post-translational modifications

Proteins corresponding to 9 of the 11 identified spots (spots 3-11) were estimated to be the products of three genes, suggesting the existence of multiple protein isoforms resulting from co/post-translational modifications of the gene products. Peptide-mass fingerprinting data were also used to identify probable co/post-translational modifications. Mass of the remaining non-matching peptides from spot 6 were subjected to search using the FindMod program in order to identify probable co/post-translational modifications. The formylation of the *N*-terminal tryptic peptide (Pro1-Arg18) of HIV-1 p24^{gag} was newly estimated by the FindMod program because the spectra of the *N*-terminal tryptic peptide (Pro1-Arg18) and the formylated form exhibited major peaks at m/z 2016.17 and 2044.19 (Fig. 5A, upper spectrum). This formylation was also detected in the corresponding spot stained with Coomassie brilliant blue R-250, suggesting that this formylation was not modified by the silver staining procedure (Fig. 5A, lower spectrum). The formylation was confirmed by sequential cleaving with 0.6 N hydrochloric acid (Fig. 5B), and mass spectrometric determination of the peptide sequence fragment (Fig. 6). The fragment ions (a-, and b-series) at m/z 239.2, 338.3, 466.5, 580.5, and 693.3 provide sufficient information for the formation of formylated Pro1, although the possibility that His12 might be formylated was considered based on protein chemistry. These data suggest that spot6 consists of two isoforms of HIV-1 p24^{gag} with free and formylated *N*-termini. Gitti *et al.* demonstrated that *N*-terminal proline residue forms a salt bridge with conserved Asp51, thereby forming the *N*-terminal beta-hairpin (Gitti *et al.*, 1996). This beta-hairpin constitutes the interface between the *N*-terminal domains of two p24^{gag} molecules (Gamble *et al.*,

1996). Furthermore, von Schwedler *et al.* demonstrated that the capsid core fails to assemble in the Asp51Ala mutant (von Schwedler *et al.*, 1998).

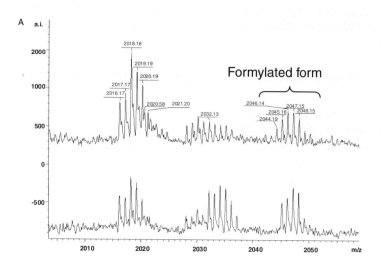

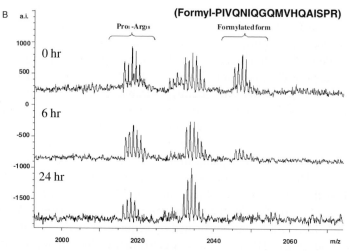

Figure. 5. Formylation of HIV-1 p24^{gag}. (A) As shown in the upper spectrum, the major peaks of $[M + H]^+$ at m/z 2016.17 represent the N-terminal tryptic peptide (Pro$_1$-Arg$_{18}$) of HIV-1$_{LAV-1}$ p24^{gag} derived from silver staining gel. The other peaks of $[M + H]^+$ at m/z 2032.13 and 2044.19 could be assigned to the methionine-oxidized form and the formylated form of the N-terminal tryptic peptide (Pro$_1$-Arg$_{18}$), respectively. As shown in the lower spectrum, these peaks were also detected in the spot corresponding to HIV-1$_{LAV-1}$ p24^{gag} derived from Coomassie brilliant blue R-250 staining gel. The molecular mass of 28.02 corresponding to the formyl group was deleted. (B) The formyl group was sequentially cleaved with 0.6 N hydrochloric acid as described in Section 2.4.1.

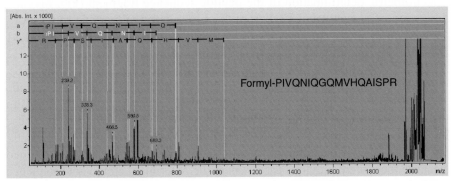

Figure 6. Post source decay spectrum of N-terminal tryptic peptide, PIVQNIQGQMVHQAISPR, of HIV-1 p24^{gag}.a-, b-, and y"-dominant fragment ions, defined according to the nomenclature of Biemann (1990), are shown, confirming the partial sequences (amino acids 1-7, and amino acids 10-18). fP is defined as a formylated proline.

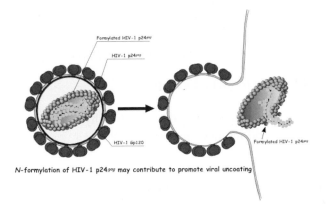

Figure 7. Proposed model for the role of N-formylated HIV-1 p24^{gag}. A fraction of the formylated HIV-1 p24^{gag} raises the possibility that N-formylation plays a promotion role in core disassembly during viral uncoating.

Therefore, a fraction of the formylated HIV-1 p24^{gag} may contribute to the weakening of the interaction between the *N*-terminal domains of p24^{gag} molecules by disturbing the formation of the salt bridge between Pro1 and Asp51 in the capsid cone architecture, suggesting that *N*-formylation of HIV-1 p24^{gag} may contribute to the promotion of viral uncoating (Fig. 7). Furthermore, the acetylation of the *N*-terminal tryptic peptide (Val1-Arg18) of CyPA$_{6.53}$ was also estimated by the FindMod program because the spectrum corresponding to the *N*-terminal tryptic peptide (Val1-Arg18) exhibited a peak at m/z 1946.30 in spot 7, and the spectrum corresponding to the acetylated form exhibited a peak at m/z 1988.26 in spot 8 (Fig. 8A, upper and middle spectra). The acetylation was confirmed by sequential cleaving with mass spectrometric determination of the peptide sequence fragment (Fig. 8B) and liberation of *N*-acetylated Val1 from the amino terminal peptide of CyPA$_{6.53}$ (Fig. 8C). The fragment ions (y"-series) at m/z 174.74, 232.21, 442.31, 571.31, 628.50, 743.24, 842.64, 1026.56, 1288.59, 1435.29, and 1732.95 provide sufficient information on the formation of acetylated Val1 (Fig. 8B). Furthermore, the peak of [M + H]$^{+}$ at m/z 1847.18 corresponded to the peptide (Asn2-Arg18) whose *N*-acetylated Val1 was cleaved by an *N*-acylamino-acid-releasing enzyme (Fig. 8C, lower spectrum). These findings demonstrated for the first time that *N*-acetylated CyPA is actually present in the virion although *N*-acetylation of CyPA has been presumed because there were previous cases where the *N*-terminal amino acid sequence of CyPA was not obtained (Ott *et al.*, 1995, Kieffer *et al.*, 1992), and that *N*-acetylation can be considered to be one of the possible causes of the difference in pI between CyPA$_{6.40}$ and CyPA$_{6.53}$.

The other modified proteins (spots 3-6 and 9-11) included the products of the *gag* and *pol* genes, but specific modifications could not be clearly established (Table III). Although 11.3-58.8 % of the total predicted sequence was recovered by proteome analysis, the remaining tryptic peptides may be either modified or not ionized well during mass spectrometry.

2.4.1   Deblocking of formyl group

The formyl group was deblocked with 0.6 N hydrochloric acid using the protocol of Ikeuchi and Inoue with modifications.

2.4.2   Identification of acetyl group

The *N*-acylamino-acid-releasing enzyme (Takara Shuzo Co., Ltd., Tokyo, Japan) liberates the *N*-terminal acylamino acid from *N*-acylated peptides (Tsunasawa and Narita, 1976). An *N*-terminal triptic peptide with a blocked *N*-terminus was dissolved in 10 μl of 15 mM sodium phosphate buffer (pH 7.2) and digested with 1.25 units of the *N*-acylamino-acid-releasing enzyme for 24 hr at 37 °C in accordance with the manufacturer's instructions. The digest was desalted using ZipTip C18 and analyzed by MALDI TOF-MS.

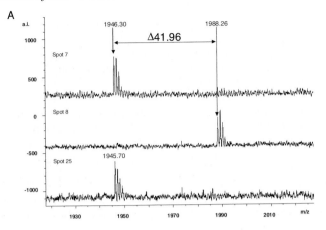

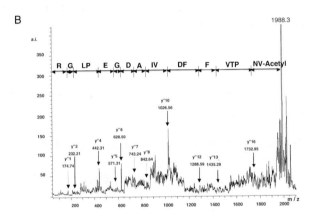

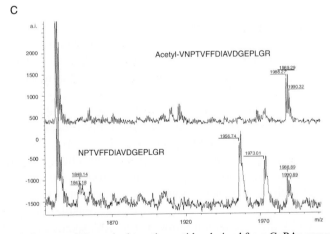

Figure 8. Spectra of N-terminal tryptic peptides derived from CyPAs corresponding to spots 7, 8, and 25, post source decay spectrum and N-acylamino-acid-releasing enzyme treatment of an N-terminal tryptic peptide derived from spot 8 protein, N-Acetyl-VNPTVFFDIAVDGEPLGR. (A) MALDI TOF-MS spectra of N-terminal tryptic

peptides derived from CyPAs corresponding to spots 7, 8, and 25. As shown in the top and bottom spectra, the peaks of $[M + H]^+$ at m/z 1946.30 and 1945.70 represent the N-terminal tryptic peptide (Val$_1$-Arg$_{18}$) of CypA. As shown in the middle spectrum, the peak of $[M + H]^+$ at m/z 1988.26 could be assigned to the acetylated form of the N-terminal tryptic peptide (Val$_1$-Arg$_{18}$). The fragment with a molecular mass of 41.96 corresponding to the acetyl group was deleted. (B) Mass spectrometric determination of the partial peptide sequence of the N-terminal tryptic peptide derived from spot 8. Y" series ions were defined according to the nomenclature of Biemann (1990). (C) The liberation of the N-acetylated Val$_1$ from the N-terminal peptide derived from spot 8 protein. As shown in the lower spectrum, the peak at m/z 1847.18 corresponding to the peptide Asn$_2$-Arg$_{18}$ (theoretical mass: m/z 1846.93) was detected after N-acylamino-acid-releasing enzyme treatment. The peak at m/z 1847.18 was not found before N-acylamino-acid-releasing enzyme treatment (upper spectrum). The unknown peaks of $[M + H]^+$ at m/z 1956.74 and 1973.01 were also found in control experiments under identical conditions, except for the omission of tryptic peptides derived from spot 8.

2.5   Characterization of the proteins outside the virion

To identify proteins outside the virion, spots that did not overlap in the protein profile of the subtilisin-untreated HIV-1$_{LAV-1}$ preparation and that of the microvesicles were also investigated. Figure 9 shows an expanded view of the same area of interest in each 2D gel image. As shown in Fig. 9A, spot 25 was observed only in the protein profile of the subtilisin-untreated HIV-1$_{LAV-1}$ preparation that was then subjected to tryptic peptide-mass fingerprinting using MALDI TOF-MS. Interestingly, the proteins corresponding to spot 25 (with a pI of 6.88) was assigned as a novel isoform of CyPA located outside the virion, which we, therefore, designated as CyPA$_{6.88}$ (Tables II and III). Indeed, spot 25 on the 2D gel of subtilisin-untreated HIV-1$_{LAV-1}$ preparation was also identified as CyPA by western immunoblot analysis with the anti-CyPA antibody (Upstate Biotechnology, Inc., Waltham, MA, USA, reference 46) (Fig. 9B). Furthermore, negative-staining transmission electron microscopy demonstrated that CyPA is outside the virion, but not outside the microvesicle (Fig. 9C).

The cause of the difference in pI between CyPA$_{6.40}$ and CyPA$_{6.88}$ is unknown, but it may be due to other types of post-translational modification because the *N*-terminal residue of CyPA$_{6.88}$ is free (Table II).

Recently, Sherry *et al.* (1998) and Saphire *et al.* (1999) have proposed that CyPA is located outside the virion and plays a direct role in the attachment of the virus to target cells. Our observation seems to be consistent with their model. However, our observation that the CyPA outside the virion is completely cleaved by the subtilisin treatment (Fig. 9A) seems to be different from that in a previous study which showed that only a small part of CyPA is removed upon proteolytic cleavage (Saphire *et al.*, 1999). In order to reconcile these data with our own observation, we propose that two isoforms of CyPA, namely, $CyPA_{6.40}$ and $CyPA_{6.53}$, are inside the viral membrane and another isoform, $CyPA_{6.88}$, is on the viral surface, and that one of the isoforms inside the virion may change to $CyPA_{6.88}$ to penetrate the viral membrane. Recently, CyPA has been reported to be a secreted growth factor induced by oxidative stress (Jin *et al.*, 2000). Although the mechanisms that permit the redistribution of CyPA on the viral surface are not yet clarified, CyPA itself may have the ability to penetrate the membrane.

We next sought to elucidate how $CyPA_{6.88}$ could redistribute on the viral surface. A chronically HIV-1$_{LAV-1}$ infected T-cell line (CEM/LAV-1) was treated with rabbit anti-CyPA polyclonal IgG and then analyzed using an EPICS XL flow cytometer. As shown in Fig. 9D, CyPA was not expressed on the cellular surface. This result suggests that CyPA is incorporated into the virion via HIV-1 Gag and redistributed from the inside to the outside of the viral membrane after budding (Fig. 10).

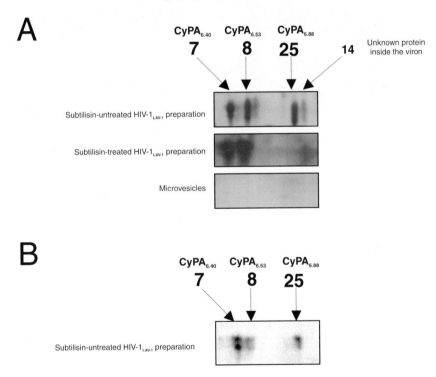

Figure 9. 2D gel image analysis, western immunoblot analysis with anti-CyPA antibody, and immunoelectron microscopic analysis. (A) After 2D gel electrophoresis and protein visualization by silver staining, the 2D gel images of subtilisin-untreated HIV-1$_{LAV-1}$ preparation, subtilisin-treated HIV-1$_{LAV-1}$ preparation, and microvesicles were expanded. The expanded image of subtilisin-untreated HIV-1$_{LAV-1}$ preparation shows four spots (spots 7, 8, 14, and 25). Spot 25 completely disappeared on subtilisin treatment. This result suggests that $CyPA_{6.88}$ exists on the viral surface. Spots 7, 8, 14, and 25 were not detected in the image of the microvesicles. The spot numbers are correspond to those in Fig. 3. (B) Spots 7, 8, and 25 on the

2D gel of subtilisin-untreated HIV-1$_{LAV-1}$ preparation were subjected to western immunoblot analysis with anti-CyPA antibody. (C) A microvesicle-contaminated HIV-1 preparation was fixed with 0.1 M sodium phosphate buffer (pH 7.4) containing 4% paraformaldehyde and 0.25% glutaraldehyde at 4° C. After brief washing, the samples were mounted on carbon coated nickel grids. Double immunolabeling techniques were used as follows. The first step was rabbit anti- cyclophilin A antibody (Upstate Biotechnology, Inc., Waltham, MA, USA) and then 15nm diameter gold-labelled as a secondary antibody, and 2nd step was murine anti gp120 IIIB monoclonal antibody (ImmunoDiagnostics, Inc., Woburn, MA, USA)) and then 5nm diameter gold-labelled as a secondary antibody. The immunolabeled samples were negatively stained with 3% uranyl acetate and examined by a transmission electron microscope. The asterisk indicates a microvesicle. (D) Grey and black lines represent staining of anti-CyPA-antibody-treated and nontreated CEM/LAV-1, respectively.

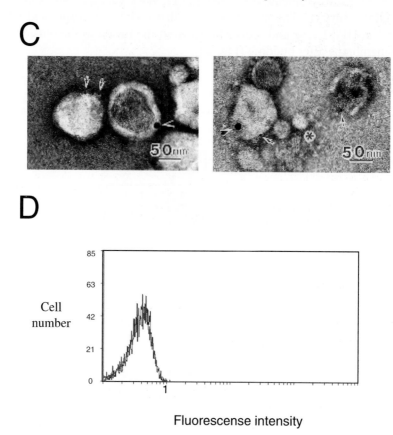

Figure 9. –Continued.

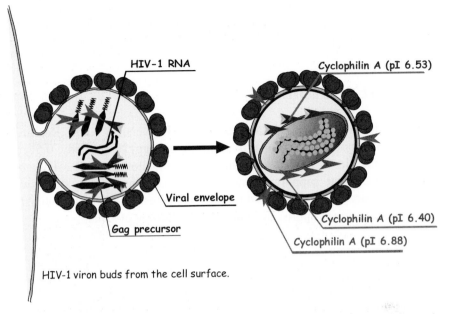

HIV-1 viron buds from the cell surface.

Figure 10. Three isoforms of cyclophilin A associated with human immunodeficiency virus type 1 and proposed model for the localization of each cyclophilin. A. Cellular CyPAs (CyPA$_{6.40}$, CyPA$_{6.53}$, and CyPA$_{6.88}$) are incorporated into progeny viruses via HIV-1 Gag precursor. Although the mechanisms that permit the redistribution of CyPA on the viral surface are not yet clarified, CyPA itself may have the characteristic of membrane penetrability.

3. CONCLUSIONS

Acid-labile formylation of *N*-terminal proline of HIV-1 p24^{gag} was found by proteomics. The role of formylation of human immunodeficiency virus type 1 p24^{gag} is unclear so far, but it is surmised that the acid-labile formylation of HIV-1$_{LAV-1}$ p24^{gag} may play a critical role in the formation of the HIV-1 core for conferring HIV-1 infectivity. Furthermore, peptide-mass fingerprinting data suggest that two isoforms of cyclophilin A (CyPA), one with an isoelectric point (pI) of 6.40 and one with a pI of 6.53, are inside the viral membrane and another isoform with a pI of 6.88 is outside the viral membrane, and that the CyPA isoform with a pI of 6.53 is *N*-acetylated. The mechanisms that permit the redistribution of CyPA on the viral surface have not yet been clarified, but it is surmised that the CyPA isoform with a pI of 6.88 may play a critical role in the attachment of virions to the surface of target cells, and both the CyPA isoforms with pIs of 6.40 and 6.53 may regulate the conformation of the HIV-1 capsid protein.

Proteome analysis of HIV-1$_{LAV-1}$ is very useful in understanding biological phenomena at the molecular level by identifying new proteins and co/post-translational modifications that are indispensable to viral replication, and in finding out attractive targets for the future AIDS-therapies.

ACKNOWLEDGEMENTS

We thank Dr. S. Matsushita (Kumamoto University, AIDS Research Institute, Kumamoto, Japan) for providing the anti-gp120 antibody, 0.5beta. We also thank Mr. T. Nirasawa (Nihon Bruker Daltonics K. K., Tsukuba, Japan) for technical support during mass sequencing. This study was supported in part by a Grant-in-Aid for Scientific Research from the Ministry of Education, Culture, Sports, Science and Technology of Japan.

REFERENCES

Brennan, R. O., and Durack, D. T., 1981, Gay compromise syndrome. *Lancet* **2:**1338-1339.
Barre-Sinoussi, F., Chermann, J. C., Rey, F., Nugeyre, M. T., Chamaret, S., Gruest, J., Dauguet, C., Axler-Blin, C., Vezinet-Brun, F., Rouzioux, C., Rozenbaum, W. and Montagnier, L., 1983, Isolation of a T-lymphotropic retrovirus from a patient at risk for acquired immune deficiency syndrome (AIDS). *Science* **220:**868-871.
Bess, J. W., Gorelick, R. J., Bosche W. J., Henderson, L. E. and Arthur, L. O., 1997., Microvesicles are a source of contaminating cellular proteins found in purified HIV-1 preparations. *Virology* **230:**134-144.
Biemann, K., 1990, Appendix 5. Nomenclature for peptide fragment ions (positive ions). *Methods Enzymol.* **193:**886-887.
Bryant, M., and Ratner, L., 1990, Myristoylation-dependent replication and assembly of human immunodeficiency virus 1. *Proc. Natl. Acad. Sci. USA* **87:**523-527.
Bukrinskaya, A. G., Ghorpade, A., Heinzinger, N. K. , Smithgall, T. E., Lewis, R. E. and Stevenson, M., 1996, Phosphorylation-dependent human immunodeficiency virus type 1 infection and nuclear targeting of viral DNA. *Proc. Natl. Acad. Sci. USA* **93:**367-371.
CDC 1982. Epidemiologic aspects of the current outbreak of Kaposi's sarcoma and opportunistic infections. *N. Engl. J. Med.* **306:** 248-252
D'Souza, M. P. and Harden, V. A., 1996, Chemokines and HIV-1 second receptors. *Nature Med.* **2:**1293-1300.
Davis, K. C., Horsburgh C. R., Hasiba U., Schocket A. L., and Kirkpatrick C. H., 1983, Acquired immunodeficiency syndrome in a patient with hemophilia. *Ann. Intern. Med.* **98:**284-286.
Dean, M., Carrington, M., Winkler, C., Huttley, G. A., Smith, M. W., Allikmets, R., Goedert, J. J., Buchbinder, S. P., Vittinghoff, E., Gomperts, E., Donfield, S., Vlahov, D., Kaslow, R., Saah, A., Rinaldo, C., Detels, R. and O'Brien, S. J., 1996, Genetic restriction of HIV-1 infection and progression to AIDS by a deletion allele of the CKR5 structural gene. Hemophilia Growth and Development Study, Multicenter AIDS Cohort Study, Multicenter Hemophilia Cohort Study, San Francisco City Cohort, ALIVE Study (see comments) (published erratum appears in Science 1996 Nov 15;274(5290):1069). *Science*

273:1856-1862.

Endrich, M. M., Gehrig, P. and Gehring, H., 1999, Maturation-induced conformational changes of HIV-1 capsid protein and identification of two high affinity sites for cyclophilins in the C-terminal domain, *J. Biol. Chem.* **26**:5326-5332.

Franke, E. K., Yuan, H. E. and Luban, J., 1994, Specific incorporation of cyclophilin A into HIV-1 virions. *Nature* **372**:359-362.

Gallay, P., Swingler, S. , Aiken, C. and Trono, D., 1995, HIV-1 infection of nondividing cells: C-terminal tyrosine phosphorylation of the viral matrix protein is a key regulator. *Cell* **80**:379-388.

Gallo, R., Wong-Staal, F., Montagnier, L., Haseltine, W. A. and Yoshida, M., 1988, HIV/HTLV gene nomenclature. *Nature* **333**:504.

Gamble, T. R., Vajdos, F. F., Yoo, S., Worthylake, D. K., Houseweart, M., Sundquist, W. I. and Hill, C. P., 1996, Crystal structure of human cyclophilin A bound to the amino-terminal domain of HIV-1 capsid. *Cell* **87**:1285-1294.

Gharahdaghi, F., Weinberg, C. R., Meagher, D. A., Imai, B. S. and Mische, S. M., 1999, Mass spectrometric identification of proteins from silver-stained polyacrylamide gel: a method for the removal of silver ions to enhance sensitivity. *Electrophoresis* **20**:601-605.

Gitti, R. K., Lee, B. M., Walker, J., Summers, M. F., Yoo, S. and Sundquist, W. I., 1996., Structure of the amino-terminal core domain of the HIV-1 capsid protein. *Science* **273**:231-235.

Gluschankof, P., Mondor, I., Gelderblom, H. R. and Sattentau, Q. J., 1997, Cell membrane vesicles are a major contaminant of gradient-enriched human immunodeficiency virus type-1 preparations. *Virology* **230**:125-133.

Görg, A., Obermaier, C., Boguth, G., Csordas, A., Diaz, J. J. and Madjar, J. J., 1997, Very alkaline immobilized pH gradients for two-dimensional electrophoresis of ribosomal and nuclear proteins. *Electrophoresis* **18**: 328-337.

Gottlieb, M. S., Schroff, R., Schanker H. M., Weisman, J. D., Fan, P., Wolf R.T. and Saxon, A., 1981, Pneumocystis carinii pneumonia and mucosal candidiasis in previously healthy homosexual men: evidence of a new acquired cellular immunodeficiency. *N. Engl. J. Med.* **305**:1425-1431.

Gulick, R.M., Mellors, J. W., Havlir, D., Eron, J. J., Gonzalez, C., McMahon, D., Richman, D. D., Valentine, F. T., Jonas, L. , Meibohm, A., Emini, E. A. and Chodakewitz, J. A., 1997, Treatment with indinavir, zidovudine, and lamivudine in adults with human immunodeficiency virus infection and prior antiretroviral therapy. *N. Engl. J. Med.* **337**:734-739.

Haendler, B., Hofer-Warbinek, R., Hofer, E., 1987, Complementary DNA for human T-cell cyclophilin. *EMBO J.* **6**:947-950.

Hammer, S. M., Squires, K. E., Hughes, M. D., Grimes, J. M., Demeter, L. M. , Currier, J. S. , Eron, J. J., Feinberg, J. E., Balfour, H. H., Deyton, L. R., Chodakewitz, J. A. and Fischl, M. A, 1997, A controlled trial of two nucleoside analogues plus idinavir in person and CD4 cell counts of 200 or less. *N. Engl. J. Med.* **337**:725-733.

Jin, Z. G., Melaragno, M. G., Liao, D. F., Yan, C., Haendeler, J., Suh, Y. A., Lambeth, J. D. and Berk, B. C., 2000, Cyclophilin A is a secreted growth factor induced by oxidative stress. *Circ. Res.* **87**:789-796.

Kieffer, L. J., Thalhammer, T. and Handschumacher, R. E., 1992, Isolation and characterization of a 40-kDa cyclophilin-related protein. *J. Biol. Chem.* **267**:5503-5507.

Kwong, P. D., Wyatt, R., Robinson, J., Sweet, R. W., Sodroski, J. and Hendrickson, W. A., 1998, Structure of an HIV gp120 envelope glycoprotein in complex with the CD4 receptor and a neutralizing human antibody. *Nature* **393**:648-659.

Laemmli, U. K., 1970, Cleavage of structural proteins during the assembly of the head of bacteriophage T4. *Nature* **227**: 680-685.

Levy, J. A., Hoffman, A. D. , Kramer, S. M., Landis, J. A., Shimabukuro, J. M. and Oshiro, L. S., 1984, Isolation of lymphocytopathic retroviruses from San Francisco patients with AIDS.

Science **225**:840-842.

Liu, R., Paxton, W. A., Choe, S., Ceradini, D., Martin, S. R., Horuk, R., MacDonald, M. E., Stuhlmann, H., Koup, R. A. and Landau, N. R., 1996, Homozygous defect in HIV-1 coreceptor accounts for resistance of some multiply-exposed individuals to HIV-1 infection. *Cell* **86**:367-377.

Maddon, P., Dalgleish, A., McDougal, J., Clapham, P., Weiss, R. and Axel, R., 1986, The T4 gene encodes the AIDS virus receptor and is expressed in the immune system and the brain. *Cell* **47**:333-348.

Masur, H., Michelis, M. A., Greene, J. B., Onorato, I., Stouwe, R. A., Holzman, R. S., Wormser, G. , Brettman, L., Lange, M., Murray, H. W., and Cunningham-Rundles, S., 1981, An outbreak of community-acquired *Pneumocystis carinii* pneumonia: initial manifestation of cellular immune dysfunction. *N. Engl. J. Med.* **305**:1431-1438.

McGrath, M., Witte, O., Pincus, T. and Weissman, I. L., 1978, Retrovirus purification: method that conserves envelope glycoprotein and maximizes infectivity *J. Virol.* **25**:923-927.

Michael, N. L., Chang, G., Louie, L. G., Mascola, J. R., Dondero, D., Birx, D. L. and Sheppard, H. W., 1997, The role of viral phenotype and CCR-5 gene defects in HIV-1 transmission and disease progression. *Nature Med.* **3**:338-340.

Misumi, S., Nakajima, R., Takamune, N. and Shoji, S., 2001b, A cyclic dodecapeptide-multiple-antigen-peptide-conjugate from undecapeptidyl arch (from Arg168 to Cys178) of extracellular loop2 in CCR5 as a novel HIV-1 vaccine. *J. Virol.* **75**:11614-11620.

Misumi, S., Takamune, N., Ido, Y., Hayashi, S., Endo, M., Mukai, R., Tachibana, K., Umeda, M. and Shoji, S., 2001a, Evidence as a HIV-1 self-defense vaccine of cyclic chimeric dodecapeptide warped from undecapeptidyl arch of exracellular loop 2 in both CCR5 and CXCR4. *Biochem. Biophys. Res. Commun.* **285**:1309-1316.

Ott, D. E., 1997, Cellular proteins in HIV virions. *Rev. Med. Virol.* **7**:167-180.

Ott, D. E., Coren, L. V., Johnson, D. G., Kane, B. P., Sowder, R. C., Kim, Y. D., Fisher, R. J., Zhou, X. Z., Lu, K. P. and Henderson, L. E., 2000, Actin-binding cellular proteins inside human immunodeficiency virus type 1. *Virology* **266**:42-51.

Ott, D. E., Coren, L. V., Johnson, D. G., Sowder II, R. C., Arthur, L. O. and Henderson, L. E., 1995, Analysis and localization of cyclophilin A found in the virions of human immunodeficiency virus type 1 MN strain. AIDS *Res. Hum. Retroviruses* **11**:1003-1006.

Popovic, M., Sarngadharan, M. G., Read, E. and Gallo, R. C., 1984, Detection, isolation, and continuous production of cytopathic retroviruses (HTLV-III) from patients with AIDS and pre-AIDS. *Science* **224**:497-500.

Ryffel, B., Woerly, G., Greiner, B., Haendler, B., Mihatsch, M. J. and Foxwell, B. M., 1991, Distribution of the cyclosporine binding protein cyclophilin in human tissues. *Immunology* **72**:399-404.

Samson, M., Libert, F., Doranz, B. J., Rucker, J. , Liesnard, C., Farber, C. M., Saragosti, S., Lapoumeroulie, C., Cognaux, J., Forceille, C., Muyldermans, G., Verhofstede, C., Burtonboy, G. , Georges, M., Imai, T., Rana, S., Yi, Y., Smyth, R. J., Collman, R. G., Doms, R. W., Vassart, G. and Parmentier, M., 1996, Resistance to HIV-1 infection in caucasian individuals bearing mutant alleles of the CCR-5 chemokine receptor gene. *Nature* **382**:722-725.

Saphire, A. C., Bobardt, M. D. and Gallay, P. A., 1999, Host cyclophilin A mediates HIV-1 attachment to target cells via heparans. *EMBO J.* **18**:6771-6785.

Scarlatti, G., Tresoldi, E., Bjorndal, A., Fredriksson, R., Colognesi, C., Deng, H. K., Malnati, M. S., Plebani, A., Siccardi, A. G., Littman, D. R., Fenyo, E. M. and Lusso, P., 1997, In vivo evolution of HIV-1 co-receptor usage and sensitivity to chemokine-mediated suppression, *Nature Med.* **3**:1259-1265.

Sherry, B., Zybarth, G., Alfano, M., Dubrovsky, L., Mitchell, R., Rich, D., Ulrich, P., Bucala, R., Cerami, A. and Bukrinsky, M., 1998, Role of cyclophilin A in the uptake of HIV-1 by macrophages and T lymphocytes. *Proc. Natl. Acad. Sci. USA* **95**:1758-1763.

Shevchenko, A., Wilm, M., Vorm, O. and Mann, M., 1996, Mass spectrometric sequencing of proteins silver-stained polyacrylamide gels. *Anal. Chem.* **68:**850-858.

Shiraishi, T., Misumi, S., Takama, M., Takahashi, I. and Shoji, S., 2001, Myristoylation of human immunodeficiency virus type 1 gag protein is required for efficient env protein transportation to the surface of cells. *Biochem. Biophys. Res. Commun.* **282:**1201-1205.

Tashiro, A., Shoji, S. and Kubota, Y., 1989, Antimyristoylation of the gag proteins in the human immunodeficiency virus-infected cells with N-myristoyl glycinal diethylacetal resulted in inhibition of virus production. *Biochem. Biophys. Res. Commun.* **165:**1145-1154.

Thali, M., Bukovsky, A., Kondo, E., Rosenwirth, B., Walsh, C. T., Sodroski J. and Göttlinger, H. G., 1994, Functional association of cyclophilin A with HIV-1 virions. *Nature* 372:363-365

Tsunasawa, S. and Narita, K., 1976, Acylamino acid-releasing enzyme from rat liver. *Methods in Enzymology*, **XLV**, 552-561.

von Schwedler, U. K., Stemmler, T. L., Klishko, V. Y., Li, S., Albertine, K. H., Davis, D. R. and Sundquist, W. I., 1998, Proteolytic refolding of the HIV-1 capsid protein amino-terminus facilitates viral core assembly. *EMBO J.* **17:**1555-1568.

Wilkins, M.R., Gasteiger, E., Gooley, A. A., Herbert, B. R. , Molloy, M. P., Binz, P. A., Ou, K., Sanchez, J. C., Bairoch, A., Williams, K. L. and Hochstrasser, D. F., 1999, High-throughput mass spectrometric discovery of protein post-translational modifications. *J. Mol. Biol.* **289:**645-657.

Wilm, M., Shevchenko, A., Houthaeve, T., Breit, S., Schweigerer, L., Fotsis, T. and Mann, M., 1996., Femtomole sequencing of proteins from polyacrylamide gels by nano-electrospray mass spectrometry. *Nature* **379:**466-469.

Wyatt, R., Kwong, P. D., Desjardins, E., Sweet, R. W., Robinson, J., Hendrickson, W. A. and Sodroski, J. G., 1998, The antigenic structure of the HIV gp120 envelope glycoprotein. *Nature* **393:**705-711.

Chapter 15

Proteomics to Explore Pathogenesis and Drug Resistance Mechanisms in Protozoan Parasites

BARBARA PAPADOPOULOU, JOLYNE DRUMMELSMITH and MARC OUELLETTE
Centre de Recherche en Infectiologie du Centre de Recherche du CHUL and Division de Microbiologie, Faculté de Médecine, Université Laval, Québec, Canada

1.   PARASITIC DISEASES

Protozoan parasites are responsible for some of the most devastating and prevalent diseases of humans and domestic animals. Protozoan parasites threaten the lives of nearly one-third of the worldwide human population, and are responsible for the loss of more than 50 million disability-adjusted life years (DALYs) and more than 2 million deaths a year. The parasite species responsible for malaria (*Plasmodium spp*) are at the forefront in terms of their ability to inflict devastating effects and mortality on the human population, but the parasites responsible for the various forms of leishmaniasis (*Leishmania spp*), African sleeping sickness (*Trypanosoma brucei*) and Chagas disease (*Trypanosoma cruzi*) are also important contributors to global morbidity and mortality figures. Many other parasitic species also contribute significantly to the world's disease burden.

Ideally, prevention would be the most efficient way to control parasitic diseases, but despite numerous efforts there are no effective vaccines against any of the clinically important parasites. For the moment, drugs are therefore the mainstay in our control of parasitic protozoans when simple prevention measures fail or prove impractical. However, a majority of the drugs directed against parasites have far from optimal pharmacological properties, with narrow therapeutic indices and limiting host toxicity. The arsenal of anti-protozoal drugs is thus limited, and this is further complicated by a rapid rise

H. Hondermarck (ed.), Proteomics: Biomedical and Pharmaceutical Applications, 367–383.

in drug resistance which is becoming a major public health problem (Ouellette and Ward, 2002). Several parasitic diseases were recently included in the World Health Organization's list of diseases where antimicrobial resistance is a major issue (www.who.int/infectious-disease-report/2000/ch4. htm).

2.   PROTEOMICS AND PARASITES

2.1   Current progress in parasite genome sequencing

The recent advances in the sequencing of microbial genomes have revitalised research in microbiology and parasitology. Indeed, since the report of the first sequence of the bacterium *Haemophilus influenzae* in 1995 (Fleischmann *et al.*, 1995) there has been an explosion of complete microbial genome sequences. As of this writing in early 2003, there are several hundred microbial genomes in the public domain that are either completed or close to completion (www.tigr.org; wit.integratedgenomics.com/gold/). The sequences of larger genomes such as those of parasites is now feasible and the sequence of at least one species of each of the major human parasites is now underway (Table 1), with the genome sequences of two *Plasmodium* species causing malaria already completed (Carlton *et al.*, 2002; Gardner *et al.*, 2002). The availability of these sequences will serve in ongoing efforts to study the parallel expression of genes and protein contents by microarray and a variety of proteomic approaches, respectively. Proteome maps of parasites should become useful reference points for studies pointing to the modulation of protein expression upon differentiation, interaction with the host or drug resistance. Proteomics, defined here as the large scale identification, quantification or localisation studies of proteins and interaction studies, has the potential to generate new therapeutic leads for the control of parasitic diseases and offers a new approach to drug and vaccine discovery. With genomes of parasites rapidly finding their place in databanks, this approach will permit the determination of how parasites interact with their hosts, respond to antiparasitic drugs and develop mechanisms to escape the immune response or to resist the action of drugs. Indeed, within these banks one might soon be able to find clues to the targets of all potential new drugs and vaccines.

Table 1. Summary of ongoing genomic and proteomic studies of the major human parasites.

Organism	Disease/ Pathology	Genome sequencing	Institute or Reference	Transcriptomic[1] and proteomic[2] studies
Brugia malayi	Filariasis	In progress	Sanger	None
Entamoeba histolytica	Amoebiasis	In progress	TIGR, Sanger	None
Giardia lamblia	Intestinal giardiasis	In progress	MBL	None
Leishmania major	Leishmaniasis	In progress	Myler *et al.*, 1999, Sanger, SBRI	El Fakhry *et al.*, 2002[2] Acestor *et al.*, 2002[2] Drummelsmith *et al.*, 2003[2]
Necator americanus	Hookworm	In progress	Sanger	None
Plasmodium falciparum	Malaria	Completed	Gardner *et al.*, 2002	Hayward *et al.*, 2000[1] Ben Mamoun *et al.*, 2001[1] Bozdech *et al.*, 2003[1] Florens *et al.*, 2002[2] Greenbaum *et al.*, 2002[2] Lasonder *et al.*, 2002[2]
Plasmodium vivax	Malaria	In progress	TIGR, Sanger	None
Schistosoma mansoni	Schistosomiasis/ Bilharzia	In progress	Sanger	Hoffmann *et al.*, 2002[1]
Toxoplasma gondii	Toxoplasmosis	In progress	TIGR, Sanger	Cleary *et al.*, 2002[1] Cohen *et al.*, 2002[2]
Trichomonas vaginalis	Trichomoniasis	In progress	TIGR	None
Trypanosoma brucei	African sleeping sickness	In progress	TIGR, Sanger	(Rout and Field, 2001)[2]

2.2 Methods applicable to the proteomic analysis of parasites

Although few papers have yet been published in the new field of proteomics for parasitic diseases, efforts are tangible and should lead to important discoveries. The most widely used technology for carrying out proteomic analyses consists of two-dimensional (2D) polyacrylamide gels, which are studied and compared by powerful new image analysis softwares.

This technique allows the separation and visualization of a large number of protein spots from within a complex mixture. The identification of proteins is carried out by various mass spectrometry techniques following protease digestion of individual protein spots and comparison of peptide fingerprint or peptide sequences with the numerous data banks available. Obviously, the identification of proteins is facilitated if the genome of the organism is completely known and we are close to this situation with at least one representative species for the majority of the important human parasites. While 2D gels are still the mainstay of most proteomic efforts, there are nonetheless a number of limitations that are associated with 2D gels. Indeed, membrane proteins, low abundance proteins, high and low molecular weight proteins and proteins with extreme pI are usually under-represented while using 2D gels. Several approaches, including a number of enrichment protocols (subcellular fractionation, chromatographic steps, membrane isolation), can be used to increase proteome representation and hence partially compensate for limitations inherent in 2D gels.

New mass spectrometry-based proteomic methods have also been developed that allow quantitative protein expression profiling. One such technique combines differential isotopic labelling with amino acid specific affinity chromatography. This isotope-coded affinity tag (ICAT) strategy has been used successfully for the accurate quantification and identification of individual proteins within complex mixtures (Gygi *et al.*, 1999). Another method is based on resolving small peptides derived from complex mixtures by high resolution multidimensional liquid chromatography with columns containing strong cation exchange and reversed phase resins for peptide separation (Carlton *et al.*, 2002). The sequence of the peptides is obtained directly by LC/LC MS/MS using sophisticated computer search algorithms against the available databases.

Both 2D gels and newer proteomic approaches have been used in the field of parasitology and we will try to give an up to date account of these findings. Work on proteomics has indicated that both 2D gels and mass spectrometry-based proteomic approaches are here to stay since they often provide different but complementary data. In an extensive review it was reported that neither 2D gels nor ICAT provided comprehensive coverage on a proteome-wide scale and this was shown eloquently by the specific example of mitochondrial multisubunit protein complexes (Patton *et al.*, 2002). In fact, 2D gels and ICAT each displayed selectivity as to the proteins detected, and this selectivity was often complementary. There are numerous innovative developments for all types of proteomic efforts and it is anticipated that a combination of techniques will be required for comprehensive coverage on a whole organism scale.

3. APPLICATIONS FOR THE COMPARATIVE PROTEOMIC ANALYSIS OF PARASITES

Whole proteome comparison of a protozoan parasite can be used to try to assess at the global level the role of the complete set of proteins under a number of conditions. As a paradigm we will describe proteomic approaches one could exploit, using as a primary example the protozoan parasite *Leishmania* (Figure 1); work carried out with other protozoan parasites will also be discussed. *Leishmania* is a parasite endemic in 88 countries that affects nearly 15 million people with 2 million new cases each year (Herwaldt, 1999). There are several species of *Leishmania* that cause different pathologies, ranging from self healing cutaneous lesions to visceral infections that are usually fatal if untreated. *Leishmania* has a digenetic life cycle that alternates between a flagellated promastigote form in the sand fly vector to a round amastigote form within macrophages. *Leishmania* can also interact with a number of antigen-presenting cells including dendritic cells. The drug of choice to treat leishmaniasis is pentavalent antimony, but resistance to this drug is now widespread (Guerin *et al.*, 2002).

A prerequisite to carrying out gel-based proteomic approaches is the generation of a proteome reference map that can then be used for comparison to find proteins whose expression is modulated under specific conditions. A number of efforts have been made to solubilise proteins suitable for 2D gel electrophoresis from parasites including *Leishmania* (Acestor *et al.*, 2002; El Fakhry *et al.*, 2002) and *Toxoplasma gondii* (Cohen *et al.*, 2002). By optimising sample preparation and using a combination of narrow pH range first dimension gels, we were able to obtain a proteome consisting of more than 3500 protein spots of *Leishmania major* (Figure 2) which represents about 35% of the expected protein complement of *Leishmania*. Several landmark proteins were identified by a combination of MALDI-TOF and LC/MS-MS techniques (Drummelsmith *et al.*, 2003) and this has led to the demonstration that despite the genome not being complete, protein identification is feasible and often successful, even when using tryptic fingerprinting-related techniques. We found the rate of identification to be commensurate with the current status of the genome. Several proteins identified were annotated as hypothetical proteins, allowing open reading frames predicted by automated algorithms to be confirmed as coding regions. Indeed, the current sequencing data of protozoan parasites suggest that ~60% of the putative gene products are annotated as hypothetical and have no known homologues.

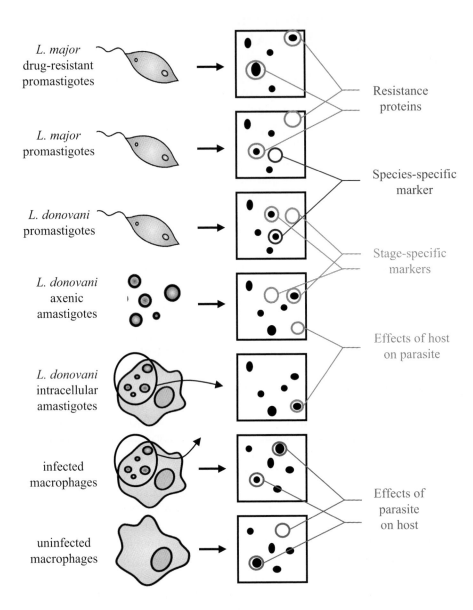

Figure 1. Use of 2D gels to study differential protein expression in the parasite Leishmania. Examples include the identification of drug-resistance proteins (green), strain or species-specific markers (blue), stage-specific proteins (orange), and the effects of host-parasite interactions on both the parasite (teal) and the host (violet).

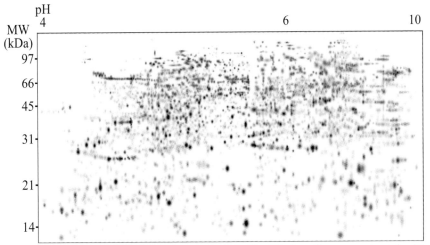

Figure 2. A 2D proteome map of L. major Friedlin. Whole cells were extracted with a lysis buffer containing urea, thiourea, and CHAPS. Samples were run using a variety of narrow-range IPG strips (4.0-5.0, 4.5-5.5, 5.0-6.0, 5.5-6.7, and 6-11; Amersham), gels were silver-stained, and the images were merged using PDQuest software (BioRad).

3.1   Strain- and Species-Specific Markers

Different species of *Leishmania* give different pathologies and this species-specific tropism is likely due to both specific microbial products and the host response. For example, *L. major* is responsible for cutaneous leishmaniasis, which is usually self-limiting, while *L. donovani* causes potentially fatal visceral leishmaniasis. Parasite species-specific products could certainly be determined by a number of genomic approaches. For example, six species of the malaria causing parasite *Plasmodium* are currently being sequenced (http://www.plasmodb.org) and sequence comparison between *Plasmodium falciparum* and *Plasmodium yoelli* revealed syntheny among metabolic genes but diversity between surface antigens and molecules enabling evasion of the immune response (Carlton *et al.*, 2002). Genomic sequence comparison of several species of the same parasite genus should thus permit the identification of species-specific products. While several genomes of parasites are currently being sequenced, the difficulty and complexity of the process makes it unlikely that several species of the same genus will be sequenced in the near future. For most parasites, the development of strategies other than comparative genomics will likely prove to be quicker and easier for the identification of species-specific proteins. A possible approach would involve the comparison of the 2D proteome maps of several representative strains of two related parasitic species. Analysis of protein spots found consistently among strains of one species and absent in the other could lead to the discovery of species-specific proteins (Figure 1).

The availability of a proteome map of *L. major* has incited us to test whether a comparison of 2D gels derived from *L. major* and *L. donovani* could lead to species-specific proteins. While several spots (~30%) between the two species are at the same position in the 2D gel, the variability between these two strains is too large to have confidence that a protein migrating differently is species-specific (data not shown). These data suggest that this approach may not lead to the discovery of species-specific proteins in this case. However, comparing the proteome of different strains of the same species could be useful to find differentially expressed proteins that should be related to specific phenotype e.g. strains more virulent than others, strains carrying a gene disruption or overexpressing a protein, or strain of the same species sensitive and resistant to a drug (see below). The availability of (almost) complete sequence data for *L. major* presents a potential mass spectrometry-based approach to the problem of high species-to-species variability and is outlined in Figure 3. By comparing the MS/MS-derived proteome profile of *L. donovani* to the *L. major* genome sequence, peptides not fitting a predicted *L. major* peptide could be subjected to *de novo* sequence analysis. Peptide sequences with homology to *L. major* proteins would be filtered out, resulting in a bank of species-specific peptide sequence tags which could then be used for further studies.

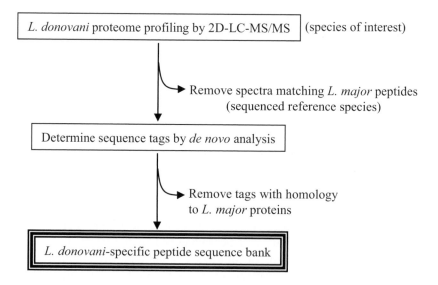

Figure 3. Example of a possible mass spectrometry-based method for the identification of protein markers used to differentiate between species when the genome sequence of one is unknown.

3.2   Stage-Specific Proteins

Leishmania has a relatively simple life cycle, switching from promastigote (sand fly) to amastigote (macrophage). Stage-specific regulation in *Leishmania* is frequently controlled at the level of translation (Boucher *et al.*, 2002), thus a proteomic approach is very well suited for isolating developmentally regulated proteins. The *Leishmania* system also has the unique advantage that both the promastigote and amastigote stages of the parasite can be cultured in an axenic fashion, eliminating possible contamination from the host cell. By analysing 2D gels from both developmental stages we found several proteins that were expressed either preferentially or exclusively in one of the two life stages of the parasite. An example of such analysis is described in Figure 4. In a preliminary analysis we estimated that more than 5% of the proteins were differentially expressed in the intracellular amastigote stage (El Fakhry *et al.*, 2002). The characterisation of two of these differentially expressed proteins by LC-MS/MS analysis revealed that they correspond to two enzymes of the glycolytic cycle (El Fakhry *et al.*, 2002). These findings are consistent with the mode of ATP production in amastigotes which depends on glycolysis. We have further pursued the analysis of soluble promastigote- and amastigote-specific proteins and have identified several proteins encoding various functions that are stage regulated not only in the amastigote, but also in the promastigote stage of *Leishmania* (unpublished observations). The ongoing proteomic approaches appear to be highly promising to find new differentially expressed proteins. As suggested previously, the use of complementary non-gel-based proteomic approaches would be a worthwhile enterprise to find further stage regulated parasitic proteins.

High-throughput in-line multidimensional liquid chromatography and tandem mass spectrometry was used to look at the protein complements of four life stages of the malaria parasite *Plasmodium falciparum* (Florens *et al.*, 2002). In total, 2415 parasite proteins were detected among sporozoites (the insect form), merozoites (invasive stage of the erythrocytes), trophozoites (multiplying form in erythrocytes) and gametocytes (sexual stage). Surprisingly, only 152 proteins were common to all four stages but known protein markers were identified in the correct life stages (Florens *et al.*, 2002). Interestingly, antigens of the *var* family present on the surface of infected erythrocytes were also found abundantly expressed in the insect stage. Several proteins identified and annotated as hypothetical had a predicted transmembrane domain, a GPI addition signal or a putative signal sequence leading to secretion or intracellular localisation. These could constitute interesting drug or vaccine targets. In a separate large scale proteomic study, 1289 malaria proteins were identified comprising 23% of the current proteome prediction (Lasonder *et al.*, 2002). Three stages of the parasite were studied, and about 50% of the proteins were found to be specifically expressed in only one stage. Interestingly, a set of peptides was obtained with significant matches in the parasite genome, but not within the open reading frames predicted by algorithms. These studies have led to the identification of additional exons and the assignment of different exon-intron boundaries which have thus led to several gene annotation changes (Lasonder *et al.*, 2002). It is also possible that further proteomic studies will pinpoint small proteins (< 75 amino acids) that have not been annotated in several genomes.

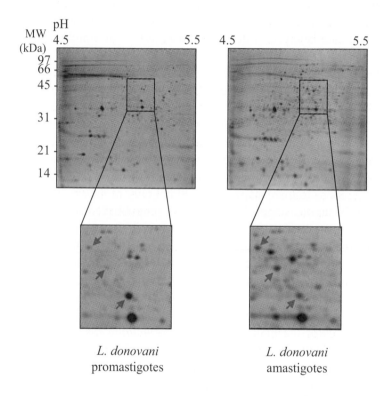

L. donovani
promastigotes

L. donovani
amastigotes

Figure 4. Comparative proteome analysis of L. donovani infantum axenic promastigotes and amastigotes. Proteins from whole cells were extracted as described in Figure 2, run on narrow-range IPG strips, and stained with SYPRO Ruby (Molecular Probes) to allow quantitative comparisons. Representative gels of each stage are shown. The expanded region contains a number of differentially-expressed spots. Spots marked with orange arrows appear to be stage-specific, while spots marked with blue arrows are differentially expressed.

3.3   Studies of Drug Response, Resistance, and Targets

The mode of action of several antiparasitic drugs, despite several years of usage, is still unknown. The possibility of a comprehensive analysis of drug response in a particular cell provided by a proteomic (and/or transcriptomic) approach should enable us to determine how these drugs act and whether they have specific targets. Drug resistance is an important problem in numerous parasitic species, and in most cases resistance mechanisms are poorly understood (Ouellette and Ward, 2002). Proteomic studies provide new strategies to try to further increase our understanding of drug resistance.

Previous proteomic studies of drug resistant bacteria have reported the differential expression of glyceraldehyde-3-phosphate dehydrogenase in erythromycin-resistant *S. pneumoniae* (Cash *et al.*, 1999) and of alkylhydroperoxide reductase in metronidazole-resistant *Helicobacter pylori* (McAtee *et al.*, 2001). While the expression of these proteins was correlated to resistance their roles in resistance remain unclear. We have applied a proteomic 2D gel approach to study drug resistance in *Leishmania* to both the model antifolate drug methotrexate (MTX) and to the clinically relevant antimonial drugs. An example of differences between sensitive and resistant parasites is shown in Figure 5. In the MTX resistant mutants we have identified a number of spots that were differentially expressed, including the overexpressed pteridine reductase PTR1 (Drummelsmith *et al.*, 2003), a known MTX resistance mechanism of the parasite (Ouellette *et al.*, 2002). This is an overwhelming validation of our proteomic strategy to isolate drug resistance markers. We have identified several other differentially expressed proteins in both antifolate and metal resistant parasites, and their roles in resistance are now being addressed experimentally using a combination of transfection protocols and functional assays.

One extensively studied new drug target in *Leishmania* is trypanothione reductase (TRYR). TRYR keeps trypanothione (Fairlamb and Cerami, 1992), a glutathione spermidine conjugate being a functional equivalent of glutathione, in a reduced form. The *TRYR* gene was transfected into wild-type *Leishmania* cells to see if we could detect the overexpressed protein in 2D gels. We had the unexpected finding that more than one spot was increased upon *TRYR* transfection. The spots had a similar apparent molecular weight, shifts in pI of about 0.08 pH units, and were identified by mass spectrometry as TRYR (Drummelsmith *et al.*, 2003). This suggests that TRYR could be post-translationally modified. TRYR is not known to be enzymatically modified *in vivo* and certainly our observation warrants further work. Protection from oxidants is essential for the survival of the parasite in macrophages and TRYR is an essential enzyme to deal with that type of stress (Dumas *et al.*, 1997). A regulation in the activity of TRYR at the post-translational level rather than at the transcriptional level could allow the parasite to quickly adjust its TRYR activity in response to the oxidative burst.

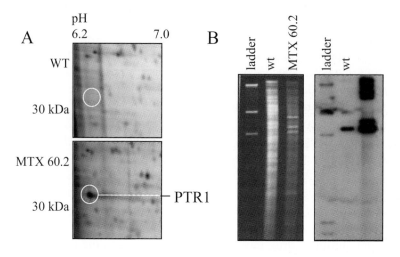

Figure 5. Overexpression of PTR1, a known methotrexate resistance protein, in a drug resistant strain of L. major (strain MTX 60.2). **A**. Comparison of 2D gel images revealed a highly overexpressed spot in the resistant mutant versus the wildtype. MALDI-TOF analysis identified the spot as PTR1. **B**. DNA amplification (visible in the ethidium bromide-stained panel to the left) is a common mechanism of protein overexpression in Leishmania. Southern hybridisation using a PTR1 probe (right panel) confirmed a substantial amplification of the PTR1 gene, providing a mechanism for the overexpression seen by 2D gels.

An interesting new drug target, the falcipain 1 of *P. falciparum,* has been characterized using a chemical proteomic screen (Greenbaum *et al.,* 2002). In this study, a biotin affinity tag cysteine protease probe was used to isolate all the parasite papain family proteases in a single step; their sequences were then determined by mass spectrometry. The falcipain-1 protease was found to have a specific role in host cell invasion, and as such is a promising new therapeutic target.

3.4   Host-Pathogen Interactions

Another application of the use of proteomics will be to study host-pathogen interactions. We have already described the use of proteomic approaches for the isolation of stage-specific proteins, several of which are likely to be important in host pathogen interactions. For example, interactions between *Leishmania* and its host are known, as determined by DNA microarray analysis, to modulate the expression of several host genes (Buates and Matlashewski, 2001) and in fact the interactions of several other parasites with their hosts were studied using DNA microarrays. However, during host-pathogen interaction in addition to changes in transcriptional activities it is highly likely that changes will occur at the level of post-translational modification. This type of regulation of activity can only be detected using proteomic approaches and would be unnoticed using transcriptomic approaches. For example, activation of the host phosphotyrosine phosphatase SHP-1 by *L. donovani* affects tyrosine phosphorylation-based signalling events (Forget *et al.*, 2001). In this manner, *Leishmania* alters the host-cell functions and can survive. Therefore, one strategy would be to put host cells in contact with the parasites for short periods, wash the cells to remove the parasites, and analyse the effects of parasite contact on the proteome of the host cell. This would be complicated by the presence of significant quantities of proteins derived from the parasite, but since the proteomes of both the host (human or mouse) and the parasite are known, it would not be difficult to resolve any cross contamination between host and parasite proteins. For example, during the analysis of stage-specific proteins of malaria by high throughput techniques, more than 1000 human proteins were also identified (Lasonder *et al.*, 2002). Global proteomic studies should lead to novel tools to study host-pathogen interactions and have the unique advantage of looking both at protein quantities, and, even more importantly in the context of host-pathogen interactions, at post-translational modifications. Interestingly, our initial analysis of the *Leishmania* proteome permitted the identification of extensive modification and proteolytic processing of the α and β tubulins of *Leishmania* (Drummelsmith *et al.*, 2003).

3.5 Organellar Proteomics

Protozoan parasites are highly divergent from the most frequently studied eukaryotic model systems, with researchers still searching for a function for a majority of their annotated proteins. One strategy to start looking at function is to determine the subcellular localisation of the proteins, one approach to which would be the isolation of specific organelles of parasites and determination of their proteome. This fractionation strategy would also have the advantage of permitting the identification of more proteins expressed at low levels, thus resulting in a better representation of the proteome. It may also be possible to characterise new proteins in organelles that are specific to protozoa such as the glycosomes of kinetoplastid parasites (Hannaert and Michels, 1994), the hydrogenosomes found in some anaerobic protozoa (de Souza, 2002), or the apicoplast in Apicomplexa parasites (Ralph *et al.*, 2001). These should all be rich in specific and novel drug targets. In one study, the nucleus of *T. brucei* was studied using this organellar proteomic approach and three proteins known to be in the nucleus were identified by mass spectrometry (Rout and Field, 2001). Using preparative free-flow electrophoresis, investigators have separated various membrane containing vesicles of the endocytic pathway (Grab *et al.*, 1998). This type of fractionation or the use of other technologies such as density gradients coupled to mass spectrometry should expedite protein discovery in parasites while at the same time providing clues to protein function. Another recent example of the use of proteomics to characterise a selected class of proteins consists of an analysis of the proteins secreted by the nematode *Haemonchus contortus* (Yatsuda *et al.*, 2003). More than 100 secreted proteins were characterised in this manner and several of these should be useful for future vaccine development strategies.

4.   CONCLUDING REMARKS

Due in part to philanthropic organisations, but also to national funding agencies with the vision to recognise the magnitude of parasitic diseases, a number of genome projects of parasites have been started. In fact, at least one species of every important human parasite is currently being sequenced. The sequences that are part of data banks are unique resources for finding the novel drugs, vaccines and diagnostic targets that are urgently needed for the treatment of parasitic diseases. Various proteomic efforts, such as the ones described in this chapter, are likely to impact greatly on a better understanding and characterization of these targets. In addition, by allowing the detection of differential expression, proteomic approaches will permit the detection of post-translational modifications, subcellular localisation, and interactions between proteins that are likely to play important roles in pathogenesis, the mode of action of drugs and drug resistance mechanisms.

ACKNOWLEDGEMENTS

This work was supported by Canadian Institutes of Health Research (CIHR) group and operating grants to B.P. and M.O. B.P. is a senior FRSQ scholar and a Burroughs Wellcome New Investigator in Molecular Parasitology. J.D. is a CIHR post-doctoral fellow, while M.O. is a Burroughs Wellcome Fund Scholar in Molecular Parasitology and holds a Canada Research Chair in antimicrobial resistance.

REFERENCES

Acestor, N., Masina, S., Walker, J., Saravia, N. G., Fasel, N., and Quadroni, M., 2002, Establishing two-dimensional gels for the analysis of Leishmania proteomes. *Proteomics* **2**:877-879.

Ben Mamoun, C., Gluzman, I. Y., Hott, C., MacMillan, S. K., Amarakone, A. S., Anderson, D. L., Carlton, J. M., Dame, J. B., Chakrabarti, D., Martin, R. K., Brownstein, B. H., and Goldberg, D. E., 2001, Co-ordinated programme of gene expression during asexual intraerythrocytic development of the human malaria parasite Plasmodium falciparum revealed by microarray analysis. *Mol Microbiol* **39**:26-36.

Boucher, N., Wu, Y., Dumas, C., Dube, M., Sereno, D., Breton, M., and Papadopoulou, B., 2002, A common mechanism of stage-regulated gene expression in Leishmania mediated by a conserved 3'-untranslated region element. *J Biol Chem* **277**:19511-19520.

Bozdech, Z., Zhu, J., Joachimiak, M. P., Cohen, F. E., Pulliam, B., and DeRisi, J. L., 2003, Expression profiling of the schizont and trophozoite stages of Plasmodium falciparum with a long-oligonucleotide microarray. *Genome Biol* **4**:R9.

Buates, S., and Matlashewski, G., 2001, General suppression of macrophage gene expression during Leishmania donovani infection. *J Immunol* **166**:3416-3422.

Carlton, J. M., Angiuoli, S. V., Suh, B. B., Kooij, T. W., Pertea, M., Silva, J. C., Ermolaeva, M. D., Allen, J. E., Selengut, J. D., Koo, H. L., Peterson, J. D., Pop, M., Kosack, D. S., Shumway, M. F., Bidwell, S. L., Shallom, S. J., van Aken, S. E., Riedmuller, S. B., Feldblyum, T. V., Cho, J. K., Quackenbush, J., Sedegah, M., Shoaibi, A., Cummings, L. M., Florens, L., Yates, J. R., Raine, J. D., Sinden, R. E., Harris, M. A., Cunningham, D. A., Preiser, P. R., Bergman, L. W., Vaidya, A. B., van Lin, L. H., Janse, C. J., Waters, A. P., Smith, H. O., White, O. R., Salzberg, S. L., Venter, J. C., Fraser, C. M., Hoffman, S. L., Gardner, M. J., and Carucci, D. J., 2002, Genome sequence and comparative analysis of the model rodent malaria parasite Plasmodium yoelii yoelii. *Nature* **419**:512-519.

Cash, P., Argo, E., Ford, L., Lawrie, L., and McKenzie, H., 1999, A proteomic analysis of erythromycin resistance in Streptococcus pneumoniae. *Electrophoresis* **20**:2259-2268.

Cleary, M. D., Singh, U., Blader, I. J., Brewer, J. L., and Boothroyd, J. C., 2002, Toxoplasma gondii asexual development: identification of developmentally regulated genes and distinct patterns of gene expression. *Eukaryot Cell* **1**:329-340.

Cohen, A. M., Rumpel, K., Coombs, G. H., and Wastling, J. M., 2002a, Characterisation of global protein expression by two-dimensional electrophoresis and mass spectrometry: proteomics of Toxoplasma gondii. *Int J Parasitol* **32**:39-51.

de Souza, W., 2002, Special organelles of some pathogenic protozoa. *Parasitol Res* **88**:1013-1025.

Drummelsmith, J., Brochu, V., Girard, I., Messier, N., and Ouellette, M., 2003, Proteome mapping of the protozoan parasite Leishmania and application to the study of drug targets and resistance mechanisms. *Mol Cell Proteomics* **15**:15.

Dumas, C., Ouellette, M., Tovar, J., Cunningham, M. L., Fairlamb, A. H., Tamar, S., Olivier, M., and Papadopoulou, B., 1997, Disruption of the trypanothione reductase gene of Leishmania decreases its ability to survive oxidative stress in macrophages. *Embo J* **16**:2590-2598.

El Fakhry, Y., Ouellette, M., and Papadopoulou, B., 2002, A proteomic approach to identify developmentally regulated proteins in Leishmania infantum. *Proteomics* **2**:1007-1017.

Fairlamb, A. H., and Cerami, A., 1992, Metabolism and functions of trypanothione in the Kinetoplastida. *Annu Rev Microbiol* **46**:695-729.

Fleischmann, R. D., Adams, M. D., White, O., Clayton, R. A., Kirkness, E. F., Kerlavage, A. R., Bult, C. J., Tomb, J. F., Dougherty, B. A., Merrick, J. M., and *et al.*, 1995, Whole-genome random sequencing and assembly of Haemophilus influenzae Rd. *Science* **269**:496-512.

Florens, L., Washburn, M. P., Raine, J. D., Anthony, R. M., Grainger, M., Haynes, J. D., Moch, J. K., Muster, N., Sacci, J. B., Tabb, D. L., Witney, A. A., Wolters, D., Wu, Y., Gardner, M. J., Holder, A. A., Sinden, R. E., Yates, J. R., and Carucci, D. J., 2002, A proteomic view of the Plasmodium falciparum life cycle. *Nature* **419**:520-526.

Forget, G., Siminovitch, K. A., Brochu, S., Rivest, S., Radzioch, D., and Olivier, M., 2001, Role of host phosphotyrosine phosphatase SHP-1 in the development of murine leishmaniasis. *Eur J Immunol* **31**:3185-3196.

Gardner, M. J., Hall, N., Fung, E., White, O., Berriman, M., Hyman, R. W., Carlton, J. M., Pain, A., Nelson, K. E., Bowman, S., Paulsen, I. T., James, K., Eisen, J. A., Rutherford, K., Salzberg, S. L., Craig, A., Kyes, S., Chan, M. S., Nene, V., Shallom, S. J., Suh, B., Peterson, J., Angiuoli, S., Pertea, M., Allen, J., Selengut, J., Haft, D., Mather, M. W., Vaidya, A. B., Martin, D. M., Fairlamb, A. H., Fraunholz, M. J., Roos, D. S., Ralph, S. A., McFadden, G. I., Cummings, L. M., Subramanian, G. M., Mungall, C., Venter, J. C., Carucci, D. J., Hoffman, S. L., Newbold, C., Davis, R. W., Fraser, C. M., and Barrell, B., 2002, Genome sequence of the human malaria parasite Plasmodium falciparum. *Nature* **419**:498-511.

Grab, D. J., Webster, P., and Lonsdale-Eccles, J. D., 1998, Analysis of trypanosomal endocytic organelles using preparative free-flow electrophoresis. *Electrophoresis* **19**:1162-1170.

Greenbaum, D. C., Baruch, A., Grainger, M., Bozdech, Z., Medzihradszky, K. F., Engel, J., DeRisi, J., Holder, A. A., and Bogyo, M., 2002, A role for the protease falcipain 1 in host cell invasion by the human malaria parasite. *Science* **298**:2002-2006.

Guerin, P. J., Olliaro, P., Sundar, S., Boelaert, M., Croft, S. L., Desjeux, P., Wasunna, M. K., and Bryceson, A. D., 2002, Visceral leishmaniasis: current status of control, diagnosis, and treatment, and a proposed research and development agenda. *Lancet Infect Dis* **2**:494-501.

Gygi, S. P., Rist, B., Gerber, S. A., Turecek, F., Gelb, M. H., and Aebersold, R., 1999, Quantitative analysis of complex protein mixtures using isotope-coded affinity tags. *Nat Biotechnol* **17**:994-999.

Hannaert, V., and Michels, P. A., 1994, Structure, function, and biogenesis of glycosomes in kinetoplastida. *J Bioenerg Biomembr* **26**:205-212.

Hayward, R. E., Derisi, J. L., Alfadhli, S., Kaslow, D. C., Brown, P. O., and Rathod, P. K., 2000, Shotgun DNA microarrays and stage-specific gene expression in Plasmodium falciparum malaria. *Mol Microbiol* **35**:6-14.

Herwaldt, B. L., 1999, Leishmaniasis. *Lancet* **354**:1191-1199.

Hoffmann, K. F., Johnston, D. A., and Dunne, D. W., 2002, Identification of Schistosoma mansoni gender-associated gene transcripts by cDNA microarray profiling. *Genome Biol* **3**:RESEARCH0041.0041-0041.0012.

Lasonder, E., Ishihama, Y., Andersen, J. S., Vermunt, A. M., Pain, A., Sauerwein, R. W., Eling, W. M., Hall, N., Waters, A. P., Stunnenberg, H. G., and Mann, M., 2002, Analysis of the Plasmodium falciparum proteome by high-accuracy mass spectrometry. *Nature* **419**:537-542.

McAtee, C. P., Hoffman, P. S., and Berg, D. E., 2001, Identification of differentially regulated proteins in metronidozole resistant Helicobacter pylori by proteome techniques. *Proteomics* **1**:516-521.

Myler, P. J., Audleman, L., deVos, T., Hixson, G., Kiser, P., Lemley, C., Magness, C., Rickel, E., Sisk, E., Sunkin, S., Swartzell, S., Westlake, T., Bastien, P., Fu, G., Ivens, A., and Stuart, K., 1999, Leishmania major Friedlin chromosome 1 has an unusual distribution of protein-coding genes. *Proc Natl Acad Sci U S A* **96**:2902-2906.

Ouellette, M., Drummelsmith, J., El Fadili, A., Kundig, C., Richard, D., and Roy, G., 2002, Pterin transport and metabolism in Leishmania and related trypanosomatid parasites. *Int J Parasitol* **32**:385-398.

Ouellette, M., and Ward, S., 2002, Drug resistance in parasites. In *Molecular Medical Parasitology* (J. Marr, T. Nielsen, and R. Komuniecki, Eds.), Academic Press,pp. 395-430

Patton, W. F., Schulenberg, B., and Steinberg, T. H., 2002, Two-dimensional gel electrophoresis; better than a poke in the ICAT? *Curr Opin Biotechnol* **13**:321-328.

Ralph, S. A., D'Ombrain, M. C., and McFadden, G. I., 2001, The apicoplast as an antimalarial drug target. *Drug Resist Updat* **4**:145-151.

Rout, M. P., and Field, M. C., 2001, Isolation and characterization of subnuclear compartments from Trypanosoma brucei. Identification of a major repetitive nuclear lamina component. *J Biol Chem* **276**:38261-38271.

Yatsuda, A. P., Krijgsveld, J., Cornelissen, A. W., Heck, A. J., and De Vries, E., 2003, Comprehensive analysis of the secreted proteins of the parasite Haemonchus contortus reveals extensive sequence variation and differential immune recognition. *J. Biol. Chem.* **278**:16941-51.

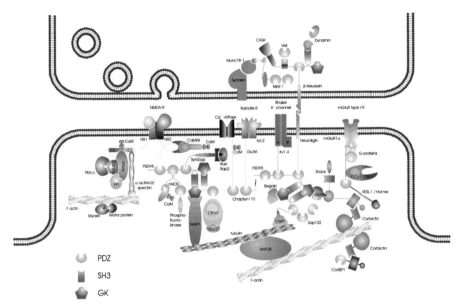

Chapter 4 Figure 5 Page 111. Schematic diagram of synaptic multiprotein complexes. Post-synaptic complexes of proteins associated with the NMDA receptor and PSD-95, found at excitatory mammalian synapses, are shown. Individual proteins are illustrated with arbitrary shapes and known interactions indicated. Proteins shown in colour are those found in a proteomic screen, whereas those shown in grey are inferred from bioinformatic studies. The specific protein–protein interactions are predicted, based on published reports from yeast two-hybrid studies. Membrane proteins (such as receptors, channels and adhesion molecules) are attached to a network of intracellular scaffold, signaling and cytoskeletal proteins, as indicated.

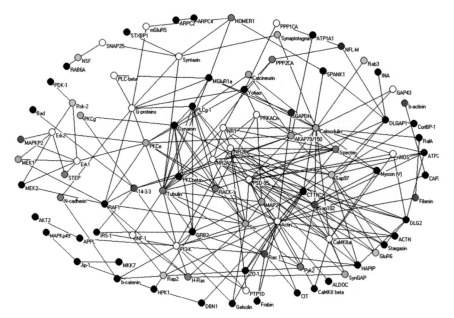

Chapter 4 Figure 6 Page 116. Network representation of MASC proteins with phenotypic annotation. 97 MASC proteins with known direct protein interactions to other MASC proteins are plotted. The NMDA receptor subunits (NR1, NR2A, NR2B) are centrally located. The association of each protein with plasticity, rodent behaviour or human psychiatric disorders is shown in the colour of the node. Key: Black – no known association; Red – Psychiatric Disorder; Green – Plasticity; Blue – Rodent Behaviour; Yellow – Psychiatric Disorders and Plasticity; Cyan – Plasticity and Rodent Behaviour; White – all three phenotypes. Orange for Psychiatric Disorders and Rodent Behaviour alone had no hits.

386

Normal Retina | Enlarged and escavated optic disc | Scattered white spots

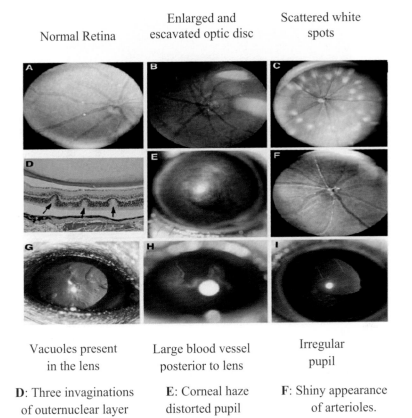

Vacuoles present in the lens | Large blood vessel posterior to lens | Irregular pupil

D: Three invaginations of outernuclear layer | **E**: Corneal haze distorted pupil | **F**: Shiny appearance of arterioles.

Chapter 5 Figure 3 Page 128. Ocular phenotypes obtained by a dominant ENU screen.Adapted fromThaung et al (2002).

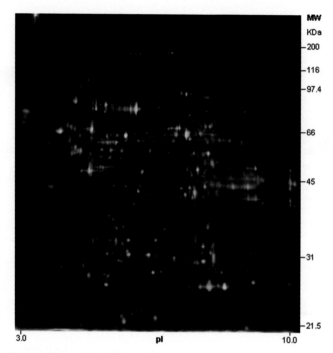

Chapter 7 Figure 6 Page 176. Differential gel electrophoresis of hepatocellular carcinoma. Cy3 (green), normal; Cy5 (red), tumorous tissue.

HIV-1 Life Cycle

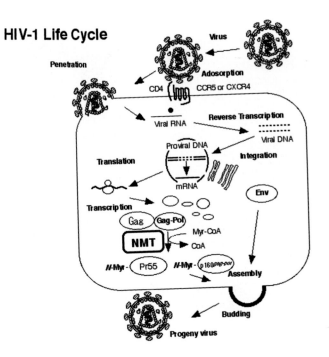

HIV-1 Gene Products

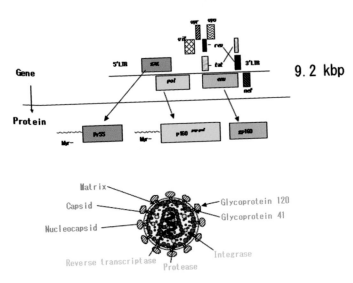

9. 2 kbp

Chapter14 Page 340. Scheme – HIV-1 Life cycle and gene products

389

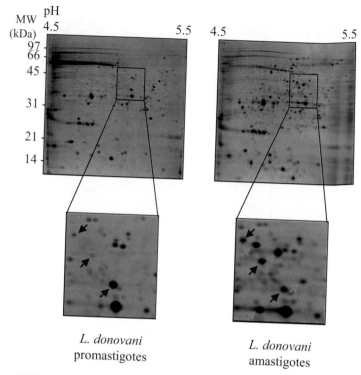

L. donovani
promastigotes

L. donovani
amastigotes

Chapter 15 Figure 4 Page 376. Comparative proteome analysis of L. donovani infantum axenic promastigotes and amastigotes. Proteins from whole cells were extracted as described in Figure 2, run on narrow-range IPG strips, and stained with SYPRO Ruby (Molecular Probes) to allow quantitative comparisons. Representative gels of each stage are shown. The expanded region contains a number of differentially-expressed spots. Spots marked with orange arrows appear to be stage-specific, while spots marked with blue arrows are differentially expressed.

AUTHORS INDEX

SUBJECT INDEX